"Wagner and Furst eloquently present the promises and perils of AI in a way that resists the tendency to over and underestimate the social, political, and economic impacts. In a world where *AI Supremacy* is more than industry domination or achieving first mover status, the authors recognize that there is no real end to this race. In their talented, critical, and in-depth geopolitical analysis of our quest to solving intelligence, the authors keep us eager to learn more about our hybrid futures and for this reason, this book is a must-read."

- **Eleonore Pauwels, Research Fellow on Emerging Cybertechnologies at the United Nations University's Center for Policy Research and Director of the AI Lab at the Wilson Center**

"Daniel and Keith help us understand social, political, and economic changes that will follow a wide adoption of AI and machine learning by adjusting norms, policies, and business models. *AI Supremacy* is a great milestone to engage us all in a critical conversation about how we want our AI-enabled future be more inclusive and sustainable."

- **Danil Kerimi, Head of Information, Technology, and Electronics Industries, World Economic Forum**

"The authors do a superb job of defining what AI is, and what it isn't, providing a comprehensive view of its use in commercial, civil, military, and intelligence applications, including the health and financial industries. *AI Supremacy* is a thorough and exhaustive review of an important, soon to be ubiquitous, technology, containing much of what is known about the subject. It is a fascinating read and highly recommended for anyone working in or interested in the subject."

- **Major General (Ret.) Robert H. Latiff, Ph.D., author of *Future War: Preparing for the New Global Battlefield***

"The AI revolution will entail vast and unpredictable disruption within societies, across countries, and among geopolitical blocks. *AI*

Supremacy is singular among the AI literature. It is a powerful, sweeping, compellingly structured and presented overview of the essential questions that the advent of AI–including the potential for superintelligence—raises. It examines the factors and forces that will drive change and define the battle lines in a new kind of quest for supremacy. This quest will determine how we experience our professional and personal lives, whether privacy and dignity– indeed ethics—will subsist, how societies are organized and stratified, how we educate our children and care for the ill and the elderly, how we fight our wars, what the global order will look like, and even how we conceive of what it means to be human. *AI Supremacy* is as thoughtful and comprehensive an exploration as you will find. It will equip readers-policymakers, corporate leaders, and ordinary citizens alike-with the knowledge needed to become informed participants in pressingly needed, high stakes societal dialogue."

- Nicolas Economou, CEO, H5, Senior Advisor of the AI Initiative at The Future Society, and chair of the Law Committee, IEEE Global Initiative on Ethics of Autonomous and Intelligent Systems

"Daniel Wagner and Keith Furst explore the problems that a radical shift toward AI utilization may present and answer pertinent questions with meticulous attention. *AI Supremacy* bridges conversations about economics and technology with comprehensive research and thorough examinations; it is a convincing resource. The book is as persuasive as it is thorough. Technology and economics professionals, and individuals concerned about the future of AI, will find the work informative and useful."

- Foreword Clarion Reviews

"Wagner and Furst provide a healthy dose of common sense to cut through the hype surrounding AI."

- Rand Waltzman, Senior Information Scientist, Rand Corporation

"In a world where there are more connected devices than people, what happens when the machines start speaking to each other with the intelligence and agency usually reserved for humanity? The specter of artificial intelligence, machine learning, and job-destroying industrial automation has far-reaching implications for business, public policy, and global economic security. In *AI Supremacy*, Daniel Wagner and Keith Furst write a considerable treatise on how AI will shape the future of work and the future of the world. This is an important work addressing one of the world's most pressing and least understood decisions – how shall we coexist with smart machines and how do we cope if Moore's Law turns against us?"

- Dante Disparte, Founder and CEO of Risk Cooperative and founding advisor to the Global Blockchain Business Council

"Wagner and Furst exhaustively explore the inner workings and implications of AI. Each chapter focuses on the current and future state of AI within a specific industry, country or society. Special emphasis is placed on how AI will shape the domestic, diplomatic, and military landscapes of the US, EU, and China. This tome is a great resource for politicians, entrepreneurs, scholars, and ethicists to research short and long-term trends in AI and where humanity fits into this brave new world."

- Russell Whitehouse, Executive Editor, International Policy Digest

"Daniel Wagner and Keith Furst have produced an inspired book that neatly unpacks the growing potential risks which accompany the race undertaken by organizations and governments alike- toward supremacy in artificial intelligence. From the globalization of AI to international espionage and the scale of China's quest for supremacy, Wagner and Furst's work is a real page turner. They raise deeply pertinent, often unanswerable questions that simultaneously highlight the enormous parameters of human fallibility, and the growing existential threat to human nature. *AI Supremacy* is a must read for individuals, businesses, and government officials who are seeking to understand what this unprecedented phenomenon means for the future of our shaky international system."

- Aditya Ramachandran, Research Associate, Hybrid Reality

"Entertaining, meaningful, and insightful, this book is a powerful revelation into the future of mankind and the world around us."

- Jennifer Ciotta, author of *I, Putin*

Also by Daniel Wagner

- *Virtual Terror*

- *Global Risk Agility and Decision-Making*

- *Managing Country Risk*

- *Political Risk Insurance Guide*

A repository of his published work (including more than 600 articles on current affairs and risk management) may be found at: www.countryrisksolutions.com.

AI Supremacy
Winning in the Era of Machine Learning

Daniel Wagner and Keith Furst

Copyright © 2018 Daniel Wagner and Keith Furst

All rights reserved.

ISBN-10: 1722113960

ISBN-13: 978-1722113964

This book contains information obtained from authentic and highly regarded sources. Reasonable efforts have been made to publish reliable data and information; however, the author and publisher do not assume responsibility for the validity of all material, nor the consequences of their use.

The author and publisher assume that both the advice and information in this book are true and accurate as of the date of publication. Neither the author nor the publisher gives a warranty (express or implied) with respect to the material contained herein, nor for any errors or omissions that may have been made.

All rights are reserved. Except as may be permitted under US Copyright Law, no part of this book may be reprinted, reproduced, transmitted or utilized in any form by any electronic, mechanical, or other means, now known or hereafter invented, including photocopying, microfilming, and recording, or in any information storage or retrieval system, without express written permission from the author.

Cover design: Anna Zayco
www.annazayco.com; aniazayco@gmail.com

Daniel Wagner

This book is dedicated to everyone with the vision, determination, and fortitude to embrace an AI-driven future with an orientation toward progress, ethics, and grace.

Keith Furst

This book is dedicated to my children, Anne-Marie, Sebastian, and Philomena, who I hope will learn to use technology as a tool to enhance their lives and the lives of others. Always remember to think for yourself and to take time to *wonder* about the mystery of being alive.

Acknowledgements

Daniel Wagner

I would first of all like to thank Keith for suggesting that we write this book and for being such an easy co-author to work with. I am confident that this book will make a meaningful contribution to the literature on AI.

I thank Devanshu Sood–formerly of Columbia University and the United Nations Development Program–for his research assistance, and Mordor Intelligence for being willing to share some of their work product with me.

Thank you to all of the reviewers who were so generous with their time, and a special thanks to Ania Zayco for her incredible design of the book cover.

Keith Furst

I would first like to acknowledge my parents, James and Karen, for their support throughout my life. I was very lucky to have parents like them, which allowed me to pursue the many opportunities that came my way.

My wife, Teresa, has always supported me throughout our lives together, from pursuing an MBA degree to starting a business and traveling overseas to write a book. Thank you for your endless patience and kind words of encouragement.

A great deal of gratitude is due to my co-author, Daniel Wagner, for acting as Editor-in-Chief for this project, taking on many of the administrative tasks associated with publishing this book, and for his guidance throughout the writing process.

Many thanks are due to my friends, Anasse and Vivek, for our intense discussions about technology and the future, just a stone's throw from Wall Street.

I would like to acknowledge my Philosophy professor, Luis E. Navia, for his earth-shattering lectures at the New York Institute of Technology. They had a profound effect on my desire to pursue more knowledge.

Also, the City of New York (CUNY) college system deserves a lot of credit as well. They offer a high quality and affordable education to swathes of children from working-class families.

Abbreviations

3D	Three-dimensional
4G	Fourth generation
5G	Fifth Generation
ABI	Activity-Based Intelligence
ACTUV	ASW Continuous Trail Unmanned Vessel
ADHD	Attention Deficit Hyperactivity Disorder
AI	Artificial Intelligence
AI2AI	AI to AI
ALIS	Autonomous Learning Investment Strategies
AML/CTF	Anti-money laundering and counter-terrorist financing
Apps	Applications
AR	Augmented Reality
ARPA-Ed	Advanced Research Projects Agency for Education
ATM	Automated teller machine
AUM	Assets under management
AUSTRAC	Australian Transaction Reports and Analysis Center
AVATAR	Automated Virtual Agent for Truth Assessments in Real-Time
AWCFT	Algorithmic Warfare Cross-Functional Team
BP	British Petroleum
CANES	Consolidated Afloat Networks and Enterprise Services
CCP	Chinese Communist Party
CFIUS	Committee on Foreign Investment in the United States
CIA	US Central Intelligence Agency
CMS	Centers for Medicare and Medicaid Services
COBOL	Common Business-Oriented Language
COiN	Contract Intelligence
COREs	Centers of Research Excellence in AI
CQW	CosmiQ Works
CRISPR	Clustered Regularly Interspaced Short Palindromic Repeats
CRM	Customer relationship management
CPTPP	Comprehensive and Progressive Agreement for Trans-Pacific Partnership

CVC	Card verification code
DAMO	Discovery, Adventure, Momentum, and Outlook Academy
DARE	Deception Analysis and Reasoning Engine
DARPA	US Defense Advanced Research Projects Agency
DL	Deep Learning
DLT	Distributed ledger technology
DMT	N,N-dimethyltryptamine
DoD	US Department of Defense
DOE	US Department of Energy
DOT	US Department of Transportation
DPAs	Data protection administrators
DTRA	US Defense Threat Reduction Agency
EDR	Endpoint Detection and Response
ENTC	Explosives National Tracing Center
ER	Extended Reality
EU	European Union
EvD	Enterprise Value of Data
FAGMA	Facebook, Apple, Google, Microsoft, and Amazon
FCC	US Federal Communications Commission
FG-ML5G	Focus Group on Machine Learning for Future Networks
FinCEN	Financial Crimes Enforcement Network
Fintech	Financial technology
FIR	Fourth Industrial Revolution
FIRRMA	Foreign Investment Risk Review Modernization Act
GAO	US Government Accountability Office
Gbps	Gigabits per second
GDP	Gross Domestic Product
GDPR	General Data Protection Regulation
GRID	Grid Resilience and Intelligence Project
HRL	Human rights law
HWCE	Hyper War Center of Excellence
IAEA	International Atomic Energy Agency
IAIA	International AI Agency
IARPA	US Intelligence Advanced Research Projects Activity
IBNS	Intelligence banknote neutralization system
ICTAIs	International Center for Transformational AI
ID	Identification

IHL	International humanitarian law
IoT	Internet of Things
IP	Intellectual Property
IQT	In-Q-Tel
ISPs	Internet Service Providers
ISR	Intelligence, Surveillance and Reconnaissance
ITTI	Intelligent Tech and Trade Initiative
ITU	International Telecommunication Network
JEDI	Joint European Disruption Initiative
Mbps	Megabits per second
MIT	Massachusetts Institute of Technology
ML	Machine Learning
MOOC	Massively open online course
MR	Mixed Reality
MSRA	Microsoft Research Asia
MST	Chinese Ministry of Science and Technology
N3	Next-Generation Non-Surgical Neuro-technology
NAFTA	North American Free Trade Agreement
NATO	North Atlantic Treaty Organization
NGA	US National Geospatial-Intelligence Agency
NGAV	Next-generation anti-virus
NGOs	Non-governmental organizations
NIC	Land and Liveability National Innovation Challenge
NLP	Natural language processing
NMI	Nature Machine Intelligence
NSA	US National Security Agency
NTSB	US National Transportation Safety Board
OBE	Out-of-body experience
OOCL	Orient Overseas Container Line Limited
P2P	Peer-to-peer
P&L	Profit-and-loss
PED	Processing, exploitation, and dissemination
PLA	Chinese People's Liberation Army
PLAN	Chinese PLA Navy
PRRS	Porcine reproductive and respiratory syndrome
PSG	Preventive Security Governance
PWC	PriceWaterhouseCoopers
QSAR	Quantitative structure – activity relationship
R&D	Research and development
RL	Reinforcement learning
RNS	Responsive neurostimulation system

SCADA	Supervisory control and data acquisition
SIF	Synthetic Identity Fraud
SMEs	Small and medium-size enterprises
SML	Supervised machine learning
SMSS	Squad Mission Support System
SSA	US Social Security Administration
SSNs	Social security numbers
SSRL	Stanford Synchrotron Radiation Lightsource
SWIFT	Society for Worldwide Interbank Financial Telecommunication
TNT	Tissue Nanotransfection
TPP	Trans-Pacific Partnership
TOEFL	Test of English as a Foreign Language
ToS	Terms of Service
UAS	Unmanned Aircraft Systems
UAV	Unmanned Aerial Vehicle
UK	United Kingdom
UML	Unsupervised machine learning
UN	United Nations
UNESCO	United Nations Educational, Scientific and Cultural Organization
US	United States
USCF	University of California San Francisco
UUVs	Unmanned underwater vehicles
VADER	Visualization and Analytics of Distribution Systems
VaR	Value at risk
VCs	Venture capitalists
VT	Virtual Reality
WTO	World Trade Organization
Zettabytes	ZB

AI-Specific Definitions

Artificial General Intelligence

A machine that can successfully perform any intellectual task that a human-being can.

Artificial Intelligence

The development of computer systems able to perform tasks that normally require human intelligence, such as visual perception, speech recognition, and decision-making. The spectrum ranges from weak AI (wherein a machine can simulate human behavior but is bound by rules and not conscious) to strong AI (wherein a machine can be conscious, think, and reason as a human would).

Augmented Reality

A technology that superimposes a computer-generated image on a user's view of the real world, providing a composite view, performing increasingly complex functions with massive amounts of data.

Deep Learning

A subset of machine learning wherein networks are capable of learning without supervision from data that is unstructured or unlabeled, performing increasingly complex functions with massive amounts of data.

Extended Reality

All real-and-virtual combined environments and human-machine interactions generated by computer technology and wearables, which includes representative forms such as augmented reality, augmented virtuality, and virtual reality.

Machine Learning

A subset of AI that often uses statistical techniques to give computers the ability to "learn" with data, based on past experiences, without being explicitly programmed. The primary types of ML are supervised learning, unsupervised learning, and reinforcement learning.

Mixed Reality

The merging of real and virtual worlds to produce new environments and visualizations, where physical and digital objects co-exist and interact in real-time (also referred to as Hybrid Reality).

Natural Language Processing

An area of computer science and AI concerned with the interactions between computers and human (*natural*) languages, and how to program computers to process and analyze large amounts of *natural language* data. The computers derive meaning from text as a human would, and can understand words, concepts, and how they are connected.

Reinforcement Learning

An area of machine learning wherein a machine learns by trial and error and the consequences of its actions. Through the allocation of rewards and punishments, a machine learns to act in a specific way in order to maximize its performance.

Supervised Machine Learning

Machine learning that pairs an input to an output through pairing (training examples). By labeling or classifying data, a machine learns the patterns in the data so that it can apply it to new data and make decisions.

Unsupervised Machine Learning

The machine learning task of inferring a function that describes the structure of "unlabeled" data (data that has not been classified or categorized). The program must find hidden patterns and relationships in data without any guidance, teaching itself by finding patterns, making mistakes, and self-correcting.

Virtual Reality

The computer-generated simulation of a three-dimensional image or environment that can be interacted with in a seemingly real or physical way by a person using special electronic equipment, such as a helmet with a screen inside or gloves fitted with sensors.

Contents

Acknowledgements i
Abbreviations iii
AI-Specific Definitions vii
Introduction 1
1. *Creeping Irrelevance* 5
2. *AI Globalization* 37
3. *Jobs and Manufacturing* 61
4. *Financial Services* 85
5. *Science, Health and Medicine* 115
6. *An Evolutionary Communications Landscape* 145
7. *The Future of Learning* 169
8. *Smarter Government* 187
9. *The AI/Cyber Nexus* 211
10. *Spying and Fighting* 251
11. *China's Quest for AI Supremacy* 283
12. *International Relations* 315
13. *Legal Safeguards* 341
14. *An AI Future to be Feared or Embraced?* 377
Notes 406
Index 458
About the Authors 498

Introduction

The idea of creating intelligent machines has ancient roots. As long ago as the 4th century B.C., the Greek mathematician Archytas of Tarentum imagined a mechanical bird propelled by steam. The Chinese, Egyptians, and a host of other civilizations from thousands of years ago envisioned machines that could assist humans, do whatever they can do even better, and make them smarter. Yet, never before in human history has the technological ability to create intelligent machines become a reality. Today, Artificial Intelligence (AI) and Machine Learning (ML) are all around us, and we are on the threshold of achievements and discoveries that were the domain of pure science fiction just a few decades ago.

The pace of progress and change is occurring at exponential rates each year and is becoming so great that many in the AI and ML community believe that in as little as a couple of decades from now human-level AI will no longer be the realm of fantasy. No wonder there is a race going on to achieve AI supremacy. Those that are not already in the race are falling further and further behind those who are already in it. But this is no ordinary race, where there is a single victor, and once the race is won, it is over. There is no single winner in this race, and any company or government that achieves supremacy in one realm at one time will not stay there for long.

The stakes are extremely high, for this is not merely a race to remain competitive; it is a race to define the contours of the future. Superintelligent AI—wherein machines possess intelligence that far surpasses human capability—is, many believe, just around the corner. Companies such as Baidu and Google, and governments such as China, Japan, Germany, Russia, and US, are devoting vast resources to create superintelligence, in an effort to seize a comparative advantage in that race.

This raises many questions. Among them, will superintelligence be used for good or evil? Will we be made extinct as a race or immortal as superintelligent AI becomes a reality? Our future may be neither dystopian nor utopian in orientation; more likely, it will be some combination of both. But are those corporate and governmental masters of the AI universe concerned with such questions, or are they simply focused on winning a race with no known boundaries and where no rules apply? Will there be boundaries and rules in the decades to come, or will it remain a free-for-all?

Regardless of whether a no-holds-barred approach to this race will continue to govern AI's future, the development of AI is quickly becoming commoditized, which is, in one sense, democratizing the production of intelligence and broadening the scope of the possible. However, it is also widening the number of participants in the race, the range of potential beneficiaries, and the possible ramifications associated with its use.

It used to be the case that humans determined machines' actions, but today, the roles are in the process of being reversed. We are gradually being reduced to *monitoring* what machines do. Soon enough, they will be determining *our* courses of action, and keeping an eye on *us*. So, the corporate and government gatekeepers have a lot at stake—not just among each other and their domains, but in terms of who oversees the machines who will "manage" us.

Experts are divided about just where the tripwire is that will enable superintelligent AI to become a reality. Some argue it is just around the corner. Others believe it is decades away. And some even suggest that there is no such thing as a tripwire at all, and that the entire notion is fanciful. While no one can know the answer until it actually happens, what we do know is that we are presently barreling down the runway at an increasingly accelerated speed into the great unknown.

If you are an airline pilot taking off on a foggy runway, you know that, very soon, you will leave the ground; the fog will clear; if there are clouds above, you will break through them, and you can reasonably expect to experience flight in a familiar manner to a known destination. That is clearly not the case with AI and ML. In this milieu, it is indeed foggy outside, and as we are careening down the runway, we have no idea what, if any, obstacles lay in our path, what awaits us upon takeoff, nor what our ultimate destination is.

This is the basic challenge organizations and governments face in considering whether to even taxi on to the runway, much less commence a takeoff roll. Many governments and organizations will not particularly care that they cannot know what conditions on the runway are, what awaits them upon takeoff, or where they are actually going. What matters to them is that they are in the race, come what may. Yet, many of them do not have a full appreciation for the near and longer-term challenges that await them. That is in large part what this book is about.

The scope of this book is deliberately broad, intended to encompass the plethora of themes surrounding the rise of AI and ML. Some of these are obvious—such as how companies and countries are competing with each other and the impact AI may have on the job market in due course. Others, you may not have thought much about —such as the impact the race of AI supremacy is having on the social fabric of societies, how wars will be fought in an AI-driven future, or whether a global legal regime governing AI may ever be effectively crafted and implemented. Every question that is asked and answered can only yield more questions; yet, if having read this book prompts organizations and governments to get into the habit of asking new questions about the implications and impacts of AI and ML, and considering the previously unthinkable, then our mission will have been at least been partially accomplished.

Achieving AI supremacy is as much about having fully embraced AI as it is about contemplating what your competitors are doing before you do, what your firm needs to do to get to the finish line in one piece, and how to still be able to look yourself in the mirror knowing that ethics and governance considerations were not abandoned in the process. The temptation to want to cut corners along the way in order to enhance the pace of change will be great. As AI and ML continue to evolve, new and ongoing challenges will arise, whether they be the continued erosion of privacy or the integration of the human body with machines.

Among the many questions AI raises are whether bias can ever be eliminated from a work product, how we can stay in control as we gradually hand the reins to machines, and whether or not we can be protected against unintended consequences. The passage of time will raise ever more questions, many of which will either not have a single answer, nor, perhaps, *any* meaningful answer. The pace of change being driven by the AI and ML arena is already breathtaking and promises to continue to upend conventional wisdom and surpass some of our wildest expectations as it proceeds on what appears at times to be an unalterable and pre-ordained course. Along the way, much of what we now consider to be "normal" or "acceptable" will change. Indeed, changing the way we think will be one of the inevitable results of racing toward AI supremacy.

1. *Creeping Irrelevance*

Cognification

AI, which many people consider embryonic and evolutionary, is already having a profound impact on the world. It is at once rewriting the rules of major industries while fueling a global arms race among the world's major powers. The stakes are high for a variety of reasons, but not being on the AI train today implies falling further and further behind the curve. Failing to get in the race for AI supremacy could lead to a company's future liquidation or an increased risk of war between nations.

Just as electricity was the driver of the Second Industrial Revolution, AI is the new electricity.[1] We are at the beginning of another Revolution that will drastically alter the manner in which we interact with our own bodies, other people, the physical world, technology, businesses, and governments. Think of it as a natural extension of how societies have developed since the 19th century, but in a previously unimaginable way and at an unfathomably rapid pace.

The First Industrial Revolution emerged from the use of machines to leverage water and steam power to mechanize production and replace or augment human labor. The Second Industrial Revolution was created from the use of electricity for mass production, which fundamentally transformed industries including transportation, defense, healthcare, food storage, and communication. It also resulted in the creation of new industries which had not previously existed before the advent of electricity.

The Third Industrial Revolution was driven by information technology and software, which automated production. The current Fourth Industrial Revolution (FIR) is an extension of the Third, but it is because of the fusion of technologies that this one is different: it is blurring the lines between the physical, digital, and biological. The FIR includes a plethora of emerging technologies such as Big Data, cloud computing, printing, AI, nanotechnology, biotechnology, and nanosatellites[2] which are transforming the world in profound ways.

One of the key aspects of the FIR is the cognification of everything—making objects smarter by connecting, integrating sensors, and building software and AI into them.[3] We are already surrounded by cognified devices such as smart speakers, watches, thermostats, cameras, computers, phones, glasses, lights, door locks, and household appliances which constitute the ecosystem of the current version of the Internet of Things (IoT). These devices enable us to craft our own personal IoT that can capture data, communicate with other devices, regulate access control, act autonomously, and provide advanced analytical insights to whatever is being monitored. While the IoT can be helpful, convenient, and even save lives, it also increases the surface area for malicious actors to conduct cyberattacks and decreases our own capacity to recall things from memory, because we become so reliant on these devices to think for us.

Imagine you are transported into the body of your future self so you can experience your daily routine 30 years from now. You are awakened by a virtual assistant which first debriefs you on world events specifically tailored to your interests and then recites your daily schedule, reminds you to take your medication and vitamins, recommends what type of clothing you should wear based on the weather, and controls the climate in your home based on who is there, taking into account individual preferences and the effect the temperature has on your health. Your home has been cognified so extensively that even mundane appliances such as a coffee maker can now be activated by voice command.

As you leave your home, there is no need for a key since one glance at a camera unlocks the door. Your self-driving car is unlocked with your fingerprints and voice command directs your Tesla (or other autonomous vehicle).

AI SUPREMACY

The first conference call of the day goes beyond voice and video, transmitting each person as a holographic image that replicates body movements and facial expressions to create identical virtual conference rooms, despite the group being dispersed across the globe.

Since it is a sunny day, you decide to get lunch with coworkers by walking to an automated store, selecting premade sushi, and simply walking out the door, since there are no cashiers present. As the day winds down, you decide it is time to buy a new sofa and begin a virtual shopping experience via an augmented reality application which transposes the holographic sofa into your living room, removing the need to actually visit the store. Your virtual assistant mentions that the color of the sofa is not an ideal style fit with the rest of your home. You thank her for the helpful advice and change the color. Finally, you decide to invite some friends over and order pizza, which is, naturally, made by robot and is delivered by an autonomous unmanned aerial vehicle (UAV). Such a "futuristic" lifestyle is currently popularized on television and in a variety of films, but we are already on our way to living exactly this type of existence with the help of AI. The "future" reality is already here.

It sounds wonderful, and it is, but there are social, economic, health, and other costs associated with the cognification of everything. First, it is important to note that, throughout the vision of the future just described, you were constantly receiving feedback from technology about decisions you would ordinarily make yourself without the need for human intervention. Is that necessarily a good thing? Even those who may prosper from the FIR may ultimately suffer from the absence of critical thinking and taking responsibility for their own actions. If we need to think less about and perform fewer mundane tasks, what will we think about or do with the extra time? Would the old adage "idle time is the devil's play thing" become a self-fulfilling prophecy?

AI embedded technology will gradually give us more time back while simultaneously eliminating wide swathes of employment sources. How will governments fill the gap? Will our culture, governments, and education systems adequately prepare us for the consequences of stripping decision-making from a significant portion of our lives? Will we gravitate toward becoming philosophers, as our leisure time increases, and ponder the meaning of life, become more actively engaged in politics, or will we simply chase fleeting pleasures in the smart hedonistic cities of the future? How will we resist the constant bombardment of information and feedback from smart objects without succumbing to the will of their masters (the companies that create and operate them)? Will the concept of personal privacy as we know it today even exist when cameras permeate every room and streams of data about our habits, pursuits, proclivities, and thoughts belong to someone else? As virtual assistants become ever more sophisticated, will we mindlessly buy into the crafted reality that AI is our friend? How will that affect our relationships with family and friends in the real world and impact our concept of ourselves?

Rising Obsolescence?

While conventional wisdom dictates that AI will result in the destruction of tens of millions of jobs over time, Gartner, a global research and advisory firm, believes just the opposite. It predicts that, starting in 2020, AI-related job creation will become a net positive, resulting in two million new jobs by 2025. It notes that many significant innovations in the past have been associated with a transition period of temporary job loss, followed by recovery, then business transformation. AI will likely improve the productivity of many jobs, eliminating millions of middle- and low-level positions, but also creating millions more new positions of highly skilled, management, entry-level, and low-skilled jobs. AI "augmentation"—a combination of human and AI—will ultimately complement each other.

Gartner thinks that, by 2022, one in five workers engaged in mostly non-routine tasks will rely on AI to do a specific job. Leveraging technologies such as AI and robotics, retailers will, for example, use intelligent process automation to identify, optimize, and automate labor-intensive and repetitive activities that are currently performed by humans, reducing labor costs through efficiency, from headquarters to distribution centers and stores. Many retailers are already expanding technology use to improve the in-store check-out process. Many consumers still prefer to interact with a knowledgeable sales associate when visiting a store. Retailers will find it difficult to completely eliminate traditional sales advisers, but they will be able to benefit from labor savings by eliminating highly repetitive and transactional jobs. Automation will lead to cost savings, while the removal of friction in value chains will increase revenue in the optimization of supply chains and go-to-market activities. By 2021, AI augmentation will generate $2.9 trillion in business value and recover 6.2 billion hours of worker productivity.[4]

As AI becomes better than us at performing narrow tasks traditionally done by humans, how will our governments create policies to protect peoples' livelihoods? Should it be up to governments to do so or will it create a new generation of entrepreneurs? Will AI create a new breed of welfare recipient or will it spur governments to dramatically reform welfare? Some have argued that, while some jobs will be displaced, new jobs will be created in their place. No one really knows exactly how all of this is going to unfold, but we can make some predictions about how different scenarios will play out, and hypothesize about what governments may need to do to proactively prepare in order to protect their nation's future (this is more fully addressed in Chapter 8).

Clearly, there are some incredibly beneficial applications of AI—such as saving lives by detecting disease earlier and more accurately than can a human doctor (which is, of course, already being done)—but it is hard *not* to imagine that AI will accelerate income inequality by reducing the number of jobs available while creating additional educational, social, and financial obstacles to acquire the skills necessary to obtain the jobs of the future.

Imagine, 20 years from now, a truck driver who has been on the job for 10 years suddenly loses his job because the trucking company he works for replaces its employees with self-driving trucks. What policies will a government have in place to ensure that this truck driver can learn new skills for the new economy, assuming there are enough jobs to go around? Even if the government were to have some sort of program for displaced workers to return to school and learn new skills, will all of them actually do it? This does not make the truck driver a bad person or necessarily lazy, but with identity and self-worth demolished, not all will react in the same way at the same time. This could lead to a number of related social issues, such as rising levels of alcoholism, drug abuse, domestic violence, obesity, or other health problems.

Ironically, truck drivers have been in demand and their salaries have been on the rise in recent years, with their wages having cumulatively increased by 10.8% from January 2014 to February 2018,[5] due to higher demand from people purchasing things online, but this will not last forever. The upward salary trajectory of truck drivers will probably create more of an incentive for companies to automate this function by adopting self-driving trucks (at least for longer routes, such as traveling on expressways between cities). Local deliveries in cities will most likely still need to be completed by human drivers, at least for now. Drones are already beginning to supplant individuals in the package delivery business for critical medical supplies,[6] especially in countries with less regulatory oversight than in Western countries, such as Rwanda. So, while AI is not likely to completely replace truck drivers, it could reduce the number of drivers needed, and consequently push wages down due to increased supply and lower demand.

Manual workers are not the only ones at risk from AI; knowledge-based workers can also be displaced. A 2018 study demonstrated that an AI platform was more skilled at identifying issues in legal contracts than twenty experienced lawyers. The participants were given four hours to review five non-disclosure agreements and identify thirty legal issues, which they were subsequently scored on. The human lawyers achieved an average accuracy rate of 85%, while the AI outperformed them by attaining 95% accuracy. While each lawyer spent an average of 92 minutes to complete the task, the AI took only 26 seconds. So, not only was the AI more accurate but, simply based on the time required to complete the task, it would take approximately 212 human lawyers working in unison to match the output of the AI.

Even at a very modest rate of $300 per hour for experienced attorneys, that level of throughput would equate to a human labor cost of $63,600. This does not mean that AI will make lawyers obsolete, since there are some tasks computer programs cannot do, such as litigating in a court room. It is more likely that AI will enable lawyers to be more efficient at their jobs by augmenting (rather than replacing) their own intelligence, so they can focus on the things they do best (such as litigation and client interaction). However, as lawyers become increasingly more efficient by leveraging AI, the number of legal jobs available will be reduced even further, in a marketplace that has been contracting for several decades.[7] (Chapter 13 will address the implications for AI on the law in greater detail).

The race for AI supremacy is well underway, given what is at stake, with the giant technology firms and others frantically acquiring companies with next generation AI algorithms, creating new products and infrastructure, in an effort to collect boundless amounts of data while constructing autonomous cyber defense and offense capabilities. In a study, PriceWaterhouseCoopers (PWC) identified more than 300 cases where AI was used in 54 economic sectors across 140 countries; the study predicted that AI could contribute as much as $15.7 trillion to the global economy by 2030, which is greater than the combined global domestic product (GDP) output of China and India in 2017.[8]

Google acquired DeepMind for $660 million in January 2014, even though the company had no customers or products at the time. Despite the high price tag, the investment seems to be paying off already as DeepMind's AI, a deep neural network based on a general framework that can understand complex dynamics, has already reduced Google's overall power consumption by 15% in their data centers.[9] DeepMind has lofty goals to "solve intelligence" and, ultimately, the firm wants to create general AI that can think as broadly and effectively as a human.[10]

The upside for economic growth and increased productivity from AI and other forms of automation is massive, but the real risk of large swathes of workers being permanently displaced is presumed. A 2017 report by the McKinsey Global Institute estimated that as many as 375 million workers globally (14% of the global workforce) will likely need to transition to new occupational categories and learn new skills in the event of rapid automation adoption. While AI is just one technology that encompasses the wider automation trend, it significantly contributes to overall job displacement concerns into the next decade.[11] That does not necessarily mean, however, that those concerns will prove to be justified in the end. Much will depend on what is done in the interim.

While many leading Silicon Valley technology titans have suggested that universal basic income[12] could be one way to address the mass displacement of workers, it misses an important point about human nature. For the truck driver (as for most workers), it is not just about money, but also about dignity, having something to do, and being able to take care of your family.

As children, many of us may have observed one or both of our parents (or caregivers) going to work every day and instilling in us the belief that hard work was a good thing and built character. The puritan work ethic was thought to be one of the foundations of what was good for the individual and society in general. This notion still pervades our collective consciousness in the United States (US) today, even outside a religious context. The working day occupies our time and minds and gives us a sense of purpose. When that is taken away, are we really ready to explore what life "could" be like doing something completely different from what we may have known our entire lives?

Doing so is not simply about income replacement. It is also about activity replacement or a completely new vision of how we want to live in society. The German philosopher Friedrich Nietzsche made the famous and controversial assertion that "God is dead". Nietzsche was highlighting that the age of the enlightenment shattered our traditional conception of God through the rise of reason and science. If reason killed God, will AI kill work?

Defining AI

Social, economic, financial, political, and other aspects of AI will be discussed throughout the book but, before we continue, let us first define AI. The term "Artificial Intelligence" was referenced in a 1956 academic proposal for a two-month Dartmouth study of how machines can simulate learning and intelligence. The goal was "to find how to make machines use language, form abstractions and concepts, solve kinds of problems now reserved for humans, and improve themselves".[13]

There were other precursors to AI, such as research by the British physician Richard Caton, who discovered the electrical nature of animal brains by recording impulses. His work led to the research of physician Hans Berger, who was the first to record human electroencephalograms, in 1924.[14] The discovery of the electrical nature of the brain inspired early work in the field of neural networks and cybernetics. The idea was that, if a brain was thought of as an electrical system, then a machine could be built to replicate it. To this end, the development of AI has been about, at its core, replicating human intelligence or biomimicry, but our definition of intelligence may need to be modified in the context of the current state of AI, to avoid anthropomorphizing it and distorting its current capabilities and limitations.

There are, naturally, differing opinions about what AI is and its impact on a plethora of sectors and issues. As Pamela McCorduck has explained, "it is part of the history of the field of AI that every time somebody figured out how to make a computer do something–such as play good checkers or solve simple but relatively informal problems–there was a chorus of critics who would say, 'that is not thinking'".[15] Skeptics and believers about AI have always existed and probably always will. They both have a point, of course, but what is true is that AI has advanced so rapidly and so significantly that there can be no doubt that it will play an even greater role in the advancement of science, technology, and daily life in the decades to come.

There seems to be some confusion around what AI can and cannot do today, and this partially stems from the use of the word intelligence. Intelligence is a nebulous concept, subject to philosophical debate, because we could think of intelligence, in a narrow human sense, in terms of an intelligence quotient–the ability to break down and solve complex problems quickly. Other interpretations of human intelligence define it more broadly as a multitude of abilities, beyond book smarts, such as the ability to "read" people or a type of emotional intelligence.

But intelligence is not limited to humans, even though humans clearly exhibit a special form of intelligence. We know that some animals can communicate with one another, coordinate attacks on a prey, and save food for the future. It could even be argued that plants exhibit a form of intelligence, as branches tend to grow away from darkness towards sunlight. Even a calculator could be considered a crude form of AI because the ability to store information and use it to perform mathematical calculations is uniquely human.

Santa Fe Institute President David Krakauer offered another definition, stating that intelligence "is making hard problems easy". Professor Krakauer goes on to explain that the greatest threat facing humanity today is not what you might expect–such as nuclear war, a bioengineered pandemic, global warming, overpopulation, or scarcity of natural resources–but rather, stupidity.[16] This implies that if we approached all of these problems in an intelligent way we could potentially address them, but nations have historically acted to protect their own interests and avoided cooperation for the greater good of humanity. This tendency for self-preservation among people and nations is part of our innate nature, but is it something that we can overcome?

In 1950, Alan Turing, one of the pioneers of computer science, proposed the idea of the Imitation Game to address the question of whether machines could think. The Game, commonly referred to as the Turing test, was to have an evaluator review the written conversation between players A and B and determine which player was human or a computer. If the computer was mistakenly taken for a human 30% or more of the time, then the test was passed. While the Turing test was an important milestone to gauge the development and capabilities of AI, it did not actually show that machines think.[17]

The phenomenon of thought remains not well understood today, so for some to claim that passing the Turing test shows that machines can think presupposes that we know what thought actually is, which we do not. While neuroscience research initiatives have made some incredible strides in understanding the human brain over the years, how consciousness is actually possible is still not clear, nor are the underlying mechanics of it. Also, there are many neurological phenomenon that are extremely bizarre and seem to fall outside the scope of something that can be explained based on our current knowledge of the brain, such as phantom limbs, psychedelic experiences, or acquired savant syndrome.

Phantom limbs give a person the sensation that a missing limb is still attached and is subject to excruciating pain. For example, Army Sergeant Nick Paupore was part of a convoy that was struck by a roadside bomb while driving through Kirkuk City during the Iraq War. The blast severely damaged his leg and he lost part of an artery, which caused a lot of blood loss. Paupore ended up losing his leg as a result of the bomb, but he still experienced excruciating pain in his phantom limb, for which powerful narcotics such as Dilaudid did not help.

A Navy neurologist, Dr. Jack Tsao, was looking for ways to help soldiers like Paupore and remembered reading a paper in graduate school about mirror treatment to help reduce phantom limb pain. A group of veterans participated in the unusual mirror therapy and many reported significant reduction of pain in their phantom limbs by using a mirror to reflect an image of the other attached limb (assuming it was still there). This created the visual appearance that the missing limb was still attached. This simple exercise tricks the brain into thinking the limb is still attached. The therapy was so effective that Paupore was eventually able to stop taking pain medication entirely.[18]

While the underlying neurological cause of phantom limb pain is not well understood, some neuroscientists believe the brain creates a body schema, which is basically a simulation of our physical body. When a limb is lost, the brain tries to compensate for the lost feedback from the missing limb and rewires itself, which leads to false sensations that the limb still exists.[19] The body schema could have evolved out of the need for our species to control our attention and limit the scope of our consciousness in everyday life.

People have reported extremely alien experiences under the influence of powerful psychedelic drugs, which raises questions about the nature of reality and self. Dr. Rick Strassman, a clinical psychologist, conducted experiments with N,N-dimethyltryptamine (DMT), a powerful psychedelic drug commonly found in nature. The source of DMT in the human body is the pineal gland, which regulates sleep cycles, among other functions. The pineal gland has a rich metaphysical history and Descartes thought it was the seat of the soul. Dr. Strassman believed that there was a potential connection between DMT and near-death experiences, psychosis, mystical experiences, and even alien abduction.[20] Terence McKenna was one of the leading intellectuals and proponents of responsible use of psychedelic drugs naturally occurring in plants. McKenna described his hallucinations on DMT, claiming to have met faceless intelligent entities or self-transforming machine elves in hyperspace. He explained that the experience was so vivid, intense, and utterly astonishing that somehow the entities he met did truly exist on some fourth dimensional plane, or, if they were really part of his own mind, there were parts of himself which he really did not know at all.

Acquired savant syndrome is very rare but it is one of the most perplexing of neurological phenomena ever studied. As an example, Derek Amato jumped into a pool to catch a football; he underestimated the depth of the pool and slammed his head extremely hard on the floor. Doctors thought he had a severe concussion and he was sent home from the hospital. After his injury, he found he was drawn to an electronic keyboard. He had never played piano before, but had acquired an uncanny ability to play as a result of his head injury. One of the theories behind acquired savant syndrome is that damage to one part of the brain results in the rewiring of other parts of the brain, releasing dormant potential, which could have been biologically suppressed by evolution.

How can a person who has never played the piano before suddenly sit down and play like a professional, which would typically take the average person years to achieve? Is it possible that Amato simply listened to or observed someone playing the piano at some point in his life and, from that memory, his brain was able to reconfigure itself in a way that released the dormant potential to play? AI is already helping us to become more capable by augmenting our intelligence synthetically. The real leap forward could be by accessing "the little rain man"[21] in all of us, via non-intrusive methods. Perhaps AI can help us hack our own biology, breaking the little drop of consciousness that evolution has granted each of us by advancing our understanding of the brain and how to activate our hidden potential.

AI Consciousness

AI algorithms have outperformed humans in certain studies when conducting specific tasks such as image classification, facial recognition, and legal contract evaluation. However, all of these tasks are considered "narrow" AI. In other words, the AI platform that can beat lawyers at identifying contract errors will not be able to drive a car autonomously. That has not stopped some from positing that AI can only lead to a scenario where we create general AI as intelligent as humans that begin to recursively improve themselves without human intervention, resulting in an intelligence explosion that evolves into superintelligence that we can no longer control.

The media is fixated on AI and greatly contributes to fears that AI is positioned to take over the world, in some type of Terminator Skynet scenario, in the not too distant future. To be sure, there are serious social, economic, and political implications to consider with the continued advancement of AI capabilities and applications, but some academic AI researchers think that it is unlikely that machines will become sentient in the near future. And a number of high-profile individuals, such as Elon Musk, Bill Gates, and the late Stephen Hawking continue to warn about the dangers of AI.[22]

The basic argument is that one day AI will outperform humans at every cognitive task. This is indeed possible, but it could be limited to solving analytical problems; it does not mean that general AI will actually experience the world as humans do, have hidden dormant potential, can dynamically adapt in hostile conditions, or have the same intellectual versatility. So, in this sense, we would argue that general AI— or the ability to perform every cognitive task as well as humans—does not actually imply or ultimately lead to sentience.

John Searle argued against general AI in his Chinese room argument. The idea was that a non-Chinese speaker could be locked in a room and be given a large batch of Chinese writing, a second batch of Chinese script, and a third batch of Chinese symbols and instructions in English to link all of the batches together. Based on this information the non-Chinese speaker would be able to formulate responses that would make sense to a native Chinese speaker, and it would not be apparent to the native speaker that the responses were coming from a non-native speaker, let alone someone who did not understand Chinese at all.[23]

The real question, if and when general AI is created, is whether we can create an AI that is truly sentient, or whether it can recreate itself. Even if we could, what would sentience actually mean for AI, and would it be drastically different from our own experience of consciousness? For general AI to truly be more intelligent than humans, we would argue that researchers would have to replicate the full spectrum of potentialities of the brain, experience, and consciousness. When you consider all of the neurological phenomena studied over the years and the vastness of the human condition, this seems like a daunting task. After all, one cannot throw a server down a flight of stairs with the result being a sudden increase in its computational power.

The real danger, at least in the near term, is that the AI programs we create will have goals that do not align with our own. Regulations, standards, and best practices for the AI industry are desperately needed, just as they are needed in the cyber-enabled world; those goals may be set by humans or, possibly, by the AI programs themselves in the future. Regulations should not stifle innovation, but the various impacts of AI need to be evaluated so that sensible policies can be put in place to protect people and their livelihoods. There is no stopping the progress of AI, given that human ambition will always push the boundaries of what is possible, but we need to find intelligent ways to regulate, monitor, and mitigate the risks associated with AI before it advances much further. The stakes only get higher with each passing year.

Machine Learning

AI is one of the most exciting and unsettling technological trends of our time, and much of the enthusiasm would not be possible without Machine Learning (ML), a subset of AI wherein machines have learned to recognize faces, classify objects in images, recommend advertisements that align with our preferences, understand and act on voice commands, drive cars autonomously, detect new types of fraud and cyberattacks, and diagnose skin cancer, among other things. It is important to make the distinction between AI and ML, which are at times referred to interchangeably. AI is, in the first instance, the science and engineering associated with intelligent algorithms, whether or not they learn from data. ML is a subfield of AI focused on algorithms that can learn automatically, without being explicitly programmed from data.[24]

One of the mathematical foundations enabling ML is linear algebra. ML algorithms have succeeded across a wide variety of domains which involve different types of data, such as text, images, voice, and video. ML solves sets of equations of all those different types of data that need to be translated into floating point numbers so they can be represented in data structures such as matrices and manipulated through various mathematical operations.[25] In this sense, everything thing we see in the world around us, and all of the data we create in cyberspace, can be processed by ML algorithms, if we can convert the data into numbers.

The discipline of AI can cover a wide variety of research and applications, including, but not limited to: ML, natural language processing (NLP), speech, expert systems, planning optimization, robotics, and computer vision. Depending on the use case, a problem could be solved with multiple AI applications. ML can also be broken down into sub-categories which, at a high level, include supervised machine learning (SML), unsupervised machine learning (UML), and reinforcement learning (RL).

SML is the most widely used high level category of ML today and has brought a tremendous amount of economic value to the Internet advertising industry by learning users' preferences based on their search and website browsing behavior and tailoring advertisements to meet their specific interests. SML has several techniques—such as decision trees, random forests, nearest neighbors, Support Vector Machines, and Naive Bayes—which can solve complex computations with hundreds of variables. SML models require data to be labeled or "supervised", meaning there is an input whose expected output is predefined, which works best when the problem is well understood and somewhat stable. For example, the problem could be to predict the market price of a used car. The input would be all of the data (also known as features) associated with the car, such as make, model, mileage, accident history, and geographic location. There could be other features not directly related to the car, such as the current strength of the economy that can also impact the market price.

To train the SML algorithm, a data set is ingested which contains all of the data on the car, current market strength, and the actual sales price. The more data the SML algorithm is given to observe, the more likely it will be to make accurate predictions on the fair market value of cars yet to be sold. As new data is introduced to the model, it can make modifications to its predications by adjusting the weights or importance of the parameters used in the algorithm. This allows it to get as close to the desired output as possible, which in this case would be an accurate prediction of the sales price.

The predicted market price of two identical cars could vary under different market conditions; the SML would be able to update its predictions based on changes in market sentiment, which could be loaded into the model automatically. At a high level, the initial learning process for an SML algorithm includes, but is not limited to: data collection, data training, feature extraction, model training, and model evaluation. Again, as new data is introduced, predictions are made based on the model's historical training.[26]

There are, not surprisingly, some limitations to SML, such as dealing with new scenarios that are not part of the existing landscape. For example, new typologies can emerge in the practice of fraud which did not exist in the past or were unknown when the model was trained. Dealing with these unknowns is one of the limitations of SML, which is why UML is sometimes used to close the coverage gap. With UML, there are no labels or supervisors, so there is no output, only input. The goal of unsupervised learning is to find regularities in the input or uncover structure in data which could be useful in several different applications.

UML leverage techniques, such as clustering, detect similarity across groups without prior knowledge of the input and expected output. In other words, no training data or labels are needed for the algorithms to identify useful insights. UML can identify networks of bad actors behaving similarly—such as bots creating fake social media accounts—allowing for the segmentation of actors into categories, which, in this example, is important for fraud prevention.

There are also situations where SML and UML can be used in conjunction with feedback from human experts to increase the overall performance of the system. In a 2017 study, researchers from MIT's Computer Science and Artificial Intelligence Laboratory and an ML startup demonstrated an AI platform that predicts cyberattacks significantly better than existing systems by continuously incorporating input from human experts. The platform was able to predict 85% of attacks, which is roughly three times better than previous benchmarks, while also reducing the number of false positives (incorrectly identifying normal activity as a cyberattack) by a factor of 5. To predict attacks, the platform would detect suspicious activity by clustering data into patterns through the use of UML. The patterns are then presented to human experts who confirm which events were actually attacks. That feedback is used as a label or supervisor for the SML algorithms to make better predictions about future attacks.[27]

The last major category of ML is RL. RL is being highlighted because of the tremendous success it has had in a variety of domains, such as self-driving cars and the defeat of a world Go champion. The idea is that an ML program is given a long-term goal, such as not crashing a car or winning a Go game, but it is not provided with immediate feedback for every action it takes. The reason is that it would require too much computational power to evaluate all of the possible board configurations of a Go game. Even if that were possible, there is no good or bad feedback to provide during the game. Despite Go's simple rules, there are 10 to the power of 170 possible board configurations, which is greater than the estimated number of atoms in the known universe.

Traditional approaches to AI programs would have machines learn complex games and construct a search tree addressing all possible positions, but the sheer number of configurations makes this method impossible. This is why researchers at DeepMind, the creators of AlphaGo, used a search tree with deep neural networks with millions of neural-like connections to mimic the human brain.[28] Ultimately, the feedback for the entire game is given at the end, after many actions have been taken. The feedback given to AI program is in the form of a win or a loss, which would reinforce or weaken a specific sequence of game play. RL has been referred to as "learning with a critic", as opposed to SML, which learns with a teacher. A "critic" differs from a "teacher" because they do not tell us what to do; rather, only how we have done in the past. Feedback does not occur as often with SML and, when it does, it arrives late, so the AI must adjust its approach to a long sequence of steps in order to improve, based on feedback from a critic.[29]

Due to its tremendous success in a wide variety of applications, the other major trend in ML is Deep Learning (DL), which we are not considering a high-level ML category, but, rather, a particular approach to ML. DL is a subset of ML wherein networks are capable of learning unsupervised from data that is unstructured or unlabeled. Facebook's DeepFace was tasked with comparing the faces from two *images* to determine if they were the same person. It reached a 97.35% rate of accuracy when compared to the average accuracy of 97.53% for *humans*.

Researchers describe the DeepFace system as a nine-layer deep neural network with a mind-boggling 120 million parameters.[30] However, this does not mean that DL neural network algorithms are infallible. Researchers in Vietnam reportedly fooled Apple's Face ID, based on a deep neural network, with a mask that cost $150 to make. Some facial recognition technology can be spoofed and cannot even tell the difference between a real face and a photo. This suggests that biometric identification (ID) could reduce different types of fraud in the short-term by raising the effort and cost to fool the software but, clearly, the criminal underworld could innovate by using 3D printers to create custom face masks or use other spoofing techniques in the future.

As discussed earlier, defining AI is far from a black and white proposition; it is subject to debate, and DL can be similarly challenging to define. One definition of DL is that it imitates the workings of the human brain in processing data and creating patterns for use in decision-making.[31] Another is that it is a neural network with more than two layers.[32] There are a number of factors that have contributed to the enormous success of DL in a relatively short period of time, which include: access to more neurons than previous artificial neural networks had, more complex ways of connecting neurons, an explosion in the amount of processing power to train models, and automatic feature extraction (labels).[33] DL is having, and will continue to have, a profound impact on how AI develops and advances in the future.

Why AI is Already So Relevant

The rise of the Internet created the foundation of an immense amount of information to be stored, created, and shared across the globe. Search companies such as Google emerged and created algorithms to search through all of that data quickly. The massive amount of data generated as a result created a need to increase computational performance beyond what was possible on a single computer. This led to new breakthroughs in parallel processing, where a task was not limited to one computer but was split across a network of computers, processed concurrently, and merged back together to create the final result. The physical limitations of one computer were erased via the use of smarter algorithms. The

Internet was one of the first steps towards Big Data; the mass adoption of smartphones was another. Through computers and smart devices, enormous amounts of data are constantly being generated, collected, analyzed, and processed. The increased digitalization of business processes, social media, peer-to-peer (P2P) sharing, and other trends have all contributed to what we call Big Data.

In 2001, 2006, and 2011 only two technology companies– Apple or Microsoft–were listed as one of the top five companies in the US by market capitalization. By 2017, all of the top five companies by market capitalization were technology companies, including Facebook, Apple, Google, Microsoft, and Amazon (FAGMA), which clearly shows the speed and scale at which technology companies have reigned supreme.[34] In 2017, *The Economist* published an article proclaiming that the world's most valuable resource was no longer oil, but data.[35] The FAGMA stocks came to dominate market capitalization for a number of reasons, including elaborate networks, broad applicability, smart devices, and sensors which collect enormous amounts of data that can be leveraged to generate huge profits quickly. And they are only getting started.

A natural byproduct of governments, companies, and people going through these various forms of digital transformations is the creation of colossal amounts of data. By 2025 the global datasphere–the sum of all digital data created, captured, and replicated on our planet–will grow to 163 zettabytes (ZB–that is, one trillion gigabytes), or ten times the 16.1ZB of data generated in 2016.[36] Dr. Anasse Bari, a New York University professor, contends that Big Data and DL are creating a new data paradigm on Wall Street. Diverse data sets such as satellite images, people-counting sensors, container ship positioning, credit card transactional data, job and layoff reports, cell phone usage, social media feeds, news articles, tweets, and online search queries can all be used to make predictions about future financial valuations. For example, data scientists can analyze the satellite images of the parking lots of major retailers over a period of time to predict future sales. Bari was part of a project that analyzed nighttime satellite images of the earth as a way to predict the GDP per square kilometer. The theory was that the greater the amount of light appearing in a given geographical area, the higher would be its implied

GDP[37]. So it is easy to tell that South Korea is a much more vibrant economy than North Korea simply by viewing the amount of lights turned on in each country at night.

Sentiment analysis is a type of NLP algorithm that can determine how people feel about a given topic. It can be a particularly powerful analytical method, which was illustrated by an anecdote in James Surowiecki's book, *The Wisdom of Crowds*[38]. In 1907, Francis Galton realized that when he calculated an average of all the guesses for people participating in the weight of an ox competition at a local fair, it would be more accurate than individual estimates, or those of supposed cattle experts. The method was not without its flaws, of course, since influence could be exerted onto each response, which could skew the accuracy of the results. However, it is a simple example of the potentially powerful insights that can be extracted from the data generated by the large numbers of participants in social media. Indeed, numerous companies now make predictions concerning their core product offerings based on data generated from social media.

During the analysis phase of a merger and acquisition transaction, social media can be a useful data source to leverage by conducting sentiment analysis on the target company. Among the insights that could be derived are the extent to which people like their products, the degree to which its customer service is perceived favorably, or whether its management is revered or reviled–all of which may have an impact on perceived validity of the purchase price. Other types of NLP algorithms attempt to extract sentiment from news articles for stock trading purposes so as to give an advantage in terms of buying or selling the subject stock.

For example, in 2011 it was observed that the mention of the actress Anne Hathaway in the news appeared to influence the stock price of Berkshire Hathaway[39], even though there is absolutely no connection between the two. This suggests that the type of NLP algorithms that automated trading strategies executed from analyzing news sentiment were somewhat unsophisticated and crude, with the algorithm not being able to distinguish between an actress and a company with one matching word (Hathaway) in their names. This brings up another key distinction about using AI-powered algorithms for different purposes: For a hedge fund, it is not about understanding how the algorithms work, but rather their impact on profit and loss. In other words, if the algorithm is profitable, then understanding it becomes secondary, but there are countless other applications where understanding needs to come first or, at the very least, the algorithm needs to be monitored and managed on an ongoing basis.

Naturally, the private sector is not the only part of the global economy in the race to achieve AI supremacy; governments are also in the race, and there is a growing amount of interconnectivity between the private and public sectors in the AI arena. As an example, Google may have inadvertently started an AI arms race between the US and China when its 2014 acquisition of DeepMind unveiled a stronger version of AlphaGo, which had defeated the world's strongest Go player. In game two, the AI program AlphaGo made an unexpected play at move 37. Many commentators at the time could not make sense of it, thought it was a mistake, and highlighted that no experienced Go Player would have made that move.[40] However, AlphaGo ended up winning game two, which brought conventional game play under question, raising the possibility that AlphaGo made the unexpected move as part of a broader game strategy that was imperceptible to even the best Go players.

In 2017, the State Council of China issued a plan to become the world leader in AI by 2030. According to two university professors, one of the major motivations for China to invest heavily in AI was AlphaGo beating a world-class champion at the notoriously complex ancient Chinese game of Go, in 2016.[41] The US was also potentially influenced by AlphaGo's historic achievement, given the amount of press coverage it received, and because the White House had issued two AI reports on the subject later that year. Many researchers had previously believed the type of AI required to beat a world-class Go champion was still 10 years away.

China has invested heavily into AI and the AI arms race— particularly between the dominant and wealthiest countries in the world–is well underway and even robust, although what is visible in the public domain in that regard appears to be somewhat benign. For example, some police forces in China use smart glasses with facial recognition capabilities to identify suspects and individuals using fake driver's licenses. The glasses are connected to a database of suspects, which allows officers to scan a crowd for possible matches. The facial recognition technology has become sophisticated enough that only one facial image is required to make the comparison. In the past, many different angles of a person's face were required to perform a comparison with a high level of accuracy. By 2018, the smart eyewear had already led to the arrest of seven people suspected of being involved in kidnapping and hit-and-run cases, and another 26 people were arrested for using fake forms of ID.[42]

China has fully stated its intention to achieve AI supremacy, which has raised concern in the US Department of Defense (DoD) that America could be falling behind China (and Russia) in an AI arms race. The Pentagon's Third Offset Strategy sought to ensure that US military supremacy, and AI had become a key aspect of its long-term strategy.[43] Russian president Putin has said that whichever country becomes the leader in AI will rule the world.[44] This statement certainly gives insight into a longer term asymmetric and information warfare strategy by Moscow, which could try to pit the US and China against each other, in theory, allowing Russia to focus on its own goals unabated. (Government-related themes are more fully explored in Chapters 10-12).

Information warfare involving AI has penetrated many aspects of the US political system. The US intelligence community believes that the Russian effort to interfere in the 2016 US presidential election had two prongs: the hacking and leaking of emails from the Democratic National Committee and Hillary Clinton's campaign chairman, and a campaign of misinformation and propaganda carried out primarily via social media. The Internet Research Agency, a shadowy entity with links to the Russian government, used some form of AI to automate its misinformation campaign on various social media platforms, but it is not entirely clear (though presumed) that their efforts were powered by ML. It is worth noting that Twitter had already announced that it will use ML to bolster its own defenses against future automated attacks. Disinformation is just one tool that will be used in the future of warfare, and ML will be a major part of offensive and defensive strategy.

AI is not only being used by foreign governments to attempt to manipulate people in other countries for their own political objectives—the candidates themselves and campaign employees are doing the same. Cambridge Analytica used proxy organizations during the 2016 US presidential campaign to feed untraceable messages on social media platforms and set up a secret email system that would self-destruct and leave no trace.[45] The firm used the personal data of 50 million Facebook users to influence the election.

Predictive Analytics

In the 1980s, psychologists developed a personality model based on five personality traits: openness, conscientiousness, extroversion, agreeableness, and neuroticism. These traits are known as the big five and are used to assess a person's needs, fears, and behavior. Accurate interpretation of data was previously the primary hurdle in developing such profiles because it required a person to fill out a lengthy questionnaire. The Internet and social media platforms changed all that.

Facebook is the world's largest repository of biometric information, with more than 2 billion members around the world. To comprehend its potential usefulness in predictive analytics, consider the following example. Michal Kosinski was a student at Cambridge University, working on his PhD at the Psychometrics Center, when he launched a small Facebook application called MyPersonality, which enabled users to answer some basic psychological questions and receive a personality profile in return. The application also allowed users to opt-in and share their Facebook profiles with the researchers. What might previously have been considered private or sensitive information, and not to be shared with third parties or strangers, is no longer the case in the era of social media and AI. It turned out that millions of people filled out the questionnaire, which provided a treasure trove of data.

Kosinski and his team worked on models that would use Facebook likes, shares, and posts to predict political affiliation, sexual orientation, intelligence; alcohol, cigarette, and drug use; and a host of other behaviors. The more data that was collected, the more accurate these models became. In 2012, Kosinski showed that, with an average of 68 Facebook likes, it was possible to predict the person's skin color with 95% accuracy, and their sexual orientation and political affiliation with 88% accuracy. As he continued to develop the model, as few as 10 Facebook likes could be used to evaluate a person better than interviewing their co-workers, and 70 likes was enough to outperform what a person's friends thought they knew about him or her. As the number of likes rose to 150, Kosinski's model could evaluate a person better than his or her own parents, and 300 likes could uncover things that their own partner did not know. Beyond 300 likes, his model could even outperform what a person thought they knew about themselves. Kosinski concluded that our smartphones are akin to a vast psychological questionnaire that we are constantly filling out–consciously and unconsciously.[46]

Aleksandr Kogan, an Assistant Professor at Cambridge University, built an application called "thisisyourdigitallife", in collaboration with Cambridge Analytica, specifically to harvest the data of Facebook users. The application Kogan created seemed similar to what Kosinski had built, where the goal was to collect data to create psychometric profiles. Hundreds of thousands of Facebook users were paid to take a personality test, under the belief that it would be used solely for academic purposes. There was a technical loophole in Facebook's platform policy that enabled Kogan and Cambridge Analytica to not only use the Facebook data of the person who filled out the questionnaire, but that of all of their friends, as well. This is how several hundred thousand completed questionnaires resulted in tens of millions of Facebook profiles being harvested. The point of collecting all of this data was to create models so that Cambridge Analytica could specifically target individuals based on the needs and fears revealed in their psychometric profiles, so as to try and influence their decision on whom to vote for by deploying personalized political advertisements.[47]

It is unclear exactly how Cambridge Analytica used this Facebook data, but the firm's data-driven insights are based on predictive analytics. The company's website stated that it had up to 5,000 data points on 250 million Americans which could be used to engage, persuade, and motivate them to act.[48] Predictive analytics use current or historical facts to make predictions about the future, which may or may not include ML algorithms. However, LinkedIn profiles of data scientists working for the company describe their role as building Big Data and ML algorithms. Hence, it is possible, perhaps even likely, that some of the psychometric profiles built by

Cambridge Analytica used ML algorithms, among its predictive analytics and statistical modeling techniques.

At the time, a variety of media outlets stated that the Facebook data Cambridge Analytica obtained actually influenced the result of the US presidential election. While it may have had an influence, it is impossible to know for certain just what its impact was, or the real extent of its influence. Professor Helmut Norpoth, of Stony Brook University, built a predictive model based on two major inputs: the results of election primaries and the "swinging of the pendulum" (or the tendency for the party occupying the White House to change after two consecutive terms).[49] On this basis, and a review of US political history since 1992, the momentum was already in favor of a Republican win and the stage had already been set by virtue of what had become the easy predictability of the US voting public. Perhaps there was no need for all the predictive analytic razzle dazzle, after all.

Given the hyper connected world we live in, the use of data to micro-target individuals cannot be surprising. Regardless of political affiliation, any campaign would be well behind the curve *not* to use some element of predicative analytics or ML to reach, and possibly influence, potential voters. Politicians have, of course, historically crafted their messages to ensure that they appeal to a target audience, as well as to the greatest number of people possible. The real issue in the Cambridge Analytica case is not that it tried to manipulate people's behavior based on data, but that it obtained the data through deceptive practices. Some individuals will continue to use social media platforms as they have in the past while others will choose to disengage—but predictive analysts will become ever more advanced, with time, to be able to predict behavior using alternative data through methods that have not yet even been imagined.

Voice Wars

Information warfare is, of course, also alive in a plethora of other areas beyond politics and social media. Technology companies are engaged in their own race to achieve AI supremacy. One of the first manifestations of their silent battle is the voice wars. All five of the FAGMA technology companies have produced their own virtual assistants.[50] Consumers have, predictably, started using voice command to conduct a growing number of Internet searches, compose emails and text, and benefit from the time and effort otherwise required to complete ordinary tasks. Some have even predicted that an estimated 50% of all searches will be voice searches by 2020.[51]

Voice-enabled computing has become the rage with nearly every new IoT and "smart" device. Electronics giant LG made all of its appliances Wi-Fi-enabled in 2017, when more than 80 LG products were already integrated with Google Home. This massive connectivity across a range of appliances is merging the physical and digital into a personal ecosystem that can be controlled by voice through a central command–the virtual assistant. While Amazon had an early lead in voice computing, it fell behind in language support because it chose to only support English, German, and Japanese. When it was launched, Google's Home assistant supported English, French, German, Italian, Japanese, Korean, Spanish, and Portuguese. Google Assistant, which enables two-way communication between the tech giant and its voice-activated users, supports 119 languages for text-to-speech conversions and searches.[52]

The holy grail of all the FAGMA companies is to connect the virtual profile they possess with each of their users' physical profiles. Your Amazon profile can be connected to Alexa so you can reorder an item simply by saying, "Alexa, reorder paper towels". Amazon considered granting access of Alexa voice recordings to third party application developers[53], which might have been legal but probably not what most consumers would either have been aware of nor have approved of. The fact is that most consumers are blissfully ignorant, either of what data is collected about them by their government and the companies they choose to engage with or what is done with it.

Most of us never read the Terms of Service (ToS) we tacitly agree to every time we use one of these companies' services, but doing so gives them the ability to basically do what they want with our information, without asking. One of the first actions Donald Trump took as president was to sign an Executive Order granting Internet Service Providers the ability to use any of our information any way they wish to without American consumers' permission. This includes selling that information to any third party for any reason at any time. At the time this action was taken, in 2017, although it was mentioned in the news, there was no public outcry. It seems we have become numb to the level of intrusiveness technology companies have assumed in our lives and we now simply take it for granted. It is as if we have collectively given up and succumbed to the seductiveness of our "smart" world.

That said, there are plenty of beneficial purposes that smart technology and voice activation are contributing to. For example, short voice samples can be used as inputs for detecting disease. A company in Berlin has focused on detecting Attention Deficit Hyperactivity Disorder (ADHD) through voice samples. Other companies have worked on detecting heart disease, manic episodes, depression, and psychotic episodes through voice recognition.[54] Every new application of technology raises profound questions. Some of the questions implied through the use of these applications include whether a user whose voice indicates a high probability of ADHD will begin to see targeted ads for ADHD medication or related books in his or her browser, whether a minor's voice will become fair game for data collection and sharing, and whether childrens' voices can or will be protected if they are recorded in the background of a virtual assistant's range.

Once the FAGMA companies begin to connect our digital profiles with our physical domain in earnest, the applications are nearly endless. It may be a good thing to be alerted about a potential illness when your virtual assistant detects something abnormal in your voice patterns, but it opens a Pandora's Box in terms of how these companies use this information. If a virtual assistant detects that a user is not getting enough sleep, will it say something like, "You seem tired. Maybe you should go to sleep early tonight"? Or perhaps, knowing your preference for avoiding habit-forming medications, it may say, "You seem tired. I have the perfect over-the-counter and non-habit-forming medication to allow you to sleep well tonight". Clearly, these assistants could make helpful (seemingly benign) suggestions, but will consumers be given the ability to configure what types of responses are given, or will the assistant "read your mind" to automatically respond without being prompted to do so? Where will a line be drawn (if at all) between subtle encouragement to purchase a product versus incessant prompting and genuine annoyance?

With just a few clicks on a smartphone or computer, our voices can access almost all accumulated human knowledge.

The sheer availability of this information, combined with the ease with which it can be obtained, has led to an overreliance on technology to remember things for us, resulting in a phenomenon known as digital amnesia. Some of us may remember a time when we could dial phone numbers from memory, without having to look them up, as we were forced to store information in our brains if we wanted ease of access. Since smartphones and other devices have become extensions of our physical bodies, many people do not even remember the phone numbers of close relatives and friends.

According Kaspersky Lab, a study of connected consumers in the United Kingdom (UK) found that many cannot recall the critical phone numbers of their children (71%), their children's schools (87%), their workplaces (57%), or their partner (49%).[55] If their phones were to run out of battery capacity, how would they contact these people? We have "outsourced" our memories to our smartphones and other devices. A phone can obviously be helpful in extreme scenarios, such as an earthquake or active shooter situation, but do we really want to wait for instructions from a virtual assistant about how to proceed in those types of situations? Have we willingly given up our ability and desire to think for sake of convenience?

Reconfiguring Ecosystems

The impacts of AI already cut across multiple domains and reconfigure complex ecosystems, one of which is the drug trade. Mexican (and other) drug cartels test and deploy new technology all the time, to successfully smuggle more drugs into the US and bring more cash back to their home countries. Mexican drug cartels have used UAVs–more commonly referred to as drones–to fly narcotics from Mexico over the southwest US border to San Diego. The cartels have even experimented with weaponizing drones to be able to drop improvised explosive devices with remote detonators.[56] Clearly, building a wall along the US-Mexico border will not eliminate the flow of drugs, guns, and money.

The Colombian cartels have used semi-submersible submarines to transport large quantities of drugs from Colombia to Mexico, but they also have developed fully submersible submarines, coated with Kevlar, which are difficult for law enforcement to detect, given the small radar signature they leave.[57] These "narco-submarines" are expensive to build, but the value of the cargo they carry justifies the cost. The cartels will undoubtedly seek to take advantage of AI to purchase (or potentially create) a fleet of unmanned underwater vehicles (UUVs) that can travel long distances via remote control. The US Navy is exploring this type of technology and has even organized an unmanned submarine competition to encourage private companies to become actively involved in its development.[58]

AI engineers may need to be careful where they travel in the future because, just as Wall Street and Silicon Valley are in a war for talent, so are the cartels. The Mexican drug cartel Los Zetas reportedly captured 36 engineers to help them build a sophisticated radio network to clandestinely communicate. One engineer was kidnapped while talking on the phone with his girlfriend outside a mechanics shop.[59] It is only logical to assume that the cartels will go to great lengths to kidnap talented AI engineers to help them customize their own fleet of UUVs and other emerging technologies.

The use of UAVs has already proven to be a significant problem for law enforcement and intelligence agencies, particularly because so many millions of people own weakly regulated drones. As will be explored in detail in Chapter 13, the law is far behind the development of cutting edge technologies. Drone hacking toolkits already exist on the Dark Web, and the legal restrictions that exist have not prevented cybercriminals and others from doing what they do best. A firm that alerts customers to unexpected drones flying in their vicinities has caught drones flying near office buildings where they can hack into Wi-Fi networks to steal proprietary information, near data centers where they can destroy massive amounts of data by (for example) disrupting cooling systems, and near prisons (presumably to deliver drugs or contraband to prisoners).

A whole host of new technologies has emerged in the AI arena in recent years, which most of us will not have been aware of, but which point to the bold new world of possibilities AI is unleashing. For example, in the energy arena the application of AI, ML, and DL is increasingly widespread. The use of AI to make distributed energy possible at scale is critical for decarbonizing the power grid, expanding the use of (and market for) renewables, and increasing energy efficiency. AI can enhance the predictability of demand and supply for renewables, improve energy storage and load management, assist in the integration and reliability of renewables, and enable dynamic pricing and trading, which creates market incentives.

AI-capable "virtual power plants" can integrate, aggregate, and optimize the use of solar panels, microgrids, energy storage installations, and other facilities. Distributed energy grids may also be extended to incorporate new sources of energy. Smart grids will also use blockchain for peer-to-peer energy trading to increase the number of distributed sources and optimize energy storage. As economies and cities move away from "heavy infrastructure" towards "smart infrastructure" with a low environmental footprint, the decentralized nature of distributed energy grids mean they have the potential to be used globally.

Precision agriculture (including precision nutrition) is expected to increasingly involve automated data collection and decision-making at the farm level–to plant, spray, and harvest crops optimally, allow early detection of crop diseases and issues, to provide timed nutrition to livestock and, more generally, to optimize agricultural inputs and returns. This promises to increase the resource efficiency of the agriculture industry, reducing the use of water, fertilizers, and pesticides. ML and DL will work in tandem with the IoT, drones, and sensors to measure crop moisture, temperature, and soil composition to provide the data needed to automatically optimize production. UAVs are increasingly being used to monitor conditions and communicate with sensors and AI-enabled systems. "Smart agriculture" has the potential to fundamentally change agriculture even more than mass-farming methods did in the 20th century.

The field of "Climate Informatics" is already blossoming, harnessing AI to fundamentally transform weather forecasting (including the prediction of extreme events) and to improve our understanding of the effects of climate change. Previously, the use of frequently updated datasets associated with climate change has required substantial high-performance computing power, with limited accessibility and applicability for the scientific and decision-making communities. AI is increasing the performance of weather and climate modelling and making it more accessible and usable for decision-making. Over time, cheaper, faster weather and climate models unlocked via AI could reduce the need for energy-hungry supercomputers, lower the cost of research, and open the field of weather and climate science to many more researchers.

Wider AI applications include simpler ML techniques combining weather models and ancillary impacts data to help predict the effects of small-scale extreme weather events (such as windstorms and floods) on human systems, allowing for better risk management. The UK Met Office has, for example, developed a chatbot application to demonstrate how "frictionless" data can be extracted from complex Big Data sets using sophisticated AI in real-time and communicating to the user through a simple interface. Another example involves artificial assistants, fed by forecast data, that can help people make decisions from what to wear to when to travel. Private companies are already working in tandem with universities and government agencies to mutually benefit from comparative advantages in the field.[60]

Data Wars, AI Supremacy, and Irrelevance

The AI wars are only beginning to heat up. The war for data can be seen as an extension of the wider AI wars, since achieving AI supremacy requires massive amounts of data. Given that data is so valuable, it is clear that not everyone will play by the rules to get it; if data cannot be collected, then it may be taken by force. The data breaches of the future may be sponsored by state actors purely to obtain data to train AI programs. Apart from voice wars, companies in the private sector will achieve AI supremacy by merging our digital and physical data profiles together, collecting and analyzing all types of data for diverse purposes and sharing it within a network of companies under the same corporate umbrella to ensure compliance, control, and the discovery of actionable intelligence.

The search for AI supremacy is also making us irrelevant in obvious as well as subtle ways. As knowledge workers use AI to become more productive, their total throughput and efficiency increases, which decreases the number of people a company needs to employ to get the same amount of work done. As jobs are displaced and business models are continually disrupted by new technology, wealth will be generated for a smaller segment of the population, making it even more difficult for the average person to maintain a consistent and sufficient income to purchase a home.

AI will present opportunities and challenges in every corner of our economy, but as AI pushes us to the edge of irrelevance, it will present opportunities and choice to individuals–as citizens, employees, employers. Governments and companies alike will seek to redefine just what our relevance is in this emerging era of machine intelligence. It is a future fraught with possibilities and peril.

2. AI Globalization

When AI and ML are combined with the interconnectedness of global supply chains, they provide a range of unprecedented opportunities and potential perils for international businesses. On one hand, rising efficiency and productivity is permitting exponential growth in some sectors and businesses. On the other hand, the gap in efficiency and productivity between those sectors and businesses that have embraced AI and ML versus those that have not is also growing exponentially, leaving those at the bottom further and further behind. The positive and negative impacts are being felt in different ways, depending on the business, sector, or nation.

AI is already poised to disrupt the globalization process, which is already in a state of near constant change. Consider this: in 2000, just 5% of Fortune Global 500 companies were based in developing countries; by 2025, emerging market companies are likely to make up more than 45% of that list. By 2035, AI could double annual rates of economic growth[1], diminishing the fortunes of some developed countries while enriching the fortunes of a small group of developing countries. Those countries that stand to gain the most are the same small group that dominates the landscape today, such as China and India, which, by virtue of their size, population, and technological capabilities have the resources necessary to take real advantage of the opportunities AI presents, and will continue to present well into the future.

The degree to which countries and populations will benefit from globalization will increasingly depend on the extent they have pursued AI successfully while managing the globalization process (to the extent that they can). Popular support for globalization has largely rested on the premise that much of the world's population would benefit from it, and that government-sponsored social safety nets would temporarily protect the disadvantaged while business focused on generating productivity, innovation, growth, and, ultimately, societal wealth. This approach worked as long as economic inequality was kept within reasonable bounds, but persistent and growing income inequality in developed and developing countries suggests that these bounds have been passed. The gradual decline of middle-class incomes in many parts of the world, despite increases in productivity, has added to a growing backlash over globalization.

Over the past decade in particular, skeptics of globalization have strengthened their case that too many are being left behind, that inequality is only rising since the advent of globalization, and that inherent societal contradictions and inconsistencies have only been exacerbated in the process. The tendency to focus on the functionality of new technologies, and the absence of an inclusive narrative that emphasizes equality of opportunity, has already stoked a backlash. The case for embracing AI may be a hard sell if some of the broader criticisms of globalization are answered with simplistic, historical analogies, making the case that equilibrium has generally resulted and that everything will work itself out in the end.

The corporate focus on maximizing shareholder value has certainly advanced productivity and created growth, wealth, and employment, but the reality is that productivity and economic growth are stalling, global trade has been shrinking in recent years, and there is a growing awareness about the unintended social and environmental side effects of globalization. Many corporations have reacted by increasing share buybacks, accumulating cash reserves, and driving dividends toward historic highs. Meanwhile, investment rates are declining, despite prevailing low interest rates. Governments are responding by printing more money and throwing more of it at social support programs intended to prop it all up. The growing sensitivity to inequality, corporate social responsibility, and climate change, among some quarters, has failed to quell much of the criticism. If AI and globalization are to integrate well, something must change.

The Slow Death of the Ancien Régime

While the trade in goods (which drove earlier phases of globalization) is stagnating, the trade in global services—particularly digitally enabled services—is growing. In 2014, services constituted 25% of total exports from OECD countries—up from 17% in 1980. This shift reflects the growing value of services in many industries, driven by the growth of digital technologies, which are blurring the boundary between products and services. The rapid growth of digital platforms has begun to make country-based business models and national borders somewhat redundant. Global market platforms such as Alibaba and Amazon, and their associated supply and delivery systems, are replacing the complex supply chains that were the foundation of prior phases of globalization, making it significantly easier for smaller companies to compete in a global market.

The familiar economic model of a single dominant economic pole, a primary technology, and a leading system of governance is slowly being replaced by multipolarity. Companies must increasingly address a plethora of paradigms, technologies, and governance rules. Data highways are becoming the new shipping routes. Cloud storage is gradually taking the place of shipping containers and warehouses. Decentralization and digitization are tearing apart conventional means of communicating and transacting.

Going forward, there is unlikely to be a single economic "pole" upon which the global economy is centered. Although there will clearly be global AI leaders (businesses and governments), their supremacy will be snapshots in time, for a given comparative advantage or superior application of technology is unlikely to endure. Other enterprises and countries will catch up and eventually surpass the leader of the moment, resulting in a perpetual race to catch up in a never-ending competition to have some time in the winner's circle.

Emerging country growth is already increasingly being determined by structural reforms designed to drive domestic demand (rather than exports) and to broaden many countries' industrial footprint, especially for commodity-producing countries. A good example of this is India, which enjoys high rates of economic growth despite falling export levels. Another example is China, which has been deliberately altering its economic model to become less reliant on export-led growth and more dependent on its massive domestic population and economic foundation to drive future growth. Future developed country growth will be driven primarily by improvements in productivity, aided by new technology and innovation.

The globally optimized value chain—a familiar feature of the current phase of globalization—will give way to value chains that *blend digital technology with older low-cost technologies, allow greater integration across products and services, and leverage the growth of independent global platforms for the exchange of goods and services*. Evolving in concurrence with this will be a more decentralized governance system with rules that are more complex, volatile, and representative of this emerging new normal. Among them will be local and regional rules that strike a new balance between national political interests and global economic realities.

We are already witnessing the rise of the most prominent newly industrialized countries in multilateral institutions. Little gets done these days without the wink and nod of China in such organizations as the United Nations and World Bank. A number of these post-war institutions are experiencing a backlash from their own national members about the division of voting powers. Countries such as China are both recipients of development aid and among the leading lending nations in the world.

Such inherent contradictions are evidence, as if any were needed, that the design of the post-war modus operandi is slowly giving way to the realities of an emerging technologically- and digitally-driven world in which new powers will rise and new institutions will take the place of the aging governance regimes created by the victors of World War II. Such institutions as the Asian Infrastructure Investment Bank and New Development Bank were established by leading emerging countries and newly industrialized countries (specifically *without* the leading developed nations of the world) and are guided by country and region-specific interests.[2] Likewise, there are calls for the creation of a global cybersecurity organization to establish clear rules of the road in our cyber-laden world (later in this book, we call for the creation of a new global organization to govern AI).

Ultimately, however, the manner in which any form of technology impacts humans depends on how we choose to develop, deploy, and control it. Microsoft's CEO, Satya

Nadella, has said that the way to create AI to augment human capabilities and enhance the human experience implies incorporating trust, transparency, the ability to assume, and infuse technology with human values and empathy.[3] That is, of course, easier said than done. While the leaders of many organizations around the world "want" to do the right thing while attempting to stay ahead of the curve, arriving at the right mix of employee and social responsibility in a hyper-competitive business landscape fixated on quarterly profits is exceedingly difficult to do.

General Electric is an example of a company whose fortunes have changed since its glory days at the top of the industrial pyramid, but whose leadership realized, in the first decade of this century, that the future was not based on the business model that made it the goliath that it used to be. Although its leadership could not have known at the time just how far the company would fall in the ensuing decade, it made a conscious decision to do a number of things to attempt to remain at the top. It is an example of a company that knew what it needed to do to *attempt* to stay ahead of the curve.

Among them was the realization that sustainable growth will require a local capability within a global footprint. The company decided to use a manufacturing strategy to open markets–producing products for the US in the US, while localizing production in large, end-use markets, such as Saudi Arabia. Its competitive advantage was digital productivity–digitizing power plants and hospitals by connecting them to the Industrial Internet, thereby improving global productivity. In Pakistan, it used analytics to improve energy efficiency and expand capacity. In India, it used the Internet to deliver healthcare to remote regions[4] and became a useful model for how to "think" about utilizing technology to innovate, grow, and prosper. That its fortunes changed dramatically had less to do with thinking outside the box than being too late in doing so.

Going forward, it will become increasingly important for global businesses to become acutely attuned to striking, and achieving, the right balance between remaining technologically savvy, being competitive, and being attuned to higher societal objectives. There is more than one way to get to the finish line– in one piece and with a clear conscience (which "should" matter). Likewise, governments cannot afford to do what they do the way they have always done it simply because that is the way it has always been done. The pot of gold at the end of the AI rainbow will, in the end, go to the organizations and governments that know and understand the difference and are not timid about doing what needs to be done to get in the game and stay in the game, in the hope of having their turn in the winner's circle.

The New Trade Regime

One of the ways AI is having a real impact on globalization is via how cross-border trade is being conducted. Until recently, data flows related to AI have been governed by World Trade Organization (WTO) rules which were drafted before the invention of the Internet. Recognizing how out of date global data governance has become, policymakers in Europe and North America have been working to link AI to trade with explicit language in bilateral and regional trade agreements. As of 2017, only one trade agreement–the Comprehensive and Progressive Agreement for Trans-Pacific Partnership (the CPTPP, formerly known as the Trans-Pacific Partnership (TPP))–included explicit and binding language to govern the cross-border data flows that fuel AI.

Specifically, the CPTPP (which is expected to be ratified and implemented in 2019) includes provisions that make the free flow of data a default provision, requires that nations establish rules to protect the privacy of individuals and firms providing data (a privacy floor), bans data localization (requirements that data be produced or stored in local servers), and prohibits all participating parties from requiring firms to disclose source code. These rules reflect a shared view among the eleven signatory nations that *nations should themselves not be allowed to demand proprietary information when facilitating cross-border data flows.*

The US (which withdrew from the TPP) wants even more explicit language related to AI in other trade agreements. US diplomats proposed that the North American Free Trade Agreement (NAFTA), which is also being re-negotiated, should include language that bans mandated disclosure of algorithms as well as source code. The US wants to ensure that its companies will not be required to divulge their source code or algorithms *even if* the other NAFTA parties consider doing so as necessary to prevent discrimination, disinformation, or the undermining of their citizens' ability to make decisions regarding their personal information.

Like most trade agreements, the CPTPP and NAFTA also include exceptions, where governments may break the rules delineated to achieve legitimate domestic policy objectives, which may include rules to protect public order, health, safety, and privacy related to data processing and dissemination. However, governments can only take advantage of the exceptions if they are proven to be necessary, are carried out in the least trade distorting manner possible, and do not impose restrictions on the transfer of information greater than what is required to achieve that government's objectives. Policymakers must delineate with greater clarity how and when they can take these steps to protect their citizens against the misuse of algorithms.

The EU is far ahead in that regard, given the 2018 implementation of the Global Data Protection Regulation (GDPR), the world's first law requiring shared governance for a digital single market that builds financial support for technological and digital innovation but also establishes clear and explicit rules protecting personal data while applying meaningful fines for non-compliance. The EU has, in addition, introduced limits in the GDPR on the use of algorithms as a human right. Article 21 allows anyone the right to opt out of ads tailored by algorithms. Article 22 enables citizens to contest legal or similarly significant decisions made by algorithms and gives them the ability to appeal for human intervention.

Some European courts have already interpreted these rules as applying to the global Internet and not just within the EU, which some companies view as a form of extraterritoriality. While the EU has taken a momentous first step in building a single digital market and encouraging public debate, its member states have not yet agreed to rules that fully govern data flows or AI in trade agreements with other countries. Policymakers are concerned that other nations will seek to dilute the European approach to protecting privacy and empowering citizens, in order to challenge the use of AI. Such measures may serve to raise the costs of obtaining data and could stifle some forms of innovation and competition.[5] Everything comes with a cost. While it is too soon to say just what costs will ultimately become associated with placing some guardrails in the data arena, few would argue that humanity would be better off without such restrictions.

International trade is already benefitting from the impact of AI which is helping proactive supply chains to thrive on data from start to finish; Online orders, warehouse packing slip information, and shipping station freight scans can all feed data to logistics programs driven by AI. Such programs can anticipate supply chain disruptions and formulate plans to compensate for them in advance. They can predict customer behavior to regulate stock, preventing order shortages or overages, and they can also calculate the fastest and cheapest shipping routes while foreseeing customer cancelations. Artificially intelligent supply chains are agile and able to alleviate the impact of inevitable disruptions.

One of the biggest challenges to achieving efficient international trade is tackling the growing burden of governance and compliance requirements. Companies that trade globally must not only know the degree to which they are complying with home country requirements, but who they are doing business with, whether any clients, suppliers, or business partners are violating any trade laws, and whether all parties also comply with international and home country laws. Software already exists to aid with compliance, but it is sometimes prone to false positives and negatives, which means that human oversight is often required. AI can improve compliance software by reducing the number of false results, reducing the amount of necessary human review, and enhancing the speed and accuracy of the trade process.

Legal-based AI programs can catalog contracts, ensuring they are being implemented properly, which helps protect companies from legal risks and ensures that they are also benefiting from a given contract by promoting on-time payments and deliveries from clients and suppliers. Some financial institutions are taking advantage of AI platforms to analyze compliance performance; they can offer more financing options to international businesses as costs drop.[6] In the future, the globalization process will be further enhanced by the growing capability and usage of AI to streamline cross-border trade.

Shipping giant Orient Overseas Container Line Limited (OOCL) joined forces with Microsoft Research Asia (MSRA) to identify, manage, and overcome a range of operational uncertainties, while creating efficiencies using AI. OOCL has, for years, fully embraced a hybrid cloud infrastructure with auto-switching and auto-scaling throughout its business. It has more than 1,000 software developers in the US, as well as Hong Kong, Manila, Shanghai, and Zhuhai. The company regards AI as critical to its future and turned to MSRA to optimize its shipping network operations, saving the shipping firm an estimated $10 million in costs annually.[7]

DDC FPO, which processes freight bills and handles other back-office functions for trucking companies, developed a product that uses AI to process bills of lading more quickly and accurately. Its intelligent data capture technology replaced a rules-based approach to billing by using an unstructured context recognition program to process bills of lading. The company processes approximately 150,000 freight bills per day, most of which is done at its offshore processing center in the Philippines. DDC uses optical character recognition technology that helps recognize freight bills and can identify the shipper, the consignee, and read bills of lading and other forms submitted in non-traditional formats. It searches for specific data as it scans bills, resulting in fewer errors and faster service.

Freight billing is just one area of the transportation world where AI and automation are being deployed. Software firms are working to automate basic processes, such as carrier selection and freight tendering. The explosion of digital data in transportation will eventually make AI a necessity. DDC's system will ultimately take the mundane data transfer and entry task away from carrier personnel, in an area of trade logistics that is largely not automated. AI has the potential to revolutionize some of the more mundane aspects of international trade, since there is not a shipment anywhere in the world that moves without a freight bill.[8] The faster they can be processed and paid, the more efficiently global trade will function.

IBM Research has introduced an AI-enabled scanner–the IBM Crypto Anchor Verifier–which will use blockchain to allow the consumers to verify the origin and authenticity of their purchases. One of its first clients was the Gemological Institute of America, which helps gem and jewelry purchasers evaluate and grade diamonds. The Verifier software can validate a product's unique optical patterns, which are sometimes undetectable by the human eye, and differentiates them from each other. In the medium-term, IBM believes that digital ledgers, blockchain, and cryptographic anchors (computers that are smaller than a grain of salt) will be able to prove the authenticity of a wide range of products.[9]

The world's first blockchain powered platform for marine insurance was launched for commercial use in 2018. Insurwave, developed in a joint venture between Ernst and Young and Guardtime, utilizes blockchain and distributed ledger technologies powered by Microsoft Azure and ACORD data standards (a global standards-setting body for the insurance market) to help businesses transform the risk management process. Insurwave was estimated to have supported more than 500,000 automated ledger transactions and to have managed risk for more than 1,000 commercial vessels in its first year.

The platform connects all participants in a secure, private network with a fixed, accurate, and irrefutable audit trail, establishing a first of its kind digital insurance value chain. A.P Møller-Maersk, the world's largest container ship and vessel supply operator, participated in the development of the platform. Migrating to Insurwave has helped Maersk automate manual processes and alleviate a range of inefficiencies and frictional costs in the way it uses trade marine insurance. The system's creators have plans to introduce it to other types of business in the future, including the marine cargo, global logistics, aviation, and the energy sectors.[10]

To encourage further innovation in the trade arena, in 2017 the WTO proposed creation of the Intelligent Tech and Trade Initiative (ITTI) to bring technology and business leaders, negotiators, and scholars together to explore ways in which blockchain technology (trust ledgers) and AI can positively impact global trade. The ITTI is an essential step designed to crush existing trade roadblocks. Via blockchain, small and medium-size enterprises (SMEs) can better access credit and link to a broader investor ecosystem, enabling them to set up new trading networks and obtain funding by sharing financial data in a security-rich and transparent public arena.

AI can also play a useful role in leveling the playing field in trade negotiations. Countries often have a difficult time preparing for trade talks, with delegations sometimes entering discussions without being fully prepared. New AI-driven technologies can enable delegations to collect and structure large amounts of information efficiently, as well as predict negotiation scenarios and model outcomes[11] (AI's ability to contribute to successful diplomatic or business negotiations in a range of contexts is almost limitless).

That is more important than ever in what can be a surprisingly unpredictable world, where countries may pull out of trade agreements or change the terms of trading arrangements with virtually no notice. The potential legal consequences and lost opportunities associated with non-compliance of trade sanctions have never been higher. Whether the US withdrawal from the TPP, the re-imposition of sanctions against Iran, or the prosecution of HSBC for money laundering, it is easy for businesses to get caught in the cross-fire.

Some compliance-oriented organizations have found a way to thrive in the chaos. Seoul-based law firm Yulchon saw trade sanctions against North Korea as an opportunity for innovation and opportunity. It developed a client-focused digital compliance system called AlgoCompliance, a multilingual algorithm that monitors government websites to identify and flag regulatory changes as they occur. By using AI, the system identifies changes that are relevant to Yulchon's clients, assessing whether a change in regulation would trigger a compliance violation. The data gathered allows the firm to produce scenarios that enable clients to focus on potential risks.

Other law firms are traveling a similar path. In 2017, Ropes & Gray produced what is essentially an in-house research unit– Ropes Risk Mitigation & Management (R2M2)– that enables its clients to better understand big picture trends, parse daily developments, and assess risks to their business operations around the globe. Law firm Ashurst developed an integrity due diligence service to help clients comply with regulatory frameworks and mitigate exposure to integrity-linked risks that can lead to reputation damage. The firm's risk assessment team scours public records across jurisdictions to develop a profile of a target entity or individual, identifying and assessing warning signs about legal, regulatory, and reputation risk-related questions.[12] As AI becomes more widely accepted and its capabilities further advanced and refined, more law firms are likely to follow Yulchon's and Ropes & Gray's path.

AI and Global Investment

According to Statista, global revenues generated from the use of AI was approximately US$5 billion in 2017, largely the result of such applications as object identification, detection, and classification. It projected that AI revenues will explode to approximately US$90 billion by 2025.[13] By contrast, KPMG believes that investment in AI is set to rocket much higher. It estimated that AI-related investment among businesses was approximately US$12 billion in 2018 and sees AI, ML, and robotic processing automation investment reaching an incredible US$232 billion by 2025.[14]

A significant portion of that growth will occur in the investment arena. AI is already disrupting the exchange-traded fund industry and it is having spillover effects into other investment vehicles. Millions of market signals, news articles, and social media posts are processed by existing funds to produce thousands of hypothetical test portfolios that are distilled into daily trade recommendations. Yet, for all its progress and promise, AI is unlikely to entirely replace human managers and analysts. What it will do is to allow them to more efficiently manage the overwhelming amounts of market data.[15] Given the trillions of dollars of assets at stake, it is a sure bet that the investment community will be embracing AI and ML.

Autonomous Learning Investment Strategies (ALIS), referred to as the 'Third Wave' of investing, has been driven by small funds that leverage recent advances in AI and ML, the explosion of data availability, and inexpensive cloud computing to generate alpha at lower costs. ALIS is about to transform the investment process by exploiting the confluence of data, data science, cheap computing power, and intelligent humans. ALIS fund managers' brains appear to be wired differently. They are often physicists, scientists, hackers, or computer gamers with a healthy disrespect for convention, and who are poised to make today's investment Ferrari look like yesterday's horse and buggy.[16]

As a result of the deployment of extreme automation and connectivity, the Fourth Industrial Revolution should be characterized by relatively low capital intensity, but a number of traditional industries will likely be disrupted by the advances that the FIR will bring to the global marketplace. Investors may be tempted to invest in the early beneficiaries of extreme automation and connectivity–including AI and robotics–but selecting winners from the new crop of technology firms may not be a straightforward process. This proved to be the case during the Dotcom boom of the late 1990s and, more recently, via the aggressive initial valuations granted to several tech "unicorns" that subsequently came under pressure.

One could argue that it may be more prudent for investors not to invest in richly valued disruptive-focused firms and to specifically avoid some traditional industries that have already been disrupted. For example, automation is mainly seen as a threat in retail, given online competition. One trip to most malls in the developed world, with typically high rates of vacancy, proves that point. By contrast, the online food business is growing quickly but still represents less than 1% in most markets. *Picking and choosing if, when, and how to invest in an AI disrupted world will remain an art rather than a science.*

Future potential beneficiaries include firms that harness Big Data to cut costs or target sales, firms that automate Big Data analysis, and firms that keep Big Data secure. As an example, large supermarkets have long been analyzing Big Data sets collected through store card use to understand customer behavior, but improvements in AI are now enabling extreme automation in the data analysis process. Customer data on physical in-store movements, collected by tracking smartphone data, may allow for more targeted sales strategies. When linked to existing AI infrastructure in supply chains, this implies a highly intelligent and demand-flexible automation for retail supply chains, with potentially significant cost savings.

In the power sector, a combination of automated Big Data analytics and extreme connectivity should produce efficiency and profit-boosting impacts over time. A year after General Electric launched its Big Data product Predix, some 23,000 wind turbines and close to 4,000 gas turbines had already been installed. Healthcare also offers a range of investment opportunities. Extreme connectivity now allows pacemaker patients to send diagnostic data from their device, via smartphone or tablet, directly to their doctors, who can assess patient health and recommend changes to the course of treatment. Increasing computing power has also been driving down the cost of analyzing the genome and the proteome (the sequence of proteins expressed by genes).

In the financial services sector, the potential effects are mixed, with widely different implications for banks, insurers, and money transfer firms. Blockchain, a shared digital ledger system that underpins cryptocurrencies, is the ultimate product of extreme connectivity since it exists *because* of the interconnection of a large number of computers. The technology works by producing a record of ownership on a shared database that is in theory impossible to edit or forge. Dozens of banks have joined R3, a global consortium of organizations in a variety of sectors working on ways to use blockchain in financial and related markets.

Blockchain could prove a double-edged sword for the banking industry, however. On one hand, it has the potential to boost profitability in several ways. It could allow banks to cut mid-level administrative labor and costs in the industry by up to US$20 billion per year. On the other hand, the ability to process transactions directly between parties, without the need for trusted central utilities that reduce counterparty risk, clearly poses a risk to traditional banking. Extreme connectivity could therefore lead to further disintermediation in an industry that some would argue is screaming for radical change.

Insurers, which tend to be stodgy, sclerotic, and resistant to change, also surely fit that category. It is possible to envision insurance processing using extreme automation, with blockchain policies that instantly pay claims based on preset information from a trusted third party. Leading insurers have been looking into ways to use blockchain to improve access to information and cut administrative costs. It seems that even the most conservative and risk-averse sectors of the global economy see the writing on the wall and are embracing their destiny with AI and ML.[17]

AI and Governments

The debate over the future potential effects of AI has been dominated by two themes: the fear that AI exceeds human intelligence and escapes human control, and the worry that a new industrial revolution will allow machines to disrupt and replace humans in part of society. There are, naturally, other concerns, such as governments having the ability to monitor, understand, and attempt to control their citizens more closely than ever before. This is a concern shared by citizens of liberal democracies as well as authoritarian governments, for both types of governments share similar intrusive capabilities as a result of the widespread utilization of AI and ML.

AI could help provide authoritarian governments with a plausible alternative to liberal democracy–a phenomenon that is already in play around the world. This could enhance the existing propensity for renewed international competition between social systems. Some countries are already moving swiftly in this direction. For example, China has begun to construct a digital authoritarian state by using surveillance and ML tools to attempt to control its potentially restive population by creating a "social credit system" (further explored in Chapter 11). *A competition between two models of government–liberal democracy and "digital" authoritarianism–may well come to define the political globalization of the 21st century.*

While liberal democracies are unlikely to be bowled over by digital authoritarianism, a plethora of leaders and governments with authoritarian tendencies are likely to see the appeal. China's "Great Firewall" approach to the Internet has already spread to Thailand and Vietnam; Chinese digital authoritarianism experts had, as of 2018, provided support for government censors in Sri Lanka and supplied surveillance or censorship equipment to Ethiopia, Iran, Malaysia, Russia, Zambia, and Zimbabwe.

China and Russia have publicly opposed the US-led conception of a borderless global Internet, and China is using its diplomatic and market prominence to influence global technical standards and normalize the notion that domestic governments should control the Internet in ways that sharply limit individual freedom. The governments that utilize such measures tend to see them as defensive in orientation and necessary to ensure domestic control. Other governments perceive them as being in fundamental opposition to their polities.[18]

Nearly two dozen countries have already jumped into the race for AI supremacy, many of them being liberal democracies, which tend to be among the most advanced and have the widest array of resources at their disposal to run in the race. There are some noteworthy exceptions, but *the extent to which any given country will become a serious contender in the race and stand a realistic chance of reigning supreme–if even briefly–will, in large part, depend on how those resources are utilized, the extent to which they are increased over time, and whether or not a meaningful, deployable strategy is part of the mix.* It will come as no surprise to learn that many contenders either do not yet have an AI strategy or are devoting paltry sums of money toward addressing their national AI development. What follows are examples of what a wide array of countries are doing to get in, and stay in, the race.

Australia does not yet have an AI strategy, but it has announced a 4-year, US$22 million allocation in the national budget to support the development of AI. Some might view this as a laughable amount of money to devote to its own future for a country as rich as Australia, but it is a place to start. The government will create a Technology Roadmap, a Standards Framework, and a national AI Ethics Framework to support the development of AI. It will also support Cooperative Research Center projects, PhD scholarships, and other initiatives to increase the supply of AI talent in Australia. The government will also prioritize AI in its Digital Economy Strategy.

Canada was the first country to release a national AI strategy. Released in its 2017 federal budget, the Pan-Canadian AI Strategy is a five-year, US$96 million plan to invest in AI research and talent that has four goals: to increase the number of AI researchers and graduates, establish three clusters of scientific excellence, develop thought leadership on the economic, ethical, policy, and legal implications of AI, and support the national research community on AI. Canada's AI strategy is distinct from others because it is primarily a research and talent strategy. Its initiatives are all geared towards enhancing Canada's international profile as a leader in AI research and training. Its overall strategy does not include investments in strategic sectors, data and privacy, or skills development. Such considerations are separate from, rather than part of, the Pan-Canadian AI Strategy.

The Strategy for Denmark's Digital Growth, released in 2018, aims to make the country a leader in the digital revolution, but rather than focusing exclusively on advances in AI, the Strategy concentrates on AI, Big Data, and the IoT. It has three goals: to make Danish businesses the best at using digital technologies, have the best conditions in place for the digital transformation of business, and ensure that every Dane is equipped with the necessary digital skills to compete in the global digital marketplace.

AI SUPREMACY

The Strategy outlines a total of 38 initiatives that include the creation of Digital Hub Denmark (a public-private cluster for digital technologies), SME:Digital (a coordinated scheme to support the digital transformation of Danish Small and Medium-Size Enterprises (SMEs)), and the Technology Pact (a nationwide initiative to foster digital skills). Approximately US$12 million was allocated for 2018, followed by US$20 million each year until 2025, and another US$12 million in perpetuity. This seems like too little money for such a grandiose vision, but the Danish government deserves credit for at least trying to position itself to punch above its weight in the AI arena.

In 2018, the EU Commission adopted the Communication on AI, a 20-page document that lays out the EU's approach to positioning European countries to compete via AI. The Commission aims to increase the EU's utilization of AI by the public and private sectors, prepare Europeans for the socioeconomic changes brought about by AI, and ensure that an appropriate ethical and legal framework is in place. Key initiatives include a commitment to increase the EU's investment in AI from €500 million in 2017 to €1.5 billion by the end of 2020, the creation of the European AI Alliance, and a new set of AI ethics guidelines to address issues such as fairness, safety, and transparency. A High-Level Group on AI will act as the steering group for the Alliance and will prepare draft ethics guidelines for member states to consider.

Finland's government appointed a steering committee in 2017 to examine how the country can become one of the world's top competitors in the application of AI technologies. The group has released two initial reports and the government has begun to incorporate its recommendations into government policy. The first report resulted in the creation of the Finnish Center for AI, an AI accelerator pilot program, and the integration of AI in the public service. A second report gave an additional 28 policy recommendations related to four aspects of the future of work—growth and employment, the labor market, learning and skills, and ethics.

Prior to releasing its formal AI strategy, Germany's federal cabinet released a paper in 2018 that outlined the goals of the strategy. In short, the government wants to strengthen and expand German and European research in AI, focus on the transfer of research results to the private sector, and create various AI applications. Proposed initiatives to achieve this include new research centers, Franco-Germany research and development collaboration, regional cluster funding, and support for SMEs and start-ups. The proposed plan is quite comprehensive and also includes measures to attract international talent, respond to the changing nature of work, integrate AI into government services, make public data more accessible, and promote the development of transparent and ethical AI. The government wants "AI made in Germany" to become a globally recognized seal of quality.

The Indian government is focusing on how India can leverage AI, not only for economic growth but also for social inclusion. NITI Aayog, the government think tank tasked with producing a strategy, calls this approach #AIforAll. The objective is to enhance and empower Indians with the skills to find quality jobs, invest in research and sectors that can maximize economic growth *and* social impact, and scale Indian-made AI solutions to the rest of the developing world. NITI Aayog provides 30+ policy recommendations to invest in scientific research, encourage reskilling and training, accelerate the adoption of AI across the value chain, and promote ethics, privacy, and security in AI.

Its flagship initiative is a two-tiered integrated strategy to boost research in AI. New Centers of Research Excellence in AI (COREs) will focus on fundamental research and act as technology feeders for the International Center for Transformational AI (ICTAIs), which will focus on creating AI-based applications in domains of societal importance. NITI Aayong identified healthcare, agriculture, education, smart cities, and smart mobility as priority sectors that will benefit the most from a social perspective by applying AI. It recommended setting up a consortium of Ethics Councils at each CORE and ICTAI, developing sector-specific guidelines on privacy, security, and ethics, creating a National AI Marketplace to increase market discovery and reduce the time and cost of collecting data, as well as a number of other initiatives to help the overall workforce acquire skills. The government wants to establish India as an "AI Garage"—meaning that if a company can deploy AI in India, it will then be applicable to the rest of the developing world.

Unlike other national strategies, which tend to focus on research and development or private sector uptake, Italy's blueprint, released in 2018, focuses exclusively on how the government can facilitate the adoption of AI technologies in public administration. The strategy devotes a significant amount of time to the challenges of integrating AI into government services and addresses the availability of skilled employees, the role of data, legal implications, and ethics. Its recommendations include the creation of a National Competence Centre and a Trans-disciplinary Centre on AI, a national platform to promote the collection of annotated data, and measures to disseminate AI-related skills through public administration. It is unclear whether Italy's government of the moment will implement and fund the recommendations on an ongoing basis, so it should be taken with a grain of salt.

Japan was the second country to develop a national AI strategy. Its Strategic Council for AI Technology was established in 2016 to develop research and development goals and a roadmap for the industrialization of AI. The resulting plan—the AI Technology Strategy—was released in 2017 and is notable for its Industrialization Roadmap, which envisions AI as a service and organizes the development of AI into three phases: the utilization and application of data-driven AI developed in various domains, the public use of AI and data developed across various sectors, and the creation of ecosystems built through connectivity. The strategy applies this framework to three priority areas of Japan's Society 5.0 initiative—productivity, health, and mobility—and outlines policies to realize the industrialization roadmap, which include new investments in R&D, talent, public data, and start-ups.

Kenya's government announced a task force in 2018 to create a strategy that encourages the development and adoption of new technologies such as blockchain and AI. It subsequently formed a task force whose goal is to provide recommendations on how the government can leverage new technologies in the following five years. The task force will also provide milestones for 2027 and 2032 and situate the strategy in the areas of financial inclusion, cybersecurity, the election process, a single digital identity, and overall public service delivery. The strategy is not strictly about AI but the technology is one of many that the government wants to harness in the coming years.

Mexico released a white paper in 2018 that laid out the foundations for a national AI strategy. The paper surveyed the current state of AI in Mexico and described potential use cases for AI at the national and regional levels. The report produced a set of recommendations grouped into five categories: government and public services, data and digital infrastructure, research and development, capacity, skills and education, and ethics. It is unclear when, if, or how the report's recommendations will be implemented.

In 2017, Singapore launched a five-year, US$110 million national program (AI Singapore) to enhance the city-state's capabilities in AI. As a government-wide partnership involving 6 different organizations, its goals are to invest in the next wave of AI research, address major societal and economic challenges, and broaden adoption and use of AI within industry. The program consists of four key initiatives: First, Fundamental AI Research funds scientific research that will contribute to the other pillars of AI Singapore. Second, Grand Challenges supports the work of multi-disciplinary teams that provide innovative solutions to major challenges Singapore and the world face (the program currently focuses on health, urban solutions, and finance). Third, 100 Experiments funds scalable AI solutions to industry-identified problems. Fourth, AI Apprenticeship is a 9-month structured program to foster a new cohort of AI talent in Singapore. It later announced an Advisory Council on the Ethical Use of AI and Data to help the Government develop standards and governance frameworks to apply ethics to AI.

The South Korean government has a 5-year plan and plans to invest approximately US$2 billion, to strengthen the country's R&D in AI. It will first secure AI talent by establishing 6 graduate schools in AI by 2022, with the goal of training 5,000 AI specialists (1,400 AI researchers and 3,600 data management specialists). It also has an initiative to train 600 people in AI to address the immediate short-term need for AI talent. It will then fund large scale projects in national defense, medicine, and public safety, and will start an AI R&D challenge similar to that begun by the DARPA. Finally, the government will invest in infrastructure to support the development of AI start-ups and SMEs, which includes funding for the creation of an AI semiconductor by 2029 and an AI-oriented start-up incubator to support emerging AI businesses.

In 2018, Taiwan's government announced the Taiwan AI Action Plan as part of a larger strategy to use Taiwan's information technology and semiconductor industries to develop new smart technologies. The Plan, which has an annual budget of about US$326 million for four years, has five key initiatives: First, the plan involves a talent program designed to cultivate 1,000 advanced AI researchers and 10,000 AI-related professionals by 2021. The government plans to actively recruit global AI talent and make it easier for them to start work in Taiwan. Second, the government plans to implement a pilot project based on the DARPA to focus R&D into niche advantages for industrial development. Third, an AI International Innovation Hub is being constructed with the aim of fostering 100 AI-related startups. Fourth, policymakers are testing open data fields to support development of intelligent applications. Fifth, the plan is to integrate AI technologies into the government's larger 5+2 industrial innovation initiative.

The British government released its AI Sector Deal in 2018, part of its larger industrial strategy that aims to position the UK as a global leader in AI. It is quite comprehensive, with policies to support public and private R&D, invest in science, technology, engineering, and mathematics education, improve digital infrastructure, develop AI talent, and lead the global conversation on data ethics. As part of the estimated £1 billion that will be spent on the project, in excess of £300 million in private sector investment is to be made from domestic and foreign technology companies and think tanks.

As US citizens, it pains us to say that, unlike a plethora of other countries, the US government does not have a coordinated national strategy either to increase AI investment or respond to the societal challenges of AI. During the final months of the Obama administration, the White House laid the foundation for a US strategy in three separate reports; however, the Trump administration has taken a markedly different, free market-oriented approach to AI. In 2018, the White House invited industry, academia, and government representatives to a summit on AI where it was announced that Trump's approach to AI has 4 goals: to maintain American leadership in AI, support the American worker, promote public R&D, and remove barriers to innovation. A Select Committee on AI was formed to advise the White House on interagency AI R&D priorities and to consider the creation of federal partnerships with industry and academia. It is unclear how much the US government intends to invest in AI R&D[19], in which sectors, or under what time frame. While much of the rest of the world seems to be barreling ahead with some bold AI initiatives, the US appears to be asleep at the wheel.

What is apparent, given this partial listing of government initiatives, is that most countries have only just begun to seriously think about their own AI future, with the majority of countries noted having only announced such initiatives in 2017 or 2018. Given that it will take most governments years to determine their path, approve funding, and execute those intentions, the two countries that appear to be the best positioned to leap forward in the coming decade are China and South Korea. Both are light years ahead of the competition in terms of intellectual capital and fiscal resources being devoted to the task on a grand scale, and only China is devoting massive, sustained resources toward achieving AI supremacy. That is why we have devoted an entire chapter to China's quest, and why China is depicted on the cover of this book as being in the clear lead in the race for AI supremacy.

A State of Metamorphosis

A new form of globalization, driven by the exponential progress of silicon, is creating massive disruptions in economies throughout the world. The world's largest media company (Facebook) has no journalists or content producers. Its biggest hospitality firm (Airbnb) has no hotel rooms. The dominant taxi company in the world (Uber) has no cars. The biggest currency repository in the world is driven by cryptocurrencies and has no buildings or physical safes. All are driven by software, which is based on knowledge and processes captured by automation. Ultimately, AI is just a more advanced type of software that is propelling us deeper into a virtual world.

The greatest near-term challenge we face is how to transition from the current economic model—driven, for instance, by conventional means of manufacturing and fossil fuels—into a new model driven by technological achievement that was until recently merely the realm of science fiction. How will we transition from our collective familiarity and comfort level with tangible, physical goods and environments to a world dominated by what cannot necessarily be seen or felt? We are already transitioning into the cyber world, where virtual reality is not only upon us, but is sought after by many of us. We are, strangely, drawn to this bold new world because it tantalizes us with possibilities. The AI world that awaits us will do much the same.

The author Wolfgang Hoeschele has given us a glimpse of how a world dominated by advanced AI may be organized. He believes that in such a world there will be a more prominent role for shared property, individual and community-level self-reliance, and participatory decision-making. Decentralized production may become the norm and marginal production costs could be reduced to near zero, creating sources of abundance. But sources of abundance can also create new forms of relative scarcity. For example, in a world of abundance, media (i.e. video, music, and literature) could produce a scarcity of attention and a predominance of robots in the work place could produce a scarcity of efficient human workers.

The "economics of abundance" refers to a mode of operating production that minimizes waste because decentralized systems are better at matching actual needs and resources. AI can provide superior coordination and planning capabilities so that value moves toward societies' edges and become the point of contact for consumers who may now be on the fringes of society, only to bring them into the center. AI may allow for the more efficient distribution of excess energy where it can be used more effectively and might reduce the cost of managing financial instruments, which should minimize the cost of running businesses.[20]

While this "vision" may seem far-fetched at this particular juncture, such a reality may not be all that far off. It seems clear that dramatic change, driven by AI and ML, is indeed coming. We appear to be at the beginning end of the runway in an AI-dominated world. The questions we must answer concern when it will formally arrive, how the change it will usher in will be managed, and what impact it will have on individuals, businesses, economies, and societies.

Conventional wisdom suggests that AI will continue to benefit higher-skilled workers with a greater degree of flexibility, creativity, and strong problem-solving skills, but it is certainly possible that AI-powered robots could increasingly *displace* highly educated and skilled professionals, such as doctors, architects, and even computer programmers. Much more thought and research needs to be devoted to exploring the linkages between the technology revolution and other important global trends, including demographic changes such as aging and migration, climate change, and sustainable development.[21] Many of these topics have either not even been broached yet or have only begun to be the subject of meaningful discussion in global fora.

While it seems clear that the growing ability of AI to autonomously solve complex problems could fundamentally reshape our economies and societies, the truth is that the impact AI may have on a whole host of issues will remain unknown for many years to come. Even when answers may appear to be apparent, they are unlikely to endure for a great length of time, for *AI is akin to an amoeba that is in a constant state of metamorphosis, forever changing its shape and adjusting to its surroundings.*

AI has the potential to dramatically change how governments and citizens interact, how businesses and consumers coexist, and how some of the world's most intractable problems are addressed and resolved. Globalization, as we know it today, will also change, but it will not disappear, for the world will only become more interconnected through the widespread utilization of AI. AI will likely play a generally positive role in its evolution, but much of how any of this transpires in a positive direction will depend on the extent to which humans have the foresight, devote the resources, and skillfully deploy strategies to cope with and embrace AI.

3. Jobs and Manufacturing

As Klaus Schwab has noted[1], the world is currently in the midst of its Fourth Industrial Revolution. The first was the advent of factory-based production, the second was the rise of the automobile, and the third was the integration of computers into our daily lives. As a result of these revolutions and their ancillary impacts, products are now manufactured faster, at lower cost, and with greater consistency. At the same time, the products being developed are increasingly complex but providing greater value to consumers. The present industrial revolution is being powered by the IoT, robotics and smart manufacturing, in a multitude of dimensions.

AI has already become an integral part of the global manufacturing process. It is enabling manufacturers to simulate the design process and the assembly line before a product is actually created, which reduces manufacturing time and cost while ensuring that the manufacturing process delivers a product in the precise manner intended. Three-dimensional (3D) printing is significantly reducing delivery time for replacement parts and transportation, cutting waste by recycling plastic, and transforming the construction business. The new generation of robotics is not only much easier to program but easier to use, and with capabilities such as voice and image recognition that re-create increasingly complex human tasks.

Smart sensors can convert data into different units of measurement, communicate with other machines, record feedback, and shut off devices if safety or performance issues arise. IoT functionality can track and analyze production quotas, consolidate control rooms, and create models of predictive maintenance. It also allows us to get the right information at the right time to make the right decisions. Cloud computing enables companies to extract and analyze information that affects the production line. Data from augmented and virtual reality is having a major impact on research and development, giving consumers more of what they want, faster than ever, and more cheaply.

Building a better manufacturing sector with AI and ML—using augmented and virtual reality, robotics, data analysis, and smart equipment—is transforming how we do what we do. Increased productivity should translate into a bright future for global manufacturing, creating vast efficiencies that will translate into greater profits, more jobs, and healthier economies.[2] It is a multifaceted landscape that is anything but black and white in terms of possibilities, outcomes, and potential costs, but there have been few times in history when transformative technologies have had as much opportunity to impact so many people and businesses in so many ways. So much remains unknown about just how, and how quickly, this revolution will continue to unfold that we are taking a bit of a collective leap into the dark in terms of imagining just what the manufacturing landscape will look like one or two decades from now, much less 100 years from now. What can be said with some certainty, however, is that this transformative process is really only just beginning, and the possible implications are both exhilarating and a source of concern.

Challenges Along the Way Toward AI Organizational Fluency

Embracing and integrating AI into the manufacturing process is a challenge for any organization but, for large companies, the culture change required to implement AI is often daunting. What some organizations may view as an essential means of remaining competitive, streamlining production processes, and cutting costs, others may view as part of a larger organizational transformation process, meant to reinvent a company. Some, of course, view it as both. There are as many potential permutations associated with embracing AI in the manufacturing process as there are potential applications for doing so.

As an example, when Airbus started production of its A350 aircraft, its plan was to increase the production rate of the aircraft faster than had been achieved with any other of its previous aircraft, so it turned to AI. Airbus combined data from previous production programs, with fuzzy matching (linking records with computer assisted translation), and a self-learning algorithm, to identify patterns in earlier production problems. The system matched up to 70% of the production disruptions in near real-time by taking into account all contextual information and providing immediate recommendations for courses of future action. Having experienced a similar issue in the past allowed Airbus to shorten the amount of time it took to address the disruptions by more than a third.

Industrial companies are anticipating the largest operational impact from AI. British Petroleum (BP), for example, chose to augment its human skills foundation with AI in order to improve field operations. Its AI-driven BP Advisor takes data produced by the company's drilling systems and generates advice for its engineers about how to adjust drilling parameters to remain in the optimum zone, while alerting them of potential operational risks. BP also automates root-cause failure analysis so that its AI system trains itself over time to be able to rapidly assess a situation and move from description to prediction to prescription.

Service providers are similarly benefitting. China's Ping An Insurance Company, the country's second-largest insurer, uses AI to improve customer service across its insurance and financial services portfolio. It offers an online loan in just 3 minutes by deploying a customer scoring tool that uses an internally developed AI-based face-recognition capability that is more accurate than humans. The tool has verified more than 300 million faces and complements Ping An's cognitive AI capabilities, including voice and imaging recognition.

Despite AI's proven applicability and successes, organizations large and small must grapple with competing investment priorities, corporate culture, and conventional modes of thinking and decision-making that stand in the way of broader adoption of AI throughout the business world. There is an inherent conflict between those encumbered by legacy business models and systems versus others who see a bold future to be had by more fully embracing AI. Some of the biggest hurdles implied in doing so include grappling with the practicalities of acquiring and developing the requisite AI talent as well as managing competing internal priorities. There is a war for AI talent, with businesses competing with each other and with government entities to identify and hire the best talent. Apart from allocating funding and resources to get the right people and integrate them into the workplace in the most appropriate and efficient manner, the need to consider how the presence of machines will impact orientation to change within an organization is an equally large challenge.

Generating business value from AI is directly connected to effective training of AI algorithms, yet few organizations understand how to do so. The starting point for many AI applications is "naked" algorithms that only become "intelligent" upon being trained with company-specific data. The ability to successfully train them depends on having well-developed information systems that can collate relevant training data. While some organizations have robust data and analytics infrastructures as well as a broad understanding of what is required to develop such data, others struggle because of limited experience with analytics or siloed data. Data collection and preparation are typically the most time-consuming activities associated with developing AI-based applications—much more so than selecting or fine tuning a model. For every new project that is rolled out, a new investment is required to prepare the data for deployment, even when previously used information can be integrated.

Inside organizations some data will be proprietary, while other data may be widely available but is fragmented across data sources, requiring consolidation and agreements with multiple stakeholders to gain complete access. In some cases, the ownership of important data may be uncertain or contested. Even if the organization owns the data it actually needs, fragmentation across multiple systems can hinder access. Many companies may come to the conclusion that they either do not have the in-house capability necessary to tackle such issues, or that it is not worth the amount of time, money, and resources necessary to manage it themselves. They may therefore choose to use external resources to get the job done.

If companies choose to rely largely on external support to create, maintain, or expand their AI platforms, they must still rely on some of their own people familiar with the issues to handle the data and remain a step ahead of evolving opportunities. External vendors may not only insist on using a company's infrastructure but their technology, which can create a conflict of interest. The tech sector is sensitive to this and has begun to offer companies technology tools to use with their own proprietary data. Many such AI algorithms and tools are already widely available in the public domain, including Google's TensorFlow, GitHub, and application programming interfaces from tech vendors.

Companies also face a variety of managerial challenges in the process of introducing AI into their organizations. Some companies have the right corporate culture to enable a smooth transition—such as leaders with the right amount of vision, a high degree of openness and ability to change, and close alignment between business and technology strategy—to allow for the type of collaboration required to integrate AI efficiently and effectively. Others will never even come close. While not every frontline manager will need to understand the difference between deep and shallow learning within neural networks, a basic theoretical understanding of the applications and what is required to roll it out is necessary. Ultimately, every manager in every organization desiring to keep pace with others in their industry needs to develop an intuitive understanding of the basic applications of AI.

Adopting AI across an enterprise places a premium on organizational flexibility that allows for new forms of collaboration among project teams that may be composed of both humans and machines. There are many potential approaches to developing AI organizational fluency, ranging from modification of the status quo to complete reinvention of a firm. In some cases, a hybrid model may make the most sense. The Teachers Insurance and Annuity Association-College Retirement Equities Fund, for example, has an analytics center of excellence, used by a variety of decentralized groups, providing expertise, guidance, and direction to internal teams. The biggest challenges may not prove to be technical but, rather, getting otherwise siloed teams to find a way to work well together and relinquish some of the control they have over their own data and processes.[3]

Making AI an integral part of an organization often requires radical thinking. It takes foresight, vision, and a willingness to step outside the organizational comfort zone to make decisions. Embracing AI may, at the outset, take guts, determination, and willpower to deploy it. However, if it is done carefully and thoughtfully, and sufficient resources are devoted at the outset to make the AI journey successful, then there is little doubt it will be a positive transformational process. That said, there are a plethora of risks involved, and there will inevitably be stumbles along the way. Yet, for many firms, there is no substitute for embracing AI. It is obviously the future.

Cobots and Paradoxes

Many of our preconceptions about the prior industrial revolutions were misplaced; the same will be true about this one. For example, when the automated teller machine (ATM) first started being used in 1969, people naturally presumed that it would translate into the death of the bank teller and dramatic shrinkage in the number of bank branches. However, contrary to what common sense and conventional wisdom dictated at the time, the number of bank tellers actually doubled, from about 250,000 in 1972 to roughly 500,000 in 2016, with approximately 100,000 of those jobs having been created since 2000. Rather than eliminating the net number of teller jobs, as many had feared and expected, the opposite turned out to be the case. While the number of tellers per branch has indeed fallen by about a third since the ATM's introduction, it became less expensive for banks to open new branches, and the number of branches in the US has risen by about 40% since 1969. The net result was more branches and more tellers. In the process, "tellers" spent less time handling money and more time selling the bank's products.

The advent of the tractor revolutionized farming, yet agricultural production levels are higher than ever and agribusiness is bigger than ever. The widespread use of assembly line and robot-enhanced production did not mean the end of manufacturing but, rather, a boom in global manufacturing. During the rise of automated means of production, the percentage of Americans participating in the workforce actually increased, from 52% in 1890 to 67% in 1990. In large part *as a result of automation, more jobs require a greater multiplicity of demands from workers*. It turns out that the automation of some jobs can make them more important and increases their economic value, while the opposite is true for other types of jobs.

A comprehensive study of 19 developed countries, addressing the period between 1970 and 2007[4], proved that a heightened degree of technology and corresponding productivity growth was not a threat to employment but actually enhanced employment. The widely held perception that industries experiencing rising labor productivity also experience falling employment is not true. While it must be said that not all sectors or industries had, or will have, similar net job gains, if productivity growth in one sector generates sufficiently large positive impacts to affect employment growth elsewhere in an economy, *productivity growth as a result of enhanced technology can indeed be employment-augmenting, even if it ends up reducing employment in the sector in which it occurs.*

Improving the value in one part of a supply or production chain increases the value of the other linkages in the chain. Yes, ATMs can dispense cash and accept deposits more efficiently than tellers, but the individuals who used to be solely tellers have become an inherently more valuable part of a bank's supply chain in the process. In 1900, around 40% of all US jobs were on farms. Today that figure is less than 2%, even though the US population was 76 million in 1900 and was 325 million in 2017. Technology can obviously eliminate jobs—by the tens of millions—but they also create jobs. Many of the products we take for granted today—such as smart phones and sport utility vehicles—either did not exist in the past or were expensive to obtain. That is no longer true today, of course.

When considering the impact of AI on jobs and manufacturing, the issue is not that there are too few jobs. Just in the US, more than 14 million new jobs were created between 2010 and 2016 and, by 2018, its unemployment rate reached low levels not seen in decades. The issue is the number of higher paying versus lower paying jobs. While technology leverages our expertise, judgement, and creativity, we are constantly adjusting our expectations and demands to the rapid pace of technological change. That change poses a real threat to economic mobility and the increasing polarization of societies.[5] Fortunately, in the developed world, we can afford to invest in ourselves. Our high level of productivity—due in large part to automation—means that we have the means to make the investments necessary to both address our economic challenges and meet our seemingly insatiable need to purchase and integrate the latest technology into our lives.

There is no way that a subject as multifaceted and complex as the impact of technology on society, the economy, jobs, and manufacturing could result in a simplistic black or white view of the world. While technology clearly boosts productivity and makes societies wealthier, it also eliminates the need for many types of jobs and leaves many workers at the lower end of the scale worse off, particularly if they are not retrained or utilized in a different way than they were previously. In the US (and elsewhere in the world), it is widely acknowledged that median income is failing to rise even as GDP soars. It is one of the greatest paradoxes of our era.

Economists are divided about whether much of the blame resides with new technologies. Many labor economists say the data is, at best, far from conclusive, and point to global trade and investment patterns, the Great Recession, or government policy. What is indisputable, however, is that computers have increasingly taken over such tasks as bookkeeping, clerical work, and repetitive production jobs in manufacturing, where many in the middle class derived their income. Yet, *higher-paying jobs requiring creativity and problem-solving skills, often assisted by computers, have proliferated. That said, demand has increased for lower skilled restaurant workers, janitors, home health aides, and others providing services that cannot be automated.* The result is a polarization of the workforce and a gradual hollowing out of the middle class. That is not the same as saying that technology is affecting the total number of jobs; the nature of the work force may change without there necessarily being a significant change in employment rates.

If new technologies are inhibiting job creation, history suggests that this may be a temporary phenomenon. As workers adjust their skills and entrepreneurs create opportunities based on the new technologies, the number of jobs will eventually rebound. Is this time different, and is long-term involuntary unemployment affecting millions of workers likely to be a net result? Since the Industrial Revolution began in the 1700s, improvements in technology have indeed changed the nature of work and destroyed some types of jobs in the process (the example of the agricultural work force in the US cited above being a prime example). Likewise, the proportion of Americans employed in manufacturing has dropped from 30% in the post–World War II years to around 10% today, due, in part, to increasing automation since the 1980s.

Still, there is no historical pattern that proves that such shifts have led to a net decrease in jobs over an extended period of time. While it often takes decades for workers to acquire the expertise needed to satisfy new types of employment, there is no identifiable long-term trend where fewer jobs are created over time. Over the long term, employment rates tend to remain relatively stable. The question is whether economic history is an appropriate guide to our bold new future. Will the job disruptions being caused by technology ultimately be temporary as the workforce adapts, or will automated processes and robots with superhuman skills make humans largely irrelevant?

Automation has historically been proven to make workers more efficient rather than replace them. Rising productivity often implies that businesses can be as productive (if not more so) with fewer employees, but it can also enable businesses to expand production with the same number of workers. For example, Kiva robot, created and sold by Kiva Systems, was a startup founded in 2002 and bought by Amazon in 2012. Its robots were designed to scurry across large warehouses, fetch racks of ordered goods, and deliver the products to humans who package the orders (sometimes referred to as "cobots" because of their interactive orientation toward humans). A warehouse equipped with Kiva robots can process up to four times as many orders as an unautomated warehouse, where workers might spend as much as 70% of their time walking around to retrieve goods. Not surprisingly for e-commerce retailers, most of Kiva's users are growing so rapidly they cannot hire people fast enough.

It could well be that cobots will become even more widely used by organizations to augment their human counterparts in the future. An example of friendlier, more flexible robots meant to work with humans is Rethink's Baxter, which requires minimal training to perform simple tasks such as picking up objects and moving them into a box. Baxter was intended to be used in relatively small manufacturing facilities where conventional industrial robots cost too much to operate profitably and pose too much danger to workers. The idea is to have the robots take care of dull, repetitive, low risk tasks that no one wants to do. Baxter is specifically manufactured to exude a desire to please. Its "eyebrows" rise quizzically when it is puzzled and its arms submissively and gently retreat when bumped. The robots created at Kiva and Rethink were cleverly designed and built to work with people and specifically designed to enhance worker productivity.[6] Far from either inhibiting human performance or making businesses less productive, the use of robots has generally done just the opposite.

Inflection Points

The installation of robots has been growing globally by between 2% and 3% annually for around a decade, roughly in line with growth in manufacturing output, and there are approximately 1.5 million industrial robots in use around the world today. Yet, their use has remained extremely limited in most industries. Even in industrialized countries such as Italy, for example, only one-half of 1% of tasks has been automated in the chemical and primary-metals industries, though it is estimated that up to 16% could be automated today. Deployment rates have also remained surprisingly low in industries that have long been at the forefront of automation. Fewer than 8% of tasks in the US transportation equipment industry are automated, compared with a potential of 53%.

What accounts for these low deployment rates? Part of the answer is that the use of industrial robots is highly concentrated. Nearly three-quarters of all robots operate in only five industrial groupings—computers and electronic products, electrical equipment, appliances and components, transportation equipment, and machinery. Approximately 80% of the robots sold each year are deployed in just five countries: China, Germany, Japan, South Korea, and the US. Since robotic systems have historically been expensive to own and operate, they exist primarily in large factories owned by corporations with big capital budgets.

So, while there have been a number of economic and technical barriers to wider adoption of robots in industry, a dramatic rise in the use of advanced robotics is imminent. Growth in installed robotics could accelerate to reach as much as 10% annually into the next decade, by which time such installations will surpass 4 million. The Boston Consulting Group has estimated that annual shipments of robots will leap from about 200,000 units in 2014 to as many as 700,000 by 2025. Even if that were to happen by that time, those installations will represent only about a quarter of all manufacturing tasks, and far less in other industries and major manufacturing economies. The growth potential over the longer term should remain immense.

Three major trends are nudging global industries toward an inflection point that will make advanced industrial robots much more commonplace in industry: greater cost-effectiveness for robots when compared with human labor, technological advances that are removing barriers to adoption in key sectors, and the arrival of systems that smaller manufacturers can more readily afford and more easily use. Let us examine each of these.

For many organizations, the biggest reason not to replace manual labor with robots is purely economic. Most organizations begin to make greater investment in automation when the cost of employing human labor rises high enough above the cost of owning and operating robotic systems to make human labor less cost-effective. Even with the ongoing rise in wages around the world, the gap has remained wide enough to prevent mass deployment. At the same time, the economics of advanced robotics are rapidly improving. For example, the total cost of purchasing and deploying a robotics system for spot welding in the US automotive industry plunged from an average of $182,000 in 2005 to $133,000 in 2014 (not adjusted for inflation). By 2025, the total cost is projected to drop by approximately another 22%, to around $100,000.

In addition, the price of robotics hardware and software, which account for only a quarter of that total cost, is about 40% lower than it was a decade ago. The cost of systems engineering—which includes installing, programming, and integrating a robotics system into a factory—has declined even more. In 2005, the average systems-engineering costs of a spot-welding robot amounted to $81,000. By 2015, those costs were around $46,000, on average, and are likely to keep dropping. The costs of peripheral equipment, such as sensors, displays, and expensive safety structures that protect workers, and that together typically cost more than the robots themselves, are plunging as well. In fact, safety barriers may not be required at all for many next-generation robots.

As costs have been declining, the performance of robotics systems has been improving by approximately 5% per year. Taken together, the changes in price and performance for spot welding have, for instance, translated into an annual improvement of 8% in the sectoral cost of deploying robotics. An investment of $100,000 in 2015 bought a robotics system capable of performing more than twice as much work as a robotics system costing the same amount as in 2005. This pace of improvement in price and performance should be sustainable for the foreseeable future. Robotics systems are thus becoming an economically viable alternative to human labor in more and more industries.

Other industries are also rapidly approaching inflection points. In 2015, the US electronics and electrical-equipment manufacturing industries had deployed about 3,300 industrial robots, many of which were relatively basic and designed to perform simple tasks. Manufacturers gradually began to add more versatile (and expensive) robots to take on increasingly complex tasks. The cost of a "generic" robotics system with a high degree of flexibility might have been about $28 per hour to operate a few years ago. *By 2020, this cost is projected to fall to less than $20 per hour, which would be below the average human worker's wage and will enable a significant increase in the number of tasks that can be automated.* As a result, the percentage of tasks that will be performed by advanced robots should reach as much as 26% by the end of this decade.

Naturally, some industries will adopt robotics more slowly than others. Tasks remain more difficult to automate in furniture manufacturing, for example, where the potential economic payoff of installing robots remains years away. The adoption of robots in that industry may not begin in earnest until the next decade, but that day is coming. The truth is, *the technical capabilities of most industrial robots remains rather limited*. Traditional robots are typically fixed in a single location and can work effectively only with objects of uniform size while moving at a predetermined speed. Most robots can process images and detect features on objects but lack logic capabilities to make decisions related to those objects.

AI SUPREMACY

Other impediments have stood in the way of higher levels of deployment. Heavy and expensive safety devices have been required to protect workers from robots in operation. If the robots are mobile, collision avoidance systems are often necessary as well. Such impediments have greatly limited the scope of robots' usefulness for many industrial applications. That said, advanced robotics systems can now perform tasks and work in environments that are far less limiting. As costs drop, companies can afford to buy robots with arms that rotate freely, giving them more flexibility to handle a wider array of objects. Robots also now have more sophisticated sensors to see and feel objects and can be more quickly and easily reprogrammed to perform different tasks.

Advanced robots are more "intelligent", as well, and can apply logic to make decisions about objects, judge quality, and provide feedback to other parts of a production system through information technology. Safety barricades are not required for many advanced robots, which means they can often work more easily along-side humans, as can some lower-cost robots. These enhancements have greatly expanded the utility of robots in a range of industries. For example, cutting and trimming meat is very challenging for traditional robots because pieces of meat come in different shapes and sizes, and their properties can vary significantly. A German company solved this problem using a 3D visual-inspection system that enables advanced robots to trim and cut.

A traditional obstacle to wider robotics use in the fabricated-metals industry has been the difficulty of translating computer-aided design drawings for complex jobs, such as grinding a gear to remove defects, into instructions for a robot. Computer-aided-manufacturing software tools such as Robotmaster integrate robot programming, simulation, and code generation to make doing so much easier. Likewise, electronics manufacturing has presented unique challenges for automation. Robots are often used to place components onto flat surfaces such as circuit boards, but it has been difficult to design a robot that can install extremely small parts, such as connectors, at odd angles with high degrees of precision and at very high speeds. As a result, the industry remains highly labor intensive. In response, a Japanese robotics firm has demonstrated a high-speed robot that rotates along six axes and has the dexterity to perform such tasks.

Most advanced robots can be quickly reprogrammed for new jobs, giving manufacturers the flexibility to produce small batches of customized products without additional capital investment. One wiring accessory manufacturer needed to increase throughput at one of its factories in Europe. The company wanted machines that could pick parts from a bin and orient them properly for assembly, a task that conventional robots could not perform at high speed, or with precision. It also wanted robots that could work on different product configurations with minimal adjustment, so it installed several robots on a new assembly line linked to a vision system equipped with digital cameras. It took only 10 minutes to adjust the system to pick and place parts for different products, which were often changed up to 30 times each week. As a result of the new system, the company boosted throughput of each shift by a factor of about nine, and only one worker was required to oversee the entire robotics assembly line.

The steady improvements in cost, performance, and functionality of robotics systems are driving another force in the next manufacturing revolution: the wider adoption of robots by small and mid-sized manufacturers. Until recently, such systems have been prohibitively expensive and unnecessarily complex for enterprises with limited budgets and engineering resources. The new generation of innovative systems is putting robots within the financial reach of small enterprises. Universal Robots, for example, markets the UR5, an industrial robot designed for material handling and assembly, with a base price of $34,000. Its two arms rotate along six axes, resulting in remarkable flexibility for a machine of that price.

The total cost of installing such low-cost generic robots generally ranges from $50,000 to $100,000, when accounting for associated costs and the project management required make it all work. Two of Rethink Robotics' robots–Baxter and the company's high-performance robot, Sawyer–are designed for precision applications such as machine tending and circuit board testing. Each costs about $40,000, including accessories, warranties, and installation. Since they can be redeployed quickly and easily across multiple product lines, the Rethink robots can be used in high-mix environments that are impractical for conventional industrial robots. In addition to their low cost, the UR5 and Rethink robots can easily be moved by hand and repurposed by workers with no programming experience. They can safely work next to humans on the same production line without expensive barricades or other protective equipment, eliminating a major cost associated with conventional industrial robots.

One small California company wanted an inexpensive automation solution that could easily be moved to perform different production tasks in the factory, so it purchased a UR5. One of the robot's first jobs was to help fill an order to make 700 valves per month. With its existing computer-numerical-control milling machine, the company could produce 400 valves per month in two work shifts. By using the UR5 robotic arm to feed parts into and remove them from the milling machine around the clock, the company was able to fill the order in 11 days while increasing production capacity by 30%. By using the UR5 for another job—to feed tubes into a bending machine—the company was able to produce 1,500 pieces in just four hours, a task that otherwise would have taken two to three days. The company estimates that it achieved a return on its investment in a few months.

As economic and technical barriers continue to fall, robots will become even more accessible to a greater variety of companies for a wider array of applications. Production efficiencies will spread beyond individual factories through entire supply chains, industries, and national economies.[7] Those companies that have already integrated robotics systems into their manufacturing processes will only continue to benefit from economies of scale, while those that hesitate will become less competitive and fall further behind over time.

The Pace of Automation and Adaptation

Automation has the potential to enhance productivity, raise throughput, improve predictions, outcomes, accuracy, and optimization, as well expand the discovery of new solutions in massively complex areas such as synthetic biology and material science. Eventually, manufacturing automation will affect all countries, sectors, and work activities in some way. Machines and algorithms are already playing a much larger role in the workplace than they did even 5 years ago. A range of factors will influence the pace and extent of automation of the work place going forward, including technical feasibility, the cost of developing and deploying solutions, labor market dynamics, economic benefits, and social and regulatory acceptance. Just how quickly those who have not yet been directly impacted by the automation process will be affected will ultimately be determined by the way each of these variables unfolds.

Full adoption across an entire sector of any technology, particularly those that are integrated into the workplace, takes a long time. Based on a review of the historical rate of adoption of previous technologies, the time from commercial availability to 90% adoption ranges from approximately 8 to 28 years (to achieve a 50% adoption rate, the range is between about 5 and 16 years). This lag applies not only to hardware-based technologies that are capital-intensive and require physical installation but also to online technologies. Those technologies with the fastest adoption rates include stents, airbags, MRIs, TVs, and online travel booking. The slower adoption category includes cellphones, personal computers, dishwashers, and pacemakers.[8]

McKinsey estimates that about half of the activities that people are paid almost $15 trillion to do in the global economy have the potential to be automated by adapting currently available technology. By their calculation, only a small proportion of all occupations (about 5%) consist solely of activities that are completely automatable today, but it found that around 30% of the activities in 60% of all occupations could be automated. This means that many workers will function alongside rapidly evolving machines, which will require that worker skills also evolve. This rapid evolution in the nature of work will ultimately affect a wide swathe of the work force.

The scale of shifts in the labor force over many decades that automation technologies will likely unleash is of a similar order of magnitude to the long-term technology-enabled shifts in the developed countries' workforces as they were when most workers transitioned from farms to factories and service jobs. One third of new jobs created in the US over the past quarter century did not previously exist (or barely existed previously) in areas such as IT development, hardware manufacturing, app creation, and IT systems management.[9]

The scope of change in manufacturing is already compelling many firms to reshape their production footprints and devise new supply chain models. A glimpse of the future of intelligent manufacturing may be found at Siemens'

Electronic Works Amberg. Production there is managed and controlled via programmable logic circuits through a virtual factory that replicates the factory floor. Through the use of bar codes, products communicate with the machines that make them, and the machines communicate among themselves to replenish parts and identify problems. Nearly 75% of the production process is fully automated, with 99.99988% of the logic circuits being defect-free.

AI is enabling some manufacturers to integrate production and client feedback in real-time to refine product design. AI-based tools can help suppliers provide better accountability throughout the supply chain, which helps aerospace manufacturers, for instance, conform to safety regulations. DL and network theory are helping engineering development teams better optimize their composition and key performance indicators in real-time. One aerospace manufacturer applied advanced algorithms to spending data in order to search for discrepancies between what it paid suppliers and what it actually owed them. The company also sought data that would help it understand differences across suppliers to develop effective procurement levers and reduce administrative costs.

In the late 2000s, GE broke new ground through its "power by the hour" service concept, which calls for operators to pay for airplane engines only when planes are actually flying. This business model is becoming increasingly dominant in aerospace manufacturing. AI technologies are ideally suited to leverage the vast amount of data collected from operating engines, especially since the manual prioritization of maintenance, repair, and overhaul sales leads is cumbersome, resource-heavy, and not always optimized. This results in unnecessarily keeping aircraft grounded, not generating revenue, and inefficiently allocating expensive engineering labor. With real-time feedback between plane and ground support facilities, ML algorithms will be able to make judgment calls about when to deploy drones and smart microrobots to conduct aircraft inspections and quality checks on intermediate and final products.

Manufacturers with the best algorithms and data will be able to offer service contracts that promise better performance at a lower cost. One firm reported a profit improvement of approximately €300 million by using ML to forecast 10 years of repair events for a fleet of more than 17,000 commercial aircraft. Advanced analytics and AI tools are also used to optimize processes around unplanned maintenance events, allowing the manufacturer to respond more effectively to disruptions. GE turned to Kaggle, a platform for predictive modeling and analytics competitions, to design new routing and ML algorithms for flight planning that optimized fuel consumption by considering variables such as weather patterns, wind, and airspace restraints. The selected algorithm showed a 12% improvement in efficiency over actual flight data.

In the future, AI tools will shift predictive analytics to cognitive assessments. The algorithms themselves will discover new rules, automatically optimizing sales and servicing for manufacturers. Preventive maintenance will be conducted with real-time feedback between plane and ground support facilities, for example, using ML and virtual assistants to identify issues, and using drones for inspections and quality checks on products. Robots the size of insects will be able to inspect airframes without removing panels and identify common defects with computer vision and ML. Smart flight systems will use predictive models based on DL technology to provide real-time feedback to flight crews, helping them to optimize energy consumption throughout flight[10], while virtual assistants will assume roles not even imagined today.

Smart Manufacturing

"Smart manufacturing"–where AI meets the IoT–is projected to grow rapidly into the next decade, with the global marketplace projected to be worth hundreds of billions of dollars in just 5 years' time. Smart manufacturing involves fully integrated manufacturing systems that respond in real-time to changing conditions and demands in a supply network, factory, or distribution system. Most of the major companies making ML tools for manufacturing are also using the same tools in their own manufacturing processes, which makes them the developer and first customers for their own tools.

The increasing prevalence of sensors in machinery, vehicles, production plants, and other hard equipment means physical equipment can be more easily digitized and monitored by AI. A variety of heavy industries (such as aviation and oil and gas) have been using GE's Predix operating system to process the historic performance data of equipment and discern a variety of operational outcomes, such as when machinery might fail. Predix can take vast amounts of information, recorded over time, to develop its forecasts through apps developed by GE and third parties. The company expects to process one million terabytes of data per day by 2020. The prognostics app enables airline engineering crews to determine how long landing gear can remain in service before a plane needs to be put into service. Doing so creates a maintenance schedule intended to reduce unexpected equipment issues and flight delays.

In essence, such predictive analytics forecast the future as a function of the past, calculating when maintenance is needed for devices, cars, trucks, and drilling machines, then scheduling repairs and upkeep before a severe failure occurs. Navistar, a manufacturer of commercial trucks, has sensors in its products that analyze brakes, lights, and engines. That adds value to maintenance services by knowing precisely when vehicles must be serviced, minimizing downtime and costs. Applying ML learning can also boost the performance of some equipment.

In 2016, Siemens launched MindSphere, designed to monitor machine fleets for service needs via machine tool and drive train analytics. It can be used by industrial companies to keep track of machine tools at plants around the world and actively monitor the performance of their assets to efficiently schedule preventive maintenance and manage how their equipment is used, to improve their operational lifespan. Like Predix, MindSphere works with machines and plants regardless of the manufacturer. The objective is to help plant operators increase the operational life of their equipment and make maintenance more efficient by assessing when a piece of machinery is expected to break down.[11]

Siemens has for years used neural networks to monitor its steel plants and improve efficiencies, has invested about $10 billion in software companies (via acquisitions) over the past decade, and has also integrated IBM's Watson Analytics into its manufacturing processes. Like GE, Siemens aims to monitor, record, and analyze everything about its manufacturing processes–from design to delivery–to identify problems and craft solutions before they impact operations. Siemens gas turbines have more than 500 sensors that continuously monitor temperature, pressure, stress, and other variables. Siemens' AI systems are learning how to continuously adjust fuel valves to create the optimal conditions for combustion based on specific weather conditions and the current state of operation of the equipment.

In 2015, GE launched Brilliant Manufacturing Suite, a holistic approach to tracking and processing everything in the manufacturing process to identify issues before they emerge, and to detect inefficiencies. The goal of the Suite is to link design, engineering, manufacturing, supply chain, distribution, and services into one globally scalable, intelligent system. Using the Suite in one of their own factories in India, with an initial $200 million investment, the company said doing so improved equipment effectiveness initially by 18%. As of 2018, GE had seven Brilliant Factories and claimed positive improvements at each. For example, its wind generator factory in Vietnam increased productivity by 5% and its jet engine factory in Muskegon had a 25% improved on-time delivery rate.

While GE and Siemens are heavily focused on applying AI to create a holistic manufacturing process, other companies that specialize in industrial robotics are focusing on making robots smarter. Japanese-owned FANUC, a leader in industrial robotics, collaborated with Cisco and Rockwell Automation to develop and deploy FIELD (FANUC Intelligent Edge Link and Drive). FIELD is described as an industrial IoT platform for manufacturing. It has also teamed up with NVIDIA to use its AI chips to create the factories of the future. FANUC is using deep RL to help some of its industrial robots to train themselves. By partnering with NVIDIA, the goal is for multiple robots to be able to learn together at the same time. What could take one robot eight hours to learn, eight robots can learn in one hour.[12]

From Productivity Gains to Innovation

A 2017 study[13] of businesses with more than 1,000 employees and annual revenue exceeding US$500 million across a plethora of industries revealed that 96% of respondents agreed that automation was the key to their organizations' digitization process, but nearly 75% said that they had yet to automate all the processes they believed should ultimately be automated. Some 98% of the respondents who used AI-supported activities to drive their digital transformation confirmed that doing so generated additional revenue for their firms. The AI applications that are the leading sources of digital transformation across industries surveyed are ML (75%), cognitive AI-led processes/tasks (57%), and institutionalization of enterprise knowledge using AI (55%). In the manufacturing and high-tech sector, use of ML was 79%, institutionalization of enterprise knowledge using AI was 66%, and cognitive AI-led processes/tasks was 60%.

Some of the reasons given for not having achieved full automation were inadequate focus on streamlining and automating wide-spanning processes, especially those deeply interconnected operational activities that supported the existing mandates of business, and the magnitude and complexity of the operational landscape. The study confirmed that automation was not just about achieving operational efficiencies and greater productivity; enterprises also desired to redirect their workers' efforts toward innovation and leveraging emerging technologies to help shape such innovations. Organizations that reinvested their people productivity gains into innovation efforts also tended to be more successful, with 63% of the respondents who said that their organizations were automating most of the tasks on their agenda also reporting that their companies were able to develop 20 or more valuable innovations in the previous 12 months.

Even though a majority of enterprises in the manufacturing sector are undergoing digital transformation via automation, few have fully accomplished their stated goals. *In the future, manufacturers' competitiveness will be measured by how well their employees are able to perform those tasks that AI cannot do, which involve human curiosity, creativity, intuition, and insight. Only those organizations highly invested in lifelong learning for their employees–to nurture all that is uniquely human in them–will have a workforce ideally suited to capitalizing on the future of business.*[14] This implies a future role for humans in all but the totally automatable functions of the manufacturing process.

The future of production raises important questions for companies well beyond the basics, however. As the AI revolution gathers real steam in the next decade, business leaders will be forced to examine a series of questions about sources of growth, innovation, competitiveness, skills, jobs, and larger questions around governance and sustainability. Speed, agility, adaptation, and adoption are the ultimate defining factors in successfully transforming businesses from the analog to the digital era. If companies cannot develop at a pace that allows them to effectively compete, they will fall further behind very quickly. Long-lasting transformation in this context requires an immediate, intense focus on understanding existing and emerging technologies, and how they can create value within a business, while developing the necessary culture and skills to execute the transformation.[15]

Just because a job can be automated does not mean that it will be, for relative costs can matter a great deal. Japanese car manufacturers rely heavily on robots while Indian textile manufacturers rely heavily on cheap labor. Even though machine capabilities are rapidly improving, seeking ever cheaper supplies of increasingly skilled labor still makes sense for the majority of businesses. Although the share of US employment in manufacturing has declined sharply since the 1950s (from almost 30% then to less than 10% today), American jobs in services have soared, from less than 50% of employment then to about 70% today.

Evidence is mounting that rapid technological progress, which accounted for the long era of rapid productivity growth from the 19th century to the 1970s, has returned. The exponential growth in chip processing speed, memory capacity, and other computer metrics is so great that the amount of progress computers will make in the next few years will surpass the progress they have made since the very beginning of the computer age. The primary innovation bottleneck is, ultimately, the time it takes society to adjust to the many combinations and permutations of new technologies and business models. At the turn of this century, technologically minded economists pointed to driving cars in traffic as the type of human accomplishment that computers were highly unlikely to master. Now, of course, autonomous vehicles are rolling off the assembly lines and being tested.

The productivity gains from future automation will be real, even if they mostly accrue to the owners of the machines. Some will be spent on goods and services, but most of the rest will be invested in firms that are seeking to expand, and presumably hire, more labor. Though inequality could soar even further in such a world, unemployment will not necessarily spike. The current doldrum in real wages may, as was the case in the early industrial era, prove to be a temporary issue, and be supplanted by higher wages for all.

The jobs of the future may look distinctly different from those they replace. Just as past mechanization freed—or forced—workers into jobs requiring more cognitive dexterity, leaps in AI and ML could create space for people to specialize in more emotive occupations, as yet unsuited to machines: a world of artists, therapists, and yoga instructors. Such work could prove to be as critical to the future as metal work was in the past. Cultural norms change slowly. Manufacturing jobs are still often treated as "better" to some than paper-pushing is to others. To some 18th-century observers, working in the fields was inherently more noble than manufacturing just about anything, and yet, the tides became completely reversed with the passage of time.

Technological progress squeezes some incomes in the short-term before making more people wealthier in the long-term, and can drive up the costs of some things even more than it eventually increases earnings. Yet, as innovation continues, automation may bring down costs in some of those areas as well. That said, the gains of the 19th and 20th centuries will be hard to duplicate. Boosting the skills and earning power of the children of 19th-century farmers and laborers took little more than providing schools where they could learn to read and write.

The transition from the analog to the digital world, from already educated lesser-skilled workers to highly-skilled workers, and from conventional manufacturing to smart manufacturing, is unlikely to be simple or painless. Modern society may find itself sorely tested if, as seems possible, growth and innovation deliver handsome gains to the skilled, while the rest cling to dwindling employment opportunities at stagnant wages.[16] Herein lies the core challenge for business leaders and policy makers alike—to find the right mix of carrots and sticks to encourage those who may not otherwise understand or be inclined to take advantage of the opportunity that is staring them in the face, to do so. Some will indeed have a choice; others will be pushed into the fire.

In that respect, the "coming" AI revolution in manufacturing will imply deploying some of the same basic management skills and orientations toward crafting high rates of employee utilization and satisfaction that have always been a requirement of profitable, thriving businesses, effective managers, and happy employees. AI will not replace the basics of business but, rather, will require that organizations become even better at perfecting efficiency, streamlining processes, and managing all aspects of the organization well. Perhaps we should be thinking about AI not as a threat to business but rather as a catalyst propelling businesses to enhance what they do well, even better.

4. Financial Services

Financial services first emerged from a need to facilitate trade in Ancient Mesopotamia, where repayment obligations for commodities were inscribed on a clay tablet. The bearer of the tablet was entitled to collect the debt inscribed on it, which became one of the earliest examples of a functioning bond market.[1] The essence of financial services has not changed all that much over the last 5,000 years; while the technology used to facilitate financial services has of course greatly evolved, the fundamental function has remained the same—to support trade and investment. Financial institutions (FIs) support cross-border transactions today just like the ancient merchants of Mesopotamia did thousands of years ago—by lending or investing money, collecting payments, and keeping records of transactions.

One could argue that every industry is information-based and that AI can impact any industry, since ML algorithms can process any type of data as long as it can be converted into numbers. However, it is important to note that financial services are distinct from other industries. In healthcare and agriculture, for example, an x-ray of a person (or images of crop fields) can be converted into numbers to draw conclusions about whether a skin growth is cancerous or make predictions about expected crop yields. Both of these industries deal with tangible things in the physical world and represent them with numbers that can be processed by algorithms. Financial services, by contrast, do not deal with tangible items; rather, they address the intangible *representation* of physical things as concepts that can be valued, bought, sold, transferred, and securitized. Money is nothing more than an idea that is transformed into physical existence by governments, but it does not actually exist in the natural world.

To illustrate the concept, Goldman Sachs acquired the personal finance startup Clarity Money[2] to bolster its online lending business in 2018. At the time, Clarity Money had more than 1 million online users, so the acquisition gave Goldman access to ML technology while increasing its portfolio of potential loan applicants and providing a lot of data about spending patterns. Goldman was purchasing data, which could be used to predict a whole range of issues indirectly relevant to finance, such as the increased probability of a user being pregnant based on the purchase of goods at a specific store, geolocation information, or the likelihood that an existing customer may be interested in making a purchase of stock. This intelligence can be fed into other parts of the organization to predict new financial sales opportunities before other FIs may become aware of it.

Depending on the nature an FI's data trove, it may choose to purchase an online lender with a large portfolio of existing customers with long payment histories. This type of data can be used to evaluate risk more effectively by, for example, forecasting default rates or servicing customers only with the best credit ratings. AI-based technology is already enabling digital lenders to originate billions of dollars in loans. The race for AI supremacy in financial services will take many forms, but some trends are already evident. Firms are using data and algorithms to make better predictions about customer behavior, offer customized product recommendations, automate routine tasks, forecast movements in the stock and bond markets, and manage risk more effectively.

FIs are positioned to experience an enormous amount of disruption as a result of the rise of AI. New risks will emerge that must be managed, further dramatic change will occur in how FIs function, and additional consolidation will likely occur in the industry. This will be prompted by the adoption of AI-driven technologies and the desire to acquire more data. In the future, the acquisition and manipulation of data will transform how FIs (and other organizations) determine their own value. This may ultimately result in creation of a new accounting standard–the Enterprise Value of Data (EvD)[3]– which could become an integral part of financial statements, capturing the value of the largest and most ignored corporate asset: data. Meaningfully defining and incorporating EvD on to balance sheets will help ensure that corporate accounting and risk management standards incorporate an AI-driven future. It will also put firms that are not specifically data-oriented on an operational equilibrium with those that are. However, those FIs that fail to embrace their own AI future will become less relevant, competitive, and profitable in time.

AI in the Financial Markets

AI already pervades the financial markets, whether triggering unanticipated (and usually unwanted) large stock sell-offs or giving stock traders an edge over their competition. The algorithms that have been built to trade are more analogous to expert systems than ML programs, with some hedge funds using AI to execute automated trading strategies. Many AI-driven trading strategies have focused on a limited number variables to evaluate entry and exit points and manage risk. The leading quantitative funds, such as Renaissance Technologies, have demonstrated incredibly high returns (15% in 2017 for its Renaissance Institutional Equities fund[4]) using various quantitative strategies. It is believed that arbitrage is a core part of its black box. Renaissance Technologies is one of the elite technology-driven quantitative funds that trades in milliseconds by using ultra-fast computers leveraging complex algorithms to take advantage of price inefficiencies in various markets.

In 2017, one of the first academic studies on the applications of ML in trading was published by Gordon Ritter, an adjunct professor at New York University. Ritter used an RL algorithm (Q-learning) to delay the payoff from a trade so that the time horizon could be extended. The algorithm discovered arbitrage opportunities and ineffective pricing of financial instruments in various markets and exploited them effectively. In the study, Ritter described why a general RL framework was needed to avoid a myopic view of trading, and why a delayed reward function was necessary in order to maximize a greater cumulative reward over time.

This conclusion was made bearing in mind that human traders are not necessarily all that good at understanding the true cost of the trade that they are engaging in. For example, if a trader submitted a large order to the market, part of the order could get executed at a specific price and part at a higher price, because the market reacted to the larger order. ML may therefore be able to understand better than a human the true transaction costs of a large trade and how the market will react.[5]

The early adopters of ML trading algorithms will reap incredible rewards because they will be competing against a limited number of other true learning trading algorithms. It is likely that ML trading algorithms will not only learn the intricacies of the market but also how other participants behave in the market, and how they can adjust their strategy to take advantage of the predicted behavior of other agents in the market. However, as the trend toward ML trading algorithms takes hold and a large percentage of funds use this strategy, it will create a new form of market-based risk unlike anything we have seen in the past. The volatility of a derivative can take on a life of its own, independent of an underlying asset. As funds deploy ML algorithms that fight against one another, extraordinary amounts of alternative data will be ingested and, slowly, the price and volatility of securities across all financial markets will take on a more extreme life of their own, completely unrelated to the security (such as the financial performance of a company).

Hedge funds are a good place to start to gauge where the future of investment and trading is headed. Hedge funds have deep pockets, some of the best minds, and are supremely motivated to win. Assets in quantitative hedge funds, many of which use AI, soared 86% between 2010 and 2017. In 2016 alone, quantitative funds received $13 billion in new investments. Such funds are proving that some conventional ideas about how financial markets operate are not always accurate. A London fund, Winton, challenged the widely held belief that major acquisitions end up hurting the buyers' shareholders. Winton analyzed data from roughly 9,000 US transactions, going back as far as the 1960s, and found that this was not always true.[6]

By 2018, the quantitative hedge fund industry was on the brink of reaching an important milestone: $1 trillion assets under management (AUM), almost double its AUM in 2010. Despite years of skepticism, AI- and data-driven strategies have taken on new life. The incredible surge in growth of the firm Two Sigma is characteristic of the shift from traditional discretionary investment strategies to AI-driven strategies. The firm's assets exploded by 733%, growing from roughly $6 billion in 2011 to more than $50 billion in 2017, which put it into a similar category as the AUM of long-established quantitative funds, such as Renaissance Technologies and D.E. Shaw. The war for technology talent has become intense, which is understandable given that many traditional hedge funds (and mutual funds) are building up their capacity to execute quantitative strategies. Quant strategies that span complex exchange traded funds have grown steadily by 15% annually since 2011.[7]

In 2016, Aidyia opened a hedge fund that makes all of the trades autonomously, without any human intervention. The AI program had several engines that processed vast amounts of data, ranging from macroeconomic to corporate account documents, with each search engine determining the best course of action for the day. This is similar to what a portfolio manager would do by getting a sense of how the market is performing before starting trading for the day, to determine how overall sentiment could impact trading decisions.

Sentient Technologies is another firm that launched trading algorithms based on AI and ML. Sentient leveraged evolutionary algorithms to generate a large number of digital stock trades based on random variables, then selected the best performers to make it to the next round. The characteristics of the winners were passed on to the next generation of algorithms—a sort of survival of the fittest among the animals of the trading world. As this process is repeated over generations of digital traders, it will eventually produce algorithms that can perform well enough to be deployed in live trading scenarios. As more firms adopt similar forms of AI-based technology, it will begin to be priced into the market.[8]

A byproduct of the digital and algorithmic revolutions is the consolidation of power among a few large companies. The investment titans Vanguard and Blackrock are expected to manage in excess of $20 trillion by 2027, which would roughly double their AUM as of 2018. This is partly due to the fact that these firms are able to aggregate massive amounts of data about their customers and use it to attract new customers (and thus, sources of capital) using robo-advisors and other algorithms. Algorithms' rise to power more generally seems inevitable as quantitative strategies invade every aspect of investment decisions and provide retail investors with access to algorithm-based financial products. Algorithm-based financial products will interact with one other and then with the irrational, unpredictable behavior of human investors, which will further complicate the decision-making process. This will increase the complexity and fragility of the financial markets to the point where sudden corrections may create huge algorithm-based sell-offs and, quite possibly, the next financial crisis.[9] It remains to be seen whether any future financial calamities caused by AI will equal or surpass those caused by humans in terms of damage done and money lost.

Rogue Algorithms

Unscrupulous traders are able to profit by manipulating the market price of financial instruments in minuscule amounts which may be barely detectable. A trader "spoofing" the market would generally place a large "bid" or "ask" price on one side of the market while placing another order he or she wanted executed on the opposite side of the market. For example, if a trader placed spoofing bid orders below the market price, it would create the impression of increased demand in the market, which could increase the price of a futures contract. Once the market moved in the spoofer's desired direction, she would cancel the phantom orders while the other, usually smaller, order would be executed at a profit.

In April 2015, Navinder Singh Sarao, a London day trader, was arrested for his alleged role in the 2010 flash crash— a 36-minute, trillion-dollar US stock market crash that resulted in wild gyrations in the market. Sarao was indicted on 22 counts of fraud, including his spoofing action—having placed deceptive orders that created the illusion of increased demand to buy or sell a particular financial instrument. He was able to reap $40 million in illicit profits from 2010 to 2014 through his market manipulation techniques[10].

Some have expressed skepticism that a single trader could truly be responsible for such an event, particularly given his limited amount of capital and the extraordinary impact it had on the market. However, it is clear that financial markets have moments of extreme volatility and that individuals like Sarao have made a lot of money by manipulating them with algorithms. It is important to note that Sarao was the trader doing the spoofing and also profiting from it. In other words, the cancelled orders and the profitable trades were all linked to the same trader. The prosecution of Sarao does not mean that the practice of spoofing will disappear, of course, but that it is likelier to be recognized when it occurs, creating a disincentive for other traders to engage in similar behavior.

That said, the next phase of spoofing markets may be the collusion between two or more independent traders not working at the same firm. Traders could agree that one person will take on the task of flooding the market with bogus orders and then cancel them when the market reacts so that the other trader can profit from the manipulation. This could in theory be repeated ad nauseam. This type of behavior is not something an individual firm monitoring its own traders could detect with certainty because the cancelled orders cannot be easily linked to the profitable trades. This is similar to how some drug deals are conducted. When large quantities of drugs are sold, the drugs and the money are usually not in the same place at the same time. If law enforcement catches a person with bags full of money, they cannot, therefore, easily prove a connection to the drug deal or prove beyond a reasonable doubt that the purchase of drugs was what the money was intended to be used for.

In 2013, traders from several large investment banks would get together to play poker and unwind from their busy schedules. It turned out that the seemingly innocent poker game was really a meeting place to collude and tip each other off about trades. The red flag was that, after every game, there was a spike in the profit-and-loss (P&L) of the traders involved in the game. The link between the poker game and P&L was not discovered until 2015, when the FI where they worked was being investigated. The connection was uncovered not by human investigators but by an AI program that leveraged ML algorithms which improved over time. The AI system built in behavioral profiles of employees and flagged anything that appeared out of character. The traders involved in the game were incredibly busy, but still found time to regularly participate in the poker game. Analysis of the P&L statements of these traders revealed that there was a link between the game and the trades.[11]

While, in this example, the AI program was able to detect traders manipulating the markets for their own benefit, who is to say that AI-based trading algorithms will not ultimately end up doing the same thing? One risk is that, just as Sarao spoofed the futures markets, an AI program could spoof markets to increase its *own* performance. As more and more funds deploy AI algorithms to trade, could the goal of the algorithms be defined so loosely that spoofing is not only likely, but becomes inevitable? The reality is that, if AI algorithmic spoofing becomes an epidemic, regulators such as the US Department of Justice and other similar law enforcement agencies across the globe will not have the capacity to prosecute all of the suspected infractions.

Ironically, identifying AI algorithmic spoofing would require another AI system to sift through the volumes of trade data available at the exchanges. What could end up happening is that, if regulators are able to train an AI algorithm to identify possible spoofing scenarios, it could send an automated enforcement action to the suspected firm. This would be analogous to running a red light and receiving a picture of your license plate in the mail, leading you to believe you had been caught. Such enforcement actions would likely follow the historical precedent of going after a low number of high value suspected offenders, then prosecuting and fining them to deter bad behavior in the future. Were that to occur, many of those inclined to misbehave would probably still believe that their chance of getting caught would be remote—and they would probably be right.

A more frightening scenario would be if two or more unrelated algorithms determine, independently of their creators, that they can increase profits by randomly placing large orders in the market and cancelling them after the price moves a small amount. Similarly, the same AI system could learn that a large swing in demand for a security it regularly trades is an indication to place a sell order to profit from the artificial price increase. What could occur is that two or more algorithms begin spoofing the market for the other algorithm, and that algorithm spoofs the market for them in return. The AI programs would not even be aware they were collaborating with another AI system or breaking any rules through their spoofing orders, because they could have been programmed not to profit themselves from excessive order cancellations. However, who can anticipate, or finds ways to prevent, algorithms from beneficially interacting with one another and learning new ways to generate greater profits over time?

In 2017, Facebook shutdown an AI experiment[12] because two chatbots created their own language, incomprehensible to their creators, to communicate more effectively. The chatbots were tasked with negotiating with each other for certain items assigned a specific value. The bartering discussions seemed to be converted into shorthand discussions, but it is odd to imagine how each chatbot understood the other based on this emergent language. Human languages tend to have rules and conventions and take many years to develop. The project was shutdown not because of any particular fear of what was going on but because the chatbots were intended to interact with people, so their utility for Facebook diminished.

The AI chatbots were instructed to negotiate and improve their bartering as they went along, but they were not told to use an incomprehensible form of English. This illustrates that AI programs can respond in unexpected ways when interacting with one another and given limited, perhaps ambiguous, instructions. While it is not realistic, at least in the near term, to imagine AI programs sitting around a virtual poker table exchanging trading tips, algorithmic collusion does not have to be anthropomorphized, and these programs do not necessarily even need to communicate directly in order to violate anti-spoofing laws or decrease market transparency and stability.

Fintech

As the transformative effects of the digital age continue to shape and reshape the world, a variety of risks are emerging. In 2017, the US Office of the Comptroller of the Currency published its Semiannual Risk Perspective report[13], which included four key areas of risk: strategic, loosening credit underwriting standards, operational, and compliance. The theme underlying all of these was the influence of digitization, which amplified the complex, evolving, and non-linear nature of risk. For example, while strategic risk and loosening credit underwriting standards were listed separately, they are clearly linked together. In other words, strategic risk was created by nonfinancial companies, including financial technology (fintech) firms, offering financial services to customers. This forced FIs to respond and adjust credit underwriting standards in an attempt to retain market share. As was evident in the aftermath of the Great Recession, when credit standards become too loose, it can be a recipe for disaster.

Fintech is broadly associated with any company using technology to more efficiently deliver financial services on digital platforms, ranging from mortgage lending to money transfer and personal finance to wealth management. Each year since 2013 has been a record for fintech globally in terms of both the number of deals and amount invested. In 2017, venture capital equity firms' investment in fintech companies totaled nearly $17 billion across more than 1,100 transactions globally.[14] It seems clear that fintech's influence on the broader FI marketplace will be felt far and wide in a short time.

Following the Great Recession, many FIs were so focused on remediation efforts and meeting new regulatory demands that innovation was not a top priority. However, at the same time, the significant technological advances in smart technology, Big Data, and ML allowed fintechs to disrupt the financial services sector, in conjunction with consumers being more open to and desiring digital alternatives. One financial service that is particularly ripe for disruption is the mortgage loan processing business, which is extremely laborious. The average mortgage costs $7,000, contains more than 400 pages of documents, requires more than 25 workers to prepare, and takes roughly 50 days to complete. This is a sector screaming for disruption and consumers have already opted for nontraditional online lenders to replace the painful mortgage acquisition experience being provided by traditional FIs.[15]

FIs' ability to collaborate is one of their greatest strengths, while simultaneously being one of their greatest sources of weakness. By agreeing to operate based on financial standards, FIs are able to interact with one another to process payments and settle security trades. But the existence of standards can make the people working in these organizations complacent. In an effort to standardize communication about cross-border payments, 239 banks from 15 countries formed the Society for Worldwide Interbank Financial Telecommunication (SWIFT) in 1973[16]. SWIFT has been the gold standard in secure message transmission about payments since that time.

Other organizations have built risk management infrastructure or data services designed to serve the financial services sector, such as Bloomberg and Mysis. In the past, innovation tended to come from within each firm. If innovation was sourced from an external company, it was designed to *support* the financial services sector as opposed to *compete* with it. Today, some fintechs are seeking to make the financial services sector more efficient via better software, but other firms are looking to take market share by offering complementary financial services directly to consumers. Fintechs deliver services digitally, so AI-driven technology is embedded into these platforms in a variety of ways. For example, NLP, powered chatbots, or ML algorithms can detect fraud in real-time. AI is not only expected of fintechs, it has become part of their standard playbook.

The fintech revolution is also impacting the world's developing countries in a positive way. In 2017, the World Bank noted[17] that the global share of adults with access to a bank account increased 7% to reach 69% between 2014 and 2017. A major contributing factor to the increased access to financial services was innovation in fintech, such as mobile money applications that have surged throughout Sub-Saharan Africa. There is still plenty of room for growth since it was estimated that 1 billion adults who are currently excluded from the financial system already have a mobile phone, with about 480 million having access to the Internet.

A major way to improve the rate of financial inclusion around the world would be for companies to pay wages digitally, as opposed to in cash, to the hundreds of millions of adults around the world without access to a bank. AI will play a key role in the future growth and inclusion of financial services globally because as mobile phones used by the unbanked and underbanked improve and Internet speeds in those countries increase, AI will be embedded into these mobile banking applications in the form of chatbots and intelligent assistants. It would not be practical to think that the unbanked and underbanked would be serviced by humans in the future because of the high labor costs and low revenue streams derived from serving poorer populations. However, AI can serve as the personal banker of the poor in the future—educating them about how basic financial services work and how to take advantage of opportunities such as a small loan or the chance to start a new business.

Fintech firms have had a laser focused approach when competing in the financial services sector, which has been described as unbundling the bank. Banks provide an array of financial services and fintechs know that they cannot compete directly with banks on all fronts. Rather, they have taken another approach and focused on a particular product or service, while creating a better user interface and promoting themselves through more effective digital marketing campaigns. Fintechs are taking additional steps to become more competitive, such as buying or partnering with other fintechs, expanding their service offerings, and integrating cryptocurrencies into their platforms.[18]

For example, Revolut is a digital banking alternative that includes a prepaid debit card and allows members to exchange foreign currencies for low fees. It also allows its members to purchase cryptocurrencies, which are considered a high-risk business given the many regulatory and governance questions remaining unresolved around these assets. However, Revolut did limit which cryptocurrencies could be purchased, and included only Bitcoin, Ethereum, and Litecoin in its portfolio. Given the demand for cryptocurrencies and many FIs' unwillingness to process payments associated with them (let alone facilitate the actual purchase of these assets), it makes sense that digital banking alternatives would step up to the plate in an effort to capture new customers and market share.

According to a 2017 report by PWC[19], large FIs across the world could lose 24% of their revenues to fintech firms by 2022. Fintechs are focusing on a wide array of applications that can be categorized as consumer facing or institutional and includes services such as business and personal online lending, payment applications, mobile wallets, and robo-advisors. Some FIs are taking an alternative approach (rather than competing directly or doing nothing), by, instead, either investing in or partnering with fintech companies, and a few of them have purchased their competitors. Since 2012, Citi has invested in 25 fintechs, while Goldman Sachs has invested in 22[20]. In 2017, JP Morgan Chase agreed to acquire the fintech company, WePay, for $220 million. Some fintechs are marketing directly to consumers and taking advantage of the power of digital platforms, even as the regulatory environment puts enormous pressure on depository institutions. This creates an opportunity for fintechs to offer snap-on financial products linked to customers' bank accounts without the regulatory burden of being an FI.

The initial phase of the fintech revolution was the complete digitization of processes and procedures. The next phase, which is already underway, is the cognification of that digital infrastructure. For example, customer service utilizing humans will continue to be deemphasized as people become more comfortable interacting with chatbots but will not disappear completely because some clients—especially high-net worth individuals—may continue to prefer personalized service. In 2016, Bank of America launched its own AI chatbot[21] that uses predictive analytics and cognitive messaging to help its customers make payments, check account balances, and pay down debt. Those customer service centers that do survive the AI revolution over the next 20 years will not be trained in how to solve customer problems; rather, they will be trained in how to interact with an AI powered chatbot on behalf of a customer. Ultimately, the human customer service provider will slowly be removed from the underlying process, to act as a catalyst to initiate the process, which will become almost completely automated.

Blockchain Rebranded

At its most basic level, blockchain is a form of mutualized record-keeping in a nearly irrevocable time-stamped ledger, designed to add security and decrease uncertainty in the administration of data and information. To understand what blockchain is, and how it applies in the AI arena, it is useful to examine the emergence of Bitcoin, which was the first successful commercial implementation of the blockchain concept.

Bitcoin propelled the idea of blockchain into the public consciousness in 2008, when Satoshi Nakamoto, a pseudonym for a person or group of people, published a paper describing how the Bitcoin protocol would allow for online payments without having to go through a trusted third party, such as an FI. Bitcoin solved the problem of trustless consensus because it provided a self-governing network through the innovative use of several different ideas, such as proof of work, cryptographic signatures, and peer-to-peer (P2P) networks.

The classic problem that Bitcoin solved is referred to as the Byzantine Generals' Problem, which is a logical dilemma of how to achieve consensus. The idea was that reliable computer systems must be able to manage malfunctioning parts that give conflicting information to different elements of a system. This concept was expressed abstractly by considering the generals of the Byzantine army, waiting outside an enemy city in anticipation of attacking it. Since the different groups under each general's command are dispersed over a wide geographic area, it makes centralized command difficult. The generals are therefore forced to relay instructions to one another through messengers so they can arrive at a consensus about when to attack the city, which will only be successful if all of the generals strike at the same time. The problem is complicated in that the generals know there is a traitor among them, so the question arises as to how to know that the message they receive is authentic and has not been tampered with. If different messages were to be sent to different generals, the cohesion of the army regarding how and when to act may break down and could cause the attack to fail.

The same idea can be applied to a P2P payment system that lacks a trusted third party validating the transactions. In a shared and distributed ledger, any payments (messages) to the ledger (coordinated attack time) must be trusted, but large distributed networks can have millions of users (generals)—so how can the payments made through an open ledger system be trusted, with no third party facilitating the activity? Blockchain essentially solved the Byzantine Generals' Problem through an innovative combination of existing concepts and applications, including cryptography, P2P networks, and game theory.

The Bitcoin blockchain is completely public (permissionless) and only necessitates minimum system requirements in order to participate. By contrast, consortium-driven blockchains are member-only, permission operated, and managed by a central authority. For that reason, consortium blockchains are best defined as distributed ledger technology (DLT). There are many use cases for DLT, including trade settlement, payment processing, and trade finance. The lack of transparency and interoperability between financial platforms—within FIs themselves and externally between other entities—creates spiraling costs and inefficiencies. DLT will drastically transform the financial services sector because it will redefine how books and records are maintained, moving from local private books to those that are globally shared.

JP Morgan Chase introduced a Contract Intelligence (COiN)[22] platform leveraging ML technology to analyze and extract important information from contract clauses. When the DLT matures, AI-based contract evaluation tools such as COiN will be repurposed to analyze a written contract between two parties and automatically generate the code to build the blockchain that will host the transaction. When AI starts to generate blockchains to conduct financial transactions, it will revolutionize contract processing to such a degree that banks will be forced to justify the continued employment of thousands of their employees. Based on this fundamentally more efficient method of leveraging AI to transform data processing, only the most digitally advanced and efficient FIs are likely to survive.

Data Lakes

It is clear that fintechs are a force to be reckoned with, being built on digital platforms, squeezing out customer insights, and predicting default rates using algorithms. One of fintech's largest online lenders, SoFI, is going after the heart of many FIs' revenue streams by offering a plethora of types of loans, including mortgages, personal, medical resident, and student loans, among others. SoFI is a unicorn (a company valued at more than $1 billion) that is able to keep costs down and service more than 350,000 members by offering all of its financial products through its digital platform, with no brick and mortar locations, which traditional FIs must have to accept cash deposits. SoFI's longer term goal is to offer credit cards and other traditional financial products in due course.[23]

Traditional FIs have some serious impediments along the road to scaling up AI innovation projects, such as being burdened by legacy systems that are still running Common Business-Oriented Language (COBOL) code that dates back to the 1960s. The common computer languages taught in universities today are Java, C, Python, and some others, but definitely not COBOL. Since COBOL is not taught in universities today, there are a shrinking number of highly skilled programmers qualified to fix mission-critical glitches in banking systems. Technology executives at FIs must sometimes scramble to hire outside contractors to fix issues as they arise.

One consulting firm, COBOL Cowboys, was created to meet the frantic demand for COBOL programmers. To put the magnitude of what is at stake into perspective, it is estimated that $3 trillion of daily commerce flows through COBOL systems each day. All of the mobile applications and digital platforms built today are based on modern computer languages which do not necessarily integrate well with antiquated COBOL systems. Other FIs have decided to bite the bullet and replace their entire core banking systems, which includes COBOL systems. For example, the Common Wealth Bank of Australia began the work to replace its core banking platform in 2012, which took 5 years and cost $750 million.[24]

Clearly, time is not on the side of FIs in terms of catching up with fintechs' futuristic orientation and the idea of attempting to stay a step ahead of them. As traditional FIs struggle to keep up, fintechs are busy accumulating dollars and creating loyal customer relationships. As fintechs build digital platforms that can accumulate data in a way that is conducive to implementing AI, traditional FIs are busy trying to keep their aging technology systems from crashing. At the very least, FIs should build more data lakes (large repositories that can hold structured and unstructured data) as an interim measure.

The idea behind the data lake is that splitting tasks across a network of computers (as opposed to relying on a single physical machine) will increase the throughput or reduce the time required to create actionable intelligence. Big Data is really about the ability to extract value for decision-making, and one of the most effective ways to do that today is by leveraging the power of parallel processing. This is a cornerstone of achieving AI supremacy because some ML methods–such as SML, UML, and RL–are extremely resource intensive and complex, so running these algorithms on one machine is not always feasible.

This raises another problem which FIs need to tackle head on if they are to take the race for AI supremacy seriously: while massive amounts of data are one key prerequisite for FIs to be viable contenders in the race of AI supremacy, the data must be in a location where it can be processed according to the rules of Big Data. FIs may therefore need to rely on external sources to harness the power of Big Data through parallel computing. Some FIs may opt to implement advanced software within their own data centers and manage the parallel computing process themselves. Others may choose to leverage cloud providers to scale up their computing capacity, but this raises questions such as, what data can be sent outside of the organization and what data cannot, and must the data first be anonymized? FIs will be grappling with such questions for many years to come.

Loosening Credit

As highlighted by the US Office of the Comptroller of the Currency, some FIs have loosened their credit underwriting standards to ensure that more business is not lost to fintechs. This is happening on a number of levels, one of which is simply that customers have access to banking alternatives such as SoFI for their lending needs. As already noted, fintech firms have generally opted not to have brick and mortar stores, which keeps their operating expenses down, enabling them to pass cost savings on to the consumer. Perhaps most concerning for the FIs is that they do not necessarily have the best models to predict default rates. FIs might mistake a low-risk customer for being high-risk (and vice versa) because the models they use to determine credit worthiness usually leverage a minimal amount of data and do not take into account other factors that could affect an individual's ability to pay back a loan.

A simple example is a young working professional applying for a mortgage which the FI could designate as high-risk because of his limited working and credit history. When the applicant receives an approval for the loan, along with a high interest rate, it could cause him to take his business elsewhere. Was this loss of revenue for the FI really necessary? Closer examination could reveal that the young professional is associated with the accounts of his parents, who happen to be affluent and valuable customers of the FI. Classic statistical methods of evaluating loan applications assume formal relationships between variables in the form of mathematical equations[25]. ML can learn from data without being explicitly programmed, which implies, among other things, not having conventional biases built into the decision-making process.

When crunching the numbers under an ML algorithm, it would become apparent that young professionals with affluent parents seldom default on mortgage loans. This is where SML can have a positive impact on more accurate credit modeling and improve revenue streams. This example is based on the assumption that algorithms are responsible for making most of the lending decisions in an automated fashion, but FIs that engage their customers through relationship managers could undoubtedly better tailor the lending needs of each specific client. AI is continuing to erode the need for humans in all sorts of processes and procedures, so that accurate credit risk modeling may eventually become tantamount to an FI's long-term success.

The loosening of credit at FIs should happen, when appropriate, but not because of increased competition from fintechs; rather, it should be driven from better credit risk modeling. One way to achieve this is for FIs to build a Big Data platform that can interact with AI algorithms. Different types of data—some possibly never considered relevant for evaluating credit risk—could be leveraged for predictive analytics. However, it should be noted that AI algorithms will be better at predicting risk than the classic statistical models when properly implemented, but neither can control the impact of randomness and extreme events (black swans), which will remain a background threat.

While FIs can apply AI algorithms to scenarios such as determining the default risk of an individual customer, it may be difficult to make similar applications to the critical credit risk modeling done at the enterprise level. For example, FIs that meet specific asset criteria are required under the US Dodd-Frank Act to conduct annual stress tests to ensure they have sufficient capital to continue operations during periods of significant economic and financial stress. The results of macroeconomic events can impact FIs differently, depending on their unique business lines, and exposure to the markets and other counterparties.[26]

One of the early models that banks used to calculate exposure to large market shocks was value at risk (VaR) methodologies. After using VaR, a bank might determine its exposure to the market, other counterparties, and the impact of negating outstanding positions; it could determine that, 99.9% of the time, the expected loss would not exceed a certain value. However, the weakness with classical statistical methods such as VaR (and even cutting-edge ML) is that, whether the underlying patterns are considered linear or nonlinear, there remains an assumption that some fundamental flaw exists.

On one hand, the data regarding extreme events will not be extensive, so ML algorithms could learn to predict future events based on more or less normal circumstances. It could be argued that extreme events such as black swans, which are very unlikely, cannot by their nature be predicted because there is no historical precedent to reference. There could be opportunities to apply AI to financial statements in order to extract meaningful insights about an FI's solvency, but taking on new data outside of that could obscure the problem and misrepresent risk drastically. So, depending on the problem, more data does not necessarily lead to better predictions. Stress testing is a clear example where AI can potentially do more harm than good, if not applied diligently.

The Speed of Fraud

As a result of increasing competition from real-time fintech payment providers such as Venmo, the major US banks responded with the creation of Zelle. Zelle allows for payments to be sent to any person with a US bank account in near real-time when, at a minimum, the sender is a member of the registered network. The real-time nature of the payments and the absence of robust protections make the network irresistible to fraudsters. PwC has noted that 90% of all Zelle transactions for one bank had been fraudulent[27], which is staggering. The methods for stealing money from bank accounts through Zelle can range from simple to complex, including gaining online access to a bank account using social engineering techniques, email phishing campaigns, and buying user names and passwords on the Dark Web.

One way to prevent fraudulent payments is to use AI algorithms to detect when something is not right after a payment has been initiated. For example, if a person changes her bank password then sends a real-time payment to someone with whom she has never done so in the past, it could raise a red flag. This is a common type of scenario that occurs during account takeovers where a fraudster will obtain access to a bank account and change account settings such as the owner's mailing address, so that the real account owner is not alerted of future activity. Account takeover fraud tends to work well with the elderly because they are often not as diligent in monitoring their account activity. There are other red flags, such as a change in the login details, where ML algorithms can add real value.

Software firms such as ThreatMetrix, Trusteer, and RSA have solutions that specialize in analyzing the physical properties, location, and behavior of devices associated with attempted logins. FIs can leverage device information as a metric used in determining whether the attempted login has a higher probability of being fraudulent and can request that the user successfully answer security questions or type a confirmation code sent via text message to the mobile phone number on file (otherwise known as two-factor authentication). By leveraging ML algorithms, more accurate predictions about potential fraud are possible. Ultimately, however, fraudsters change tactics and are constantly finding new ways to hide or obfuscate their login details. SML can be used to cover well-known fraud trends and UML can be leveraged by identifying clusters of abnormal activity against emerging fraud trends that do not fit predefined patterns or signatures.

Some FIs are implementing biometric-based authentication to access their accounts, products, and services. In 2006, for example, Bradesco became the first bank in Brazil to leverage biometric verification at ATMs by validating customer hand vein patterns to confirm that they match those on file. The bank claimed that ATM fraud had been almost completely eliminated using biometric authentication. There are other benefits to using biometric ID, which is especially useful in a country such as Brazil, given that Brazilian IDs can be forged fairly easily and there is an illiteracy issue that leads to people forgetting their ATM pins when trying to withdraw their monthly government benefits.[28] The core of the problem is that no sooner does an FI find a solution that works then the fraudsters think of new ways to break the security system. It is a virtual game of cat and mouse.

Biometrics can be well addressed by AI and even ML, depending on the algorithms used. Matching algorithms will continue to leverage advances in AI to ensure they are spoof proof. A fraudster should not be able to print a picture of someone's palms and pass the biometric authentication procedure, but that will not stop a thief from chopping off another person's hand to gain access to an account via an ATM (which has actually happened). That is why biometric authentication systems should be obligated to support features such as liveness detection, because a "dead hand" will not have active blood flow and a severed eyeball cannot blink.

US law enforcement agencies have been known to use the fingerprints of a corpse to unlock their phone in an effort to find clues to a crime. The reason they can do this without a warrant is that a dead person does not have privacy rights, and the relatives and friends of the deceased cannot claim privacy rights over information they sent to the deceased, since it was shared voluntarily. Law enforcement agencies have even been exploring how the Apple Face ID feature can be accessed for investigation purposes. In 2017, Vietnamese security researchers claimed that the Apple Face ID feature was hacked using a custom-created mask.[29]

Synthetic Identity Fraud (SIF) is an emerging trend in the US which, on the surface at least, seems like it should not even be possible, but it is. SIF involves the use of stolen social security numbers (SSNs) which are combined with other, often fabricated, personal information, such as a name, date of birth, and address. The key for fraudsters is to find SSNs of young children or others, such as immigrants, who do not yet have a file with the credit bureaus. By using SSNs without a credit history, fraudsters are able to trigger the creation of a synthetic credit file. Over time, the file is further developed in an effort to increase a credit rating and history so that larger credit lines can be extended and maxed out, which are then never paid back. The child will eventually grow up and realize that her SSN was compromised and used as part of various fictitious identities to commit fraud.

The US Social Security Administration (SSA) developed a way for FIs to verify an SSN, but it requires a physical signature which is not practical in the digital age and has not kept pace with consumer expectations. Hence, financial products such as credit cards that support online applications are not able to leverage the information stored by the SSA. While FIs can leverage the latest and greatest in AI technology, sometimes simply getting access to the most basic but important data elements associated with correct names (such as SSNs) can have a critical impact on fraud reduction.[30]

In 2017, the US Government Accountability Office (GAO) published a report[31] on SIF and estimated that it is responsible for annual losses of between $50 and $250 million per year. FIs were not the only ones affected by SIF, of course—government agencies also faced substantial losses by paying benefits to fictitious people. It has been estimated that one state paid $200 million in fraudulent unemployment claims to synthetic identities that were nurtured over time. The report also described the inefficiencies associated with current SSA procedures to verify a person's SSN and name. The GAO report contributed to the introduction of bipartisan legislation to address SIF. In 2018, the Protecting Children From Identity Theft Act was introduced in the US Congress, requiring the SSA to develop a database to facilitate the verification of consumer information upon the request of a certified FI.

AI can help identify possible fraud by detecting inconsistencies in a customer's information and behavior, but there are often limitations to its potential usefulness because of a lack of data. The proposed Act requiring the SSA to provide a database to validate names was therefore the right decision. The US government needs to pass additional laws to provide access to data to strengthen its ability to protect its citizens' privacy and identify fraudulent activity. In this sense, we have truly entered the age of data wars; Mechanisms to share information through public and private partnerships will be critical to preventing mass scale fraud and promote national security.

FIs' use of AI and ML can help to substantially reduce identity fraud by creating labels to determine whether activity was truly fraudulent or not. These labels can be fed into SML algorithms to identify trends in fraudulent activity that could be specific to a given FI or part of a broader industry fraud trend. As new forms of fraud emerge, UML algorithms can be used to identify anomalies that can be investigated. If they are identified as fraud, such scenarios could be used as labels to be fed into SML algorithms. This type of fraud prevention occurs only after a transaction actually occurs, as a second line of defense.

There is a very delicate balance to be strived for by FIs because they do not want to create unnecessary friction for customers using real-time payment services such as Zelle, but at the same time they need to protect against increasingly serious losses. A first line of defense that is already widely used comes in the form of additional verifications, such as providing answers to pre-determined questions or two-factor authentication. This must clearly be expanded into biometric ID that requires verification of a customer's face, voice, iris, or fingerprint. This raises the stakes in creating spoof proof algorithms, but also raises the long-term risk that not only will our SSNs become exposed, but so will the very biometric information being used to verify it. And, once that biometric information is released into cyberspace, it cannot be replaced. That is the real danger, for it is only a matter of time until large data breaches of biometric data begin to occur.

There is a lot of excitement surrounding RL and DL (neural networks), but these types of algorithms tend to be even more complex to put into place correctly and effectively than SML and UML. SML has produced extraordinary amounts of economic value by allowing for the personalization of ads based on a user's preferences and browsing history. Getting SML and UML right will be a big step forward in the fight against financial fraud. That being said, as more sophisticated forms of fraud emerge, and fraudsters begin using AI to conduct fraud in the form of intelligent bot attacks, FIs will have to leverage more advanced types of ML techniques to address the complex dynamics of a learning agent, which is still years away.

A 2017 report[32] estimated that $19 billion (or 9%) of the total digital advertising spend in 2018 would be lost to ad fraud, much of which was conducted by malicious bots programmed to repeatedly *click* on ads hosted by websites to drive revenue away from the advertiser and toward the host website. These bots will gradually get smarter over time and better at imitating human mouse movements and clicks. Ad fraud is heavily automated because the environment they operate in is completely digitized. These types of fraud bots will eventually make their way into financial services, as the sector itself becomes more digitized, years from now. Until then, many types of financial fraud will remain labor intensive.

Money Laundering

Money laundering is the process of concealing or disguising the existence, source, movement, destination or illegal application of illicitly-derived property or funds, to make them appear legitimate[33]. The estimated amount of money laundered each year is between 2% and 5% of global GDP (or between US$800 billion and US$2 trillion)[34]. Most developed countries have robust anti-money laundering and counter-terrorist financing (AML/CTF) regulatory frameworks that FIs operating in those countries must comply with. Historically, FIs have complied with AML/CTF regulations because they have to, not because it was necessarily part of their corporate mission. FIs are slowly becoming more publicly cognizant of the fact that the financial services provided by their organizations can be used for nefarious purposes. This is evident because some large FIs (such as Standard Chartered) have established dedicated web pages[35] explaining what financial crime is and how their organization is leading the fight against it. While this could be interpreted as merely a public relations campaign, these organizations are taking some long overdue responsibility for their role in the fight against money laundering and financial crime.

Money laundering is actually quite different from fraud, even though the two may overlap at times. Many cases of fraud, especially at FIs, can be identified with a reasonable degree of confidence after the fact. For example, when you receive a text from your FI asking to confirm whether you made a recent purchase at a local store or online, this is intended to prevent fraud. There may have been something about that transaction that appeared unusual, based on your historical profile, so you were asked to confirm it. By confirming you did or did not conduct the transaction, you are helping train an SML algorithm (assuming the FI is using ML). The transaction can be labelled as fraud or non-fraud based on your response, which helps the algorithm improve its performance over time and reduces customer friction by only requesting confirmation for transactions with the highest probability of being fraud. Money laundering is different because, while there are labels identifying a transaction as such, the FI cannot call up the customer and ask whether she is laundering money, which is against the law.

When FIs investigate money laundering, at the end of the process a decision is made about whether to file a suspicious transaction report or suspicious activity report. If a report is filed with the appropriate regulators, it can be used as a label for SML algorithms. The problem is that money laundering is based on a spectrum of suspicion rather than a concrete "yes" or "no", as in some fraud investigations. FIs must have a reasonable suspicion that money laundering is occurring, or, at the very least, they must be unable to identify a reasonable apparent business purpose underlying the recurring transactions.

Also, regulators such as the Financial Crimes Enforcement Network (FinCEN) in the US do not provide feedback to FIs on what types of reports they have found to be useful. This is understandable for a regulator such as FinCEN because feedback may be difficult to provide from a logistical standpoint. If no feedback is provided on a certain set of reports, FIs may also consider them to be less useful. However, in the hypothetical scenario where feedback was given, it would be erroneous to believe that only suspicious reports that were identified end up being useful to law enforcement, since the repository of reports as a whole serve as a massive confidential informant, which can be referenced on demand, providing useful pieces of financial information.

FinCEN also needs to evolve with the financial services industry. As FIs begin to more regularly use AI-driven technology to detect and report money laundering activities, FinCEN should also be using AI to evaluate those reports. FinCEN already leverages AI with its FinCEN AI system [36], but the way money laundering is reported may need to be rethought because AI can find large networks of nefarious actors working together and there could be a better way of reporting this activity to regulators. This should not stop FIs from using AI to combat money laundering, but FinCEN should not exempt itself from the AI revolution. AI will surely be one of the most useful resources for law enforcement agencies to extract desperately needed intelligence to combat crime and terrorism going forward.

There are plenty of other challenges that FIs can tackle with AI-driven technology, such as the problem of shell companies. Unfortunately, some countries (including within the US, such as in case of the state of Delaware) allow the formation of legal entities with little or no information being provided on the actual beneficial owners of those entities. This issue was highlighted by the Panama Papers data leak, which embarrassed many of the world's elite in 2015. In 2018, two bills were presented to the US Congress–the True Incorporation Transparency for Law Enforcement Act (S. 1454) and the Corporate Transparency Act (H.R. 3089/S. 1717)–to attempt to address the beneficial ownership transparency issue. Although the bills had bipartisan support, neither proposed penalizing states for noncompliance. Delaware, and states like it, are unlikely to comply, given their economic interest in keeping the barriers to opening a legal entity in their states low.

Regardless of whether countries and states in the US derive beneficial ownership information at the time legal entities are formed, there remain opportunities for AI to aid FIs in identifying the people involved. For example, there are companies such as Bureau Van Dijk that have large databases of ownership information regarding complex and global corporate structures that can be leveraged to link a company in one country to a company in another. In some cases, a company can own other companies in countries that require beneficial ownership to be filed, so knowing the owners of one company could infer ownership in another, assuming there are links between the two entities. NLP algorithms can also process massive amounts of information in corporate databases and via public sources to link companies, based on news articles or common addresses. This is important because, when investigating money laundering, FIs struggle with understanding the nature of the business transaction, who is really behind an opaque company, and what additional risks they may or may not pose.

Additionally, identifying potential red flags associated with money laundering is incredibly complex but boils down to extracting the intent underlying a series of transactions, supplemented by other related information. But how can one understand the financial intent of a person or legal entity based on data? It is actually quite difficult, but AI can help the money laundering investigation process in many ways, such as by analyzing the massive amounts of news about people and companies to determine whether any of that media is negative. Adverse media and negative news can increase the money laundering risk associated with the person or company in question. Analyzing media articles is also a complex NLP problem. First, the person or company being screened needs to be matched to negative news, but doing so based on a first and last name can lead to false positive matches. As investigators work with tools that screen for adverse media they will provide feedback to the system verifying whether the negative news article pertains to the person or company of interest. Again, this label can be used in SML algorithms that leverage NLP as a way for the system to get better with time.

Regulatory Sandboxes

FIs are under severe scrutiny for money laundering infractions and the US regulatory regime is notoriously punitive, based on the amount of fines levied against institutions which have failed to comply. The FI industry has traditionally responded by throwing bodies at the compliance challenge, as opposed to experimenting with new technology. One exception has been HSBC, which was fined $1.92 billion in 2012 by US authorities for allowing cartels to launder drug money through Mexico, and for other compliance failures. HSBC Holdings engaged an AI firm to help reduce the number of false positive alerts generated by the bank's transaction monitoring system. During the pilot of the technology, HSBC saw a 20% reduction in the number of investigations, without losing any of the cases referred for additional scrutiny.

HSBC's pilot program took place in the UK. In 2015, the UK's Financial Conduct Authority announced its Regulatory Sandbox[37], which described its intention to support disruptive innovation through an appropriate regulatory framework to help ensure continued economic growth. Other countries' regulatory authorities are doing the same. In 2017, Singapore announced that it would use technology and data analytics to combat transnational crime and money laundering.[38] Singapore also established a regulatory sandbox that year, when it was reported that local bank OCBC engaged an AI firm that helped it reduce its transaction monitoring system alert volume by 35%. The Australian Transaction Reports and Analysis Centre (AUSTRAC) has also collaborated with researchers at RMIT University in Melbourne to develop a system capable of detecting suspicious activity for large volumes of data. AUSTRAC had found it increasingly difficult to stay on top of their data[39], so it, too, created an innovation hub and sandbox to explore the prospects of applying AI to fintech, while doing a better job of combating financial crime, terrorist financing, and the other forms of organized crime.

The OCC opened its own Office of Innovation in 2017 to support the financial services sector. The OCC has preferred to use different terminology to ensure that FIs and fintechs are held responsible for their actions. It prefers the term "bank pilot" as opposed to "regulatory sandbox", which could be misinterpreted as experimenting without consequences.[40] By not adopting the language of regulatory sandboxes–which is remarkably similar across other jurisdictions, such as Singapore, the UK, the United Arab Emirates, and Australia– the US regulators could be putting US FIs at a global disadvantage in terms of competitiveness. While one could interpret this as nothing more than a word game, it could be a major step in the right direction for the US to adopt the language of innovation already in use across the globe. At the very least, US regulators such as FinCEN should offer FIs some form of safe harbor rule[41] to allow them to innovate their AML/CTF programs without fear of regulatory sanction.

Conclusion

AI will have an incredible impact on financial services in the future because it is essentially an information-based business. In order for FIs to achieve AI supremacy, they must transform antiquated technology into modern digital architectures. However, large FIs have a natural advantage over their smaller brethren since data is the fuel that powers AI, and the bigger institutions have more of it. One of the major goals for FIs leveraging AI-driven technology should be customers–focusing on bringing a better experience and more value to end users via predicted needs and personalized recommendations.

AI SUPREMACY

If FIs simply focus on cutting costs by deploying more AI, they could miss a major opportunity to connect with customers. Such a mistake could prove very costly. Fintechs are being built from the ground up and are not shackled by legacy technology. This allows them to design systems and processes that are customer-centric and deliver value to the end consumer. Given their openness to technology, as millennials accumulate more income they will be more inclined to interact with FIs or fintechs that can provide financial services in the most hassle-free way. Fintechs will continue to dig in to FIs' market share in the retail space while moving into the commercial side of the business, which will ultimately lead to more consolidations through mergers and acquisitions and future FI failures.

The increased digitization of banks is a double-edged sword which needs to be handled with care. On one hand, FIs must digitize all of their operations as a matter of survival in the new age of the digital economy because consumers demand it. However, when a FI digitizes its end-to-end operations it opens itself up to an even greater extent to new and fast-moving fraud and cyber risks. This is why AI should be embedded in customer-facing applications, but also in the back-end, to protect against fraudulent activity and cyberattacks.

Other parts of the AI revolution in financial services will be focused on reducing costs and assessing credit default risk more accurately. Compliance functions within FIs can benefit from AI by providing more efficient ways to comply with the expanding complexity of the regulatory landscape. FIs are burdened by legacy systems, motivated regulators, and complex and compartmentalized departments, which make achieving AI supremacy difficult. FIs should strive for AI *transformation* for their business operations.

FIs can achieve AI supremacy in the financial markets by deploying algorithms that process alternative data and use the latest ML methodologies to outperform their competitors. There will likely be a wave of fortunes made in the financial markets driven by AI and ML. The quantitative fund industry is only beginning to heat up, and AI traded securities will be the hot new product on Wall Street and beyond. The financial markets are already extraordinarily complex, with hidden feedback loops that are impossible to sufficiently model. When

AI–a complex learning agent–disrupts an already complicated system, it will impact the overall ecosystem in ways we cannot even begin to fathom.

One thing is certain–regulators must maintain a watchful eye on the impact of AI on the financial markets, which will diminish the importance of evaluating the fundamental financial performance of companies and gradually increase market volatility. Extreme volatility will, however, accrue quietly in the background, waiting patiently to trigger the next financial crisis. The role AI may or may not play–whether it will either diminish the impact of that crisis or make it worse than it might otherwise have been–will remain unknown.

5. Science, Health and Medicine

We are living in the Anthropocene Era, in which man-made risk clashes with natural risk (and vice versa) as never before in human history.[1] Since the dawn of the nuclear age, human activities have had a profound and impactful influence on the natural world, most notably through global warming and climate change. While a subject of considerable debate and discussion, particularly over the past 20 years, the results of our collective actions are being felt in the environment with ever greater clarity. In 2018, the World Economic Forum predicted[2] that the technological innovations that will emerge from the Fourth Industrial Revolution will produce six critical environmental challenges that demand humanity's profound attention in the 21st century: climate change, biodiversity and conservation, healthy oceans, water security, clean air, and weather and disaster resilience. AI is poised to help humans tackle these critical challenges, as well as make dramatic advances in the realms of science, health, medicine, and other areas in ways most of us have not yet even imagined.

For example, AI can decipher our diverse opinions and options through swarm intelligence (by getting humans to communicate like honeybees and other self-organizing organisms in nature) and help us collaborate to achieve common objectives. While AI-driven technology can detect bacteria in water with the aid of some basic electronic components and a Neural Stick, AI cannot, of course, prevent the pollution of water sources or make contaminated water drinkable. We must be careful not to think of AI as a panacea. There is a risk, with all of the hype surrounding AI, that we may mistake those issues for which AI and ML can assist humans versus those they cannot. As the futurist Ben Hammersley has noted, we cannot eat AI because we still need to do the farming.[3] It is, therefore, important that we acknowledge that much of what AI and ML do today, and much of what they will do in the future, will be to improve underlying processes rather than reinvent them.

Mining Scientific Knowledge

Despite its obvious limitations in the areas of science and medicine, there is a lot of potential for AI to transform both for the better. One important area that can be revolutionized by AI is knowledge hidden and too often obscured from use in scientific journals. A 2015 report[4] estimated there were more than 28,000 active scholarly peer-reviewed English-language journals and in excess of 6,400 non-English-language journals publishing approximately 2.5 million articles annually. Some studies have estimated[5] that a good portion of the scientific papers published are generally not read by anyone apart from the author, referees, and journal editors; Even fewer are cited by other researchers. The exact number of readers and citations of scientific papers is naturally contested but, clearly, only a small fraction of all of the research produced is actually read.

AI can help democratize scientific knowledge by making it more easily available to more researchers, but also by optimizing search engines. AI-driven search engines are attempting to rebuild the landscape and make discovering knowledge easy and intuitive. For example, Iris.ai developed an interactive visual interface[6] that displays keywords relevant to a search and underlying documents. The system leverages NLP and UML to create documents, fingerprints, and semantic clusters based on extracting keywords, contextual synonyms, and hypernyms by accessing a database of 83 million open access research papers. By using UML for this purpose, clusters of words form around certain research papers- even though the context of the paper itself is unknown-there is a greater likelihood of identifying a noteworthy relationship between the words. The system's main objective is to reduce the amount of time people spend sifting through information in order to find the research documents they need, raising the profile of authors and their work in the process.

Building better search engines with sleek visual interfaces is an important step in the direction of making scientific knowledge more accessible, but it is only the beginning. The next step in the evolution of scientific research and AI is extracting *meaning* from papers. In 2017, a ScienceIE competition was held with the goal of extracting keywords from scientific papers and identifying relationships between them. The competition in question extracted only keywords, as opposed to an entire hypothesis, but keyword extraction is an important component of the process. One of the world's leading publishers of scientific researchers–Elsevier, which publishes more than 2,500 journals–has taken notice of how AI can revolutionize their massive collection of manuscripts by engaging an AI researcher. Ultimately, Elsevier wants to be able to provide capabilities for a person to ask a question about a paper and get an AI-driven answer. This could result in the creation of a sophisticated chatbot or virtual assistant specifically designed to incorporate the nuances of natural language in scientific papers. This aspiration implies that the next milestone will be to actually extract the problem, solution, evaluation, and result from scientific papers quickly and efficiently.[7]

When AI gets better at reading journals taking into account nuances and inconsistencies, will researchers change the way they publish to integrate the use of AI on their own accord, or will they need to be prompted by publishers to do so? Since financial institutions have standard protocols governing information exchange such as SWIFT, what is to prevent the scientific community from arriving at standards for presenting numerical research and important aspects of a paper to make it easier for AI algorithms to process? This goes beyond the standards and conventions that already exist for formatting research papers. The idea would be to digitize the scientific publishing process and arrive at standard codes and formats to make it easier for algorithms to process and understand. While we are undoubtedly a long way from AI being able to conduct peer reviews, it is not difficult to imagine that day will come sooner than we may think.

Science publishing is estimated to be an $8-10 billion a year industry, with subscription fees steadily increasing by 6% annually. It was estimated[8] in 2017 that the University of California San Francisco (USCF) spends roughly $60 million per year on subscriptions to closed-access research publications. Although USCF is considered a tier-1 research medical school, the rising costs are unsustainable, which is a primary motivation for it having committed to the OA2020 initiative—a global open access research publishing model.

Many in the AI research community believe that, given how swiftly the pace of change is occurring today, scientific research should be made more open, especially with the advent of AI. In 2018, in excess of 3,000 people (mostly academics), signed a petition[9] promising not to submit, review, or edit articles for Nature Machine Intelligence (NMI). The purpose of the boycott was to highlight that closed and subscription-based journals such as NMI prevent the advancement and public scrutiny of AI, and a dialogue could be better served with open access and low-cost journals. Given that AI is a dual-use technology and has the potential to impact lives in significant ways, the petition made sense.

Clearly, the scientific publishing companies will push back from allowing open access research, given how it will impact their profit margins, however, as shown by OA2020, scientific publishing companies stand to be disrupted in major ways over the next 5 to 10 years. The scientific publishing industry could respond to the open publishing model by granting access to some of its platform to the public, in addition to offering subscription-based models to provide access to more data or advanced analytical tools. Even if all of the major universities were to endorse the open access publishing model tomorrow, the scientific publishing companies will still own a treasure trove of historical research, which acts like building blocks of knowledge that researchers need not necessarily replicate, but can build upon. It is, therefore, more than likely that the publishers that embrace AI the earliest and fastest will be the ones who survive the pending open access revolution.

The NMI boycott could be about something much deeper; perhaps it is not really about *people* accessing scientific papers at all. AI researchers are obviously not capable of reading every single paper written, but they want their own AI to be able to do so. Perhaps, therefore, one of the motivations for open AI research, although no one may be inclined to admit it, is to gain access to large data sets—for the same researchers supporting the boycott—to ultimately train their own AI programs. Although allowing AI researchers' greater access to scientific research will benefit humanity, it could also allow for more public debate, based on quantifiable results and studies. Non-profit organizations such as OpenAI[10] have stated that their mission is to discover and enact a path toward Artificial General Intelligence, wherein a machine can perform any intellectual task a human can.

Some may argue that reducing the pace at which AI exponentially builds knowledge on itself (not unlike compound interest) could be a good thing, but it would be naïve to think that the relentless instinct of humans to push the limits of what is possible is repressible. Using AI to cluster similar scientific papers together and extracting meaning and hidden connections from them could encourage scientists to embrace new lines of thought and possibly make new discoveries. AI could also help revitalize the idea of the genius to the 21st century. Scientific history is littered with notable figures considered geniuses, but this has become less common. Perhaps part of the reason is because much of the low hanging fruit of scientific discovery has already been claimed. Researchers are now teaming up with high powered computers to make significant new scientific breakthroughs. When AI is able to categorize, analyze, and extract meaning from scientific papers, it will have the ability to formulate new hypotheses based on knowledge drawn from multiple studies that do not already exist in the body of scientific literature. When a scientist and AI program team up in the future, their collective intelligence may just unleash the power of a new genius.

AI Puts on a Lab Coat

Another major challenge that AI can tackle is the successful reproduction of experiments. A 2016 Nature survey[11] found that 70% of scientists failed to reproduce the experiments of their peers and more than 50% failed to reproduce their own experiments. There are multiple reasons for this, given the diversity and complexity of scientific research, such as pressure on researchers to publish research and selective reporting. However, other factors point to lack of standardization of processes and poor statistical analysis. Both of these areas could potentially be improved through the integration of AI in the lab and during the publication process. The net result of this reproduction crisis is that researchers are spending too much time chasing false leads. Data about the reproducibility of scientific experiments could help AI mine through the most relevant and reputable research. This is another process that should be incorporated into the scientific publishing process to ensure that scientific papers of lower quality do not lead researchers down rabbit holes.

Although AI is only beginning to scratch the surface of mining scientific knowledge, robots have already entered the lab and put on a coat. In 2016, a robotic lab assistant created a Bose-Einstein condensate (hyper-cold gas) experiment that had won three scientists the Nobel Prize in 2001. The experiment involved using focused radiation that slows down a cluster of atoms. The process is sensitive to fluctuations in energy so the AI was designed to monitor change in different variables and adjust settings in real-time to maintain the optimal conditions for the experiment to continue successfully.[12]

AI is not only being used for obscure physics experiments, but in labs to create useful things for the real world. In 2018, a group of scientists from the US Department of Energy's SLAC National Accelerator Laboratory, the National Institute of Standards and Technology, and Northwestern University used AI to discover and improve metallic glass 200 times faster than was previously possible. The team leveraged an AI system–Stanford Synchrotron Radiation Lightsource (SSRL)–that combined ML and laboratory automation by creating and screening hundreds of sample materials concurrently. Metallic glass with the right blend of metals can be stronger, lighter, and more corrosion-resistant when compared to traditional alloys such as steel.

In the last *century* scientists have investigated approximately 6,000 combinations of materials that form metallic glass, but by leveraging the SSRL system researchers can complete 20,000 combinations in *one year*. The team used the data from historical experiments of metallic glass as a starting point for training their models, but also built a feedback loop by using an SSRL x-ray beam to scan two sets of alloys and pass the results of the scan back into the ML model to make new predictions for what combination of alloys would maximize their desired result. The group discovered three new combinations of materials to create metallic glass, two of which had never been used before. Scientists noted that ML removed the need to develop elaborate theories because the research was data-driven, which can lead to research being redirected in unexpected ways since algorithms make inferences and draw conclusions which could, at times, move beyond current scientific understanding and biases.[13]

AI is tackling other even more complex problems, such as drug production. The process of creating new drugs is more complex than new alloys because of the number of components in human biochemical systems, and the dynamic interactions among those parts. In 2015, a research paper[14] described how a robotic lab assistant known as Eve discovered compounds that could be used against drug-resistant malaria. A computer scientist from the University of Manchester in the UK and his team revealed that Eve had discovered a chemical called TNP-470 which essentially targets an enzyme that is key to the growth of Plasmodium vivax, one of the parasites that causes malaria. The challenge that pharmaceutical companies and scientists have long faced is how to identify specific drug molecules that can fit well into protein or enzyme locks such as receptors to trigger a signal or biological response.

After scientists can establish that a drug molecule can trigger a signal (which could be based on thousands of compounds) they need to chisel the composition of the compound through a process called quantitative structure–activity relationship (QSAR). QSAR is mathematical model based on regression that predicts how new compounds will impact biological and chemical processes based on the molecule's composition, shape, and other properties such as electrical charge. Eve was given an initial data set of 5,000 molecules, then predicted which remaining molecules had the highest probability of solving the challenge the best, and only tested those. There is obviously real benefit in what Eve does, but also, perhaps, an indication of the advantage of what she does not do. Commercial compounds used in screening cost an average of $15 per milligram, which means that testing compounds randomly and by brute force costs a lot of money and takes too much time. Eve saves both time and money.

What connects all of the use cases of AI in the lab is not simply process automation, but predicting which variations of parameters in a given experiment will yield the best results. The data used to train ML models are crucial in saving time and avoiding the need to test every possible permutation of a given experiment unnecessarily. While AI can outperform humans in many domain-specific tasks such as facial recognition, when it does fail, it is generally not done gracefully, and AI fails in ways a human never would.

As AI becomes more integrated with standard experimentation processes, the publication of results could evolve to include details about how a model was trained, algorithms were used, the size of the data set, and all of the AI's predictions (including those that were incorrect). Despite the risks, the introduction of AI in the lab is exponentially increasing the output of scientists, which may lead to a surge of new discoveries across a range of disciplines. AI can potentially help drug manufacturers by automating the drug development process and leveraging ML and Big Data to streamline R&D resources, focusing on the most promising molecules to address a particular disease while reducing wasteful overhead costs.

Medicine

Most medical doctors have well established models for dealing with common illnesses that occur frequently, but rare illnesses can lead to misdiagnoses and medical errors. Doctors use their knowledge and experience to treat patients, but their tools are at times rudimentary and they may propose generalized forms of treatments that do not work for all patients. The implications of AI in the areas of disease diagnosis and precision medicine are profound and are already beginning to bear fruit.

In 2015, a patient was admitted to a hospital[15] affiliated to the University of Tokyo's Institute of Medical Science in Japan. Doctors initially diagnosed the patient as having acute myeloid leukemia, a type of blood cancer. Based on this diagnosis, the patient underwent a round of chemotherapy, but the recovery period post-remission was abnormally slow, so the doctors thought it might have been the result of another form of leukemia. The doctors decided to use IBM's Watson and compared the patient's genetic data with its own large database to identify thousands of genetic mutations and focus on those that could be relevant to the disease. Watson responded in 10 minutes with a conclusion that would have taken human scientists at least two weeks, and helped doctors conclude that the patient had another rare form of leukemia caused by Myelodysplastic Syndrome.

According to a 2016 study by Johns Hopkins[16], experts calculated that more than 250,000 deaths per year are due to medical errors, making it the third leading cause of death behind heart disease and cancer in the US. Researchers have advocated updating the medical coding system to explicitly identify medical errors on death certificates. The medical coding system was developed to be consistent with the operational procedures of insurers, but fails to properly identify deaths resulting from treatment gone wrong. Various conditions contribute to fatal medical errors in hospitals, such as failure to follow standard protocols, doctors working while sleep deprived, inaccurate medical diagnosis, fragmented insurance coverage, and the cognitive biases of hospital staff.

Researchers have demonstrated that DL algorithms can diagnose disease at least as well as physicians in the fields of cardiology, dermatology, and oncology. In 2016, the International Symposium on Biomedical Imaging held a competition[17] to identify DL algorithms that detect metastatic breast cancer in images of lymph node biopsies. The winning algorithm—a form of DL based on a neural network—demonstrated a 92% success rate, compared with a pathologist's 96% success rate. However, it is worth noting that the doctor outperformed the AI when both were working autonomously; when the pathologist joined forces with the AI his accuracy rate increased to 99%. While DL algorithms have shown incredible promise across a range of medical fields, the main method for a pathologist to diagnose disease remains looking at a sample of cells through a compound light microscope, which is one of the critical barriers to scaling up the technology. The reason is that these algorithms require a digital representation of the cells being examined in order to work.

The applications of AI in healthcare are not only limited to small data sets and one-off competitions, such as when Google's DeepMind launched a cooperative project in 2016 with the Moorfields Eye Hospital NHS Foundation Trust to improve eye treatment. Moorfields shared one million eye scans and other related anonymized information with DeepMind so that its DL algorithms could go to work. In 2018, Google researchers described[18] how a modified microscope could democratize the adoption of DL tools for pathologists around the world. The Augmented Reality Microscope platform included a modified light microscope that supports real-time image analysis and highlights the results of the ML algorithms into the pathologist's field of view. In other words, as the pathologist looks at the cells through the microscope the ML algorithms are dynamically working to highlight possible areas of interest based on what a researcher is viewing at any given time.

Google may not necessarily produce significant revenue from this initiative, but technology companies often develop products not to necessarily make money, but to collect data. *The race for AI supremacy in the medical field is likely to quickly evolve into a data war*. Technology companies competing in the healthcare field will accelerate their efforts to lay the foundation for a global pipeline to move massive amounts of medical data in real-time to satisfy the insatiable appetite to produce more accurate predictions from their AI platforms. The infrastructure will be built through strategic partnerships with hospitals, medical associations, research universities, and even with users providing biomedical information via their smartphones, once the regulatory hurdles can be overcome.

However, some medical organizations will naturally view the idea of large technology companies accessing their massive medical archives as a threat, will fight to keep their data private, and build their own AI in house. In other words, the nature of healthcare-related information could be fundamentally transformed as it becomes more accessible to the average person. The inequality of information that currently exists between healthcare provider and patient could be reconfigured as AI accumulates knowledge greater than what is retained by a single physician or hospital. This shift of knowledge will have significant economic implications for the healthcare industry but will also shift power to the smartest medical AI platforms. Perhaps the greatest benefit of AI appearing on the medical scene is that it could reduce the need for routine doctor visits if machines can examine and diagnosis us in real-time.

Machines could be developed that can be operated by a registered nurse in a local pharmacy or health clinic to perform all of the vital checkup functions, leaving the ultimate diagnosis to AI rather than a human doctor. The nurse could point an apparatus with a light at the ears, eyes, nose, and throat, and send images of the patient to AI for diagnosis. This, of course, raises several legal concerns about data privacy and other forms of liability based on misdiagnosis, but, on the other hand, not developing powerful AI that can diagnose people in a clinical setting could be seen as a national security risk. A 2017 study[19] by the Association of American Medical Colleges estimated that there will be shortage of more than 100,000 physicians in the US by 2030, the result of a plethora of factors, including the high cost of becoming and remaining a doctor, and the draw of other professional sectors. Populations that cannot be cared for domestically may seek to reside elsewhere. This is becoming an increasingly important issue for a range of countries in the developed and developing worlds.

So, the general AI doctor will be built; the question is by whom, and will it necessarily lead to more accessible healthcare for all? The large technology firms are in the best position to build the brains of a general AI doctor that can be stationed in pharmacies and clinics across the world, but it will take colossal amounts of data to get it right. Since some medical organizations in developed nations could perceive the rise of AI not under their explicit control as a threat to their sovereignty, it could make sourcing medical data from developing nations an important long-term strategy. It is not hard to imagine a future version of Google's ARM platform being Wi-Fi-enabled to link to the cloud and provide more accurate predictions in real-time, which raises cybersecurity issues. In exchange for a more accurate prediction it is more than likely that the image of the cells being reviewed will be sent to the cloud, stored, and catalogued as input for another version of AI to learn from.

Although precision medicine will clearly be a focal point of the healthcare practices of the future, there are times when meticulous care is not as important as comfort. As people grow old and suffer from more ailments, quality of life becomes increasingly important to many. According to a 2017 report by the UN[20], the number of people aged 80 or over is expected to triple by 2050, from 137 million in 2017 to 425 million in 2050. Based on longer life expectancies, there will be an increased demand for home healthcare. However, it is not only the aging population that will drive this increasing demand, it is the unsustainable way many nations let their elderly die.

In the US, the end of life is a big business since 80% of the 2.5 million Americans who died in 2011, for example, received Medicare benefits which cost the federal government approximately $85,000 per patient for their last six months of life.[21] Apart from being exorbitant in terms of cost, patients' quality of life is often reduced by being subjected to unnecessary tests and aggressive treatment under the presumption that they wish to continue living, when many people may, in fact, prefer to die comfortably at home. Many of the challenges associated with healthcare and end of life are shifting the focus of medical practice to a more palliative approach in some countries, if only because not having elderly patients in hospital ready for treatment could prompt a greater number of medical malpractice lawsuits.

AI can help improve the quality of how we die if we use it intelligently. One of the fears of keeping elderly people inside their homes and alone is that, when they do need help, no one may be able to assist them. At some point in the future, robots may help to physically care for the elderly in their homes in a manner similar to how they were portrayed in the film iRobot, but the high level of dexterity required will take many more years to perfect, and their cost could make deploying them in large droves prohibitively expensive, quite apart from the possibility that many elderly may not accept them as caretakers.

In the near-term, home healthcare is more likely to be transformed by AI-driven monitoring systems installed throughout the house of an elderly person to interact via voice commands, and by monitoring the person's movements around the home. An elderly patient could easily wear smart devices that track their vital signs and other bodily functions to ensure everything is normal. If the AI detects that a patient has fallen or is having trouble getting out of bed, a human health aide could be alerted or, in extreme circumstances, the AI could call for an ambulance on behalf of the patient. AI can help keep elderly patients in their homes while providing real-time health updates and alerts to nurses, doctors, or family members, depending what type of information is being disclosed.

Death is one of the most personal of all experiences, which raises the question, do we really want the voice of AI to be the last voice we hear before we die? Some of the dying will not care, or perhaps even notice. There is a strong argument to be made that AI should be used to improve the lives of the elderly while creating clear boundaries regarding where and how it should and should not be deployed. As is the case with many aspects of AI and ML, its promise can only be as good as its application. Surely, this credo has the most importance when applied to healthcare.

Early Detection

One of the prevailing practices in modern medicine is the annual physical exam with a primary care physician, which is supposed to monitor one's current state of health while identifying any changes or symptoms that could be indicative of disease and require treatment or strategies to mitigate its development via preventive measures. Some diseases can develop between annual physical exams; identifying them too late could have severe consequences. Furthermore, some diseases are more insidious and the symptoms do not appear immediately or necessarily in dramatic form, so discovering them too late can be catastrophic.

It is worth bearing in mind that the physical exam is limited to the physician's knowledge and the types of tests that are at his or her disposal. Patients may visit their physician as their bodies are developing cancer, but they may not be feeling sick nor may they be exhibiting any physical symptoms. In a 2018 study, researchers from Johns Hopkins University described how a blood test could be used to detect the early signs of cancer, even before a person starts feeling symptoms, which could be a big win for liquid biopsy technology. The researchers used the blood of more than 1,000 patients who were previously diagnosed with different types of cancer to identify specific proteins and gene mutations. Biotechnology startup Grail Bio raised over $1 billion to develop a single blood test capable of detecting the early stages of many different cancers.[22]

While these blood tests can detect traces of tumor DNA in a person's bloodstream, it is not guaranteed that the cancer will develop aggressively enough to pose a real risk to the patient or require any treatment. Hence, overemphasizing potential signs of cancer can lead to misdiagnosis and unnecessary treatment. Therefore, large data sets and ML algorithms are needed, not only to detect signs of cancer but also to determine which patients will need treatment. As the cost of these types of blood tests decline, a person may need to give blood multiple times to track the progression of the cancer. Velocity patterns could be indicative that the trend will continue and the cancer is on a likely trajectory to become malignant.

AI is not only helping to detect disease early when no symptoms are present but is also more accurately predicting which patients are at risk of a heart disease, even when they are examined by cardiologists with conventional methods. Out of 60,000 heart scans completed each year, some 12,000 (or 20%) are misdiagnosed. Researchers from Oxford Hospital in the UK have developed an AI system that diagnoses heart scans more accurately than specialists. The AI system extracts more than 80,000 data points from a single echocardiogram image to increase the accuracy of heart disease diagnosis from 80% to more than 90%.[23]

AI is not only being used to examine our blood, DNA, and hearts but also how we move around our homes, as an early indicator of Alzheimer's disease. Researchers from the Massachusetts Institute of Technology (MIT) developed a fall detector for the elderly and quickly realized it could be repurposed for other uses, such as detecting the early signs of Alzheimer's disease. A small rectangular device is placed into a person's home and reflects radio waves off of all objects (including human bodies) in a 30-foot radius to detect movements. The device uploads all of the raw data into the cloud so that ML algorithms can map out a person's gait, determine whether or not they have fallen, are waking up at night and walking around, or display other behavior that may be either indicative or out of character.

The type of ML algorithms being used are DL-based because each person and fall is unique, so the AI must be trained to identify which patterns constitute a fall and still be able to detect it. The objective of the study was to identify certain behaviors that occur in many different people and circumstances so that they may be generalized as an indicator of Alzheimer's. Treatment can then be given and good candidates identified for experimental drug studies. From 2002 to 2012, 99% of experimental Alzheimer's drugs failed because of our limited understanding of the disease. By identifying and monitoring potential candidates for experimental drugs, AI can use a small window of time to enable drugs to have a greater probability of positive impact.[24]

Wearables such as Fitbit will eventually evolve into implantables wherein we no longer wear devices but actually embed them into our bodies and they monitor much more than basic vital signs. They can already monitor the gait of our movements, sweat patterns, and other indicators many of us are not aware of. All of the streams of data that are collected about our bodies can already be sent into the cloud to enable AI to make predictions and highlight areas of risk. We are still years away from the development of powerful wearable devices or implantables that can collect large amounts of health data in real-time and have access to enough Internet bandwidth to upload it into the cloud, however. The first sophisticated real-time health monitors are likely to first appear in our homes. In 2016, Google applied for a smart bathroom patent[25] where sensors are embedded in the toilet, mirror, and tub. The patent suggests that the "smart bathroom" would go beyond simple health monitoring to potentially track the activity of nervous, endocrine, and muscular systems.

We face a similar series of questions related to allowing AI into our homes and lives as we do with data generation and dissemination more generally. Most people do not tend to think very much about data collection from the Alexa or "smart" appliances sitting in their kitchen, yet they have wittingly or unwittingly made a decision to allow the maker of these devices to collect sometimes personal data about them, not necessarily knowing or caring about what happens to that data. Cyber risk has muscled its way into most peoples' lives, and there is little reason to believe that, as AI develops, it will not do the same. Some would say it already has. But the stakes are particularly high when it comes to the data produced from our health and medical data, for there is nothing more personal.

Data privacy will become an even stronger point of contention in the future because third party companies could lobby for access to real-time health data. If third parties were to gain access to this type of data, it could lead to unintended consequences, such as targeted marketing based on a person's unique genetic makeup, health history, and bodily functions. For example, perhaps if someone were at greater risk for heart disease, he or she would automatically see advertisements for health clubs or healthy foods when they surf the Internet. To a limited degree, this is already the case, because, as we all know, once we have purchased something on Amazon, searched for a hotel room, or looked for a doctor online, we already receive unsolicited advertisements, courtesy of Internet service providers.

Health-related data need not necessarily be shared with third parties, however, because, in the future, smart health "mirrors" could be installed in different places across a city. Simply passing in front of a smart mirror would trigger a scan of your body and a specific product recommendation would be made to you on the screen based on the results. If this seems far-fetched, bear in mind that facial recognition cameras already exist in cities and countries throughout the world. Based solely on your face, these systems can identify exactly who you are and access a data base about you ranging from your driving record to your social media platforms and financial information.

We could end up in a world where every decision will be scrutinized by AI. Our genetic predisposition to certain diseases could lead to ongoing smartphone-based recommendations to alter our eating habits, purchase a gym membership, or see the doctor for an exam. Will our governments allow us to opt-out of real-time health scans and will we choose to sacrifice our privacy in exchange for early disease detection? These are among the many questions that come to mind as we contemplate an AI-driven future in the healthcare arena. No doubt, the answers to such questions will be as different as the choices to be made. The real question is, to what extent will we *have* a choice in the matter in the future? Will governments, insurers, or health care providers decide what is necessary, or what is good or bad for us?

Genetic Sequencing

Another major piece of precision medicine that is still in early stages of development is genetic sequencing. The effort to sequence the first human genome—3 billion base pairs of DNA that reside on 23 chromosome pairs—officially began in 1990, cost $2.7 billion to create, and took nearly 15 years to complete. By 2015, the cost plummeted and someone could have her entire genome sequenced and analyzed for about $1,400. Still, only about 1-2% of people who do so get actionable results[26] because interpreting genetic information remains an embryonic field and environmental factors can influence the final outcome.

By sequencing a genome, geneticists can identify mutations that could identify predisposition to certain illnesses or manifest themselves in a developing disease. Simply identifying gene mutations does not necessarily lead to a definitive prognosis that a person who is susceptible to a certain disease will be develop it at some point in his life. For example, in 2011, scientists identified 50 genetic hot spots[27] that could contain important information about how multiple sclerosis develops. There have been documented cases where genome sequencing has become a matter of life and death.

In 2003, Dr. Lukas Wartman, a young oncologist working at Washington University, was diagnosed with adult acute lymphoblastic leukemia. He went through intense chemotherapy but, after going into remission for a time, the cancer returned. After his final relapse, he was again treated with intense chemotherapy, but the cancer did not go into remission. As Wartman was undergoing treatment, Washington University was sequencing on an ongoing basis the genome of his cancer, along with healthy cells, via 26 sequencing machines and a supercomputer. When the results came back, researchers identified one normal gene had been producing large amounts of a protein and fueling the aggressive growth of the cancer. Fortunately, a Food and Drug Administration (FDA)-approved drug (Sutent) was available to treat his advanced kidney cancer. Although the drug was not within the scope of standard treatment protocols for adult acute lymphoblastic leukemia, having identified the malfunctioning gene through sequencing allowed researchers to conclude that an existing drug could treat it. After Wartman took the drug for a few weeks and completed a lower dose of chemotherapy, his cancer went into remission.[28]

Wartman is probably only alive today because he happened to work for a lab that had access to supercomputers that could sequence his genes, and researchers who could interpret the results. As the cost of sequencing has dropped exponentially, it is becoming more accessible to the average person, but who will interpret the results and what will be the cost? While the interpretation of genetic information is still in its early stages, it can clearly be catapulted forward with the help of AI.

Advances in genetic sequencing and in identifying the potential causes of disease progression could rapidly outpace the drug development process. According to a 2014 study,[29] it cost an astronomical $2.6 billion (in 2013 dollars) to develop a drug and bring it to market over a decade. Based on the high cost of drug development, there has been a flood of investment into AI startups.[30] In 2017, Deep Genomics, which focuses on researching the genetic cause of disease, got into the drug development business.[31] Deep Genomics is focusing on oligonucleotide therapeutics, which is being called the third major drug-development platform, following small molecules and biologics. Oligonucleotide therapies require the analysis of tens of billions of compounds to achieve effective, targeted combinations. Deep Genomics wants to tackle these with DL and high-performance computers.

The future of medicine will be driven by computing because the human body is basically a large program and we are only beginning to model pieces of it. The first phase of AI in medicine will be to identify genetic anomalies that are linked to specific diseases while targeted drugs will be created to manipulate those genes. The next phase will be to create targeted drugs that are optimized to reduce negative impact to other parts of the body. As more people get their genomes sequenced, a pool of information will be created that can be referenced with other data such as medical records and lifestyle. At some point in the future, a person may be able to get their genome sequenced and have their medical records and lifestyle indicators screened through AI that predicts the likelihood of developing specific diseases, based on what happened with other people.

It is not only the sequencing of the human genome that can lead to breakthroughs in healthcare–several US biotechnology firms are harnessing AI and genetics to fight antibiotic-resistant bacteria, which is one of the greatest healthcare threats humanity faces. In the US, each year, approximately 2 million people are infected with resistant bacteria and at least 23,000 of them die from those infections.[32] These firms are exploring how to identify which bacteriophages or bacterial viruses can be used to invade bacterial cells and cause them to dissolve.[33]

Adaptive Phage Therapeutics wants to create custom therapies, based on the bacteria that have infected patients, and has a lab test to make these predictions. The company is training an ML algorithm with the genomes from bacteria that can predict the best bacteriophage to fight a particular infection by reducing the amount of time required to make an analysis.[34] In 2017, Veritas Genetics bought an AI company[35] to help make sense of the genetic information it is sequencing. If companies can sequence the genome of large numbers of individuals, gather information on their lifestyle, and monitor the development and progression of disease, ML can do what it does best and find links in the genes without being explicitly programmed to do so.

If genetic and lifestyle information is the raw data and the development of disease is the outcome, then different types of ML can be leveraged to find what types of genetic mutations lead to certain diseases in a given timeline. Such analysis obviously requires huge computing power, given that a single person's genome takes up about 150 gigabytes of storage space (roughly 9.7 million average pages of text in a Microsoft Word document)[36]. Finding a sufficient amount of storage space is the starting point of the genome sequencing process, which is well under way.

The next exciting area in genetics is editing. Clustered Regularly Interspaced Short Palindromic Repeats (CRISPR) refers to a natural defense system that bacteria use to protect themselves from infection against viruses. When the CRISPR-Cas9 system in a bacterium detects an invading virus, it creates two types of short RNA, one of which contains a genetic sequence that matches the virus. The two RNA strands form a complex with a protein called Cas-9, which is a type of enzyme that can cut DNA. Scientists discovered they could reprogram this natural system to cut the DNA of any living cell, including those of humans, at specific locations by changing the guide RNA to match the target gene. Scientists can even cut out a mutant gene and replace it with a healthy copy.[37]

Another way to think about the CRISPR-Cas9 system is that it is a nano-sized biological sewing kit that can alter and cut DNA at a specific point in a specific gene.[38] The CRIPSR-Cas9 system can also target many different genes simultaneously, which is useful for researchers given that some diseases arise from the complex interactions of many mutated genes, as opposed to a single one. The ramifications of CRISPR-Cas9 gene editing technology almost sounds like science fiction, for it can potentially allow us to control our evolution by ending genetic diseases, curing cancer, bioengineering designer babies, ending aging, and modifying our bodies for space travel and extraplanetary colonization. Regardless of how the CRISPR technologies develop, AI will play a central role in its advancement as a treatment that can be used in humans.

Gene editing technology is still in the earlier stages of development, but some successful studies point to a world of future possibilities. In 2017, researchers successfully removed the human immunodeficiency virus DNA from the genome of living animals, preventing further infection.[39] There are roughly 50,000 known genetic mutations linked to disease in humans, approximately 32,000 of which are caused by a single mutation in a base pair. This makes gene editing technology powerful because it can be made reliable for use in humans and it has the potential to eliminate more than half of the genetic diseases currently known.

However, gene editing is not without its risks and, while it has shown to be somewhat reliable in certain scientific studies, the technology risks making off-target edits or unintended changes in the genome which could be permanent and/or difficult to reverse.[40] China, the UK, and the US have conducted genetic editing experiments of human embryos; however, in the UK and US, researchers must destroy the embryos after a few days and are not permitted to implant them into a womb. Based on looser regulations, China is on course to be the first country to produce a genetically edited child.

The implications of gene editing technology are far reaching. If this technology can truly be made viable in humans, genetic disease could become a thing of the past. Genetic editing could also have a profound impact on the economy. If we move towards a society where only the most desirable traits are selected in humans of the future, what will be the basis of competition? We already live in a world where genetic information influences life and death, when, for example, pregnancies are terminated because a genetic disease is detected. The potential misuse of gene editing is also a source of great concern. Terrorists or rogue states could decide to use gene editing to develop evenly deadlier pathogens[41] than exist today, which could potentially cause global epidemics of biblical proportions. This is already a concern among the world's intelligence agencies. The threat of next generation biological weapons could also revitalize a biological arms race among nation states, which is why it must be handled with extreme care.

Man and Machine Merge

The FIR is characterized by the blurring of lines between things physical, digital, and biological. This is already being seen with unbelievable medical procedures that are literally merging a human being with a machine through implantable devices. Kimberly Bari is a young woman who began suffering from frequent and violent seizures in 2010. Bari had hundreds of seizures because of epilepsy. Her condition was particularly debilitating because it was drug-resistant. Fortunately, neurologists found that she was a good candidate for a procedure that fewer than 2,000 people had attempted, involving implanting the world's first responsive neurostimulation system (RNS) (or computer) into her brain.

According to NeuroPace, the creators of the RNS, the technology is only possible because of ML and a lot of data. The NeuroPace engineers trained an ML algorithm that powers the RNS by using more than 2 million recordings of brain activity from other patients. The system was not explicitly programmed to detect abnormal activity, but programs itself with the labels of normal and seizure brain activity it is given from researchers. In this way, as it monitors new brain activity of existing patients using the RNS (such as Bari), it can predict what type of patterns are likely to lead to a seizure, based on how it was trained, and intervene with electrical stimulation completely autonomously.[42]

The RNS includes a small neurostimulator connected to leads (tiny wires with electrodes) that are placed in up to two of the onset seizure areas. The RNS is similar to a pacemaker but, instead of monitoring the heart, it monitors and responds to brain activity in real-time. Patients use a device to scan the neurostimulator to collect data about their brain activity and upload it to the cloud, where doctors can access the information and plan changes as needed. The RNS may or may not allow for neurologists to make minor adjustments to parameters that alter the way the system responds to their specific patients. However, the knowledge framework about what the system considers abnormal brain activity is not likely to be drastically changed. If the RNS keeps predicting that a certain type of brain activity is a seizure, when neurologists provide feedback that indicates that a prediction was a false positive, this label will be used by the AI to make more accurate predictions in the future.

When medical procedures are conducted to implant a digital device into the human body, there are usually significant health factors contributing to that decision, particularly when other non-invasive treatment methods have been determined to be unviable. However, some people are simply motivated by the ability to enhance their senses by embedding technological devices into their bodies. Neil Harbisson is an artist who was born with a rare genetic disorder–achromatopsia–which causes complete color blindness. For him, the world was viewed in greyscale and he had never been able to perceive colors through his vision. He began exploring what types of alternative perceptions of color might be possible and discovered he could install an antenna in his head to interpret different light frequencies–from infrared to ultraviolet–through a fiber-optic sensor hovering in front of him. The sensor transferred that data to a chip in the back of his head, which was converted into vibrations in his skull and, finally, sound. For Harbisson the perception of color was completely distinct from his vision and hearing, so it was an entirely new sense.

Harbisson became widely known as the world's first official cyborg after the UK government allowed him to take his passport photo with his antenna attached, which, prior to 2004, would have been rejected because of a prohibition of electronic equipment in photographs.[43] He argued that the antenna is not electronic equipment but a part of his body that extends his perception of color through sound. In a 2012 Ted talk[44], Harbisson described his unique experience of color and some of his most stimulating environments, which turned out to be supermarkets because of their fluorescent lights and brightly colored products, which were the equivalent of nightclubs for him. Even his experience of eating food and choosing what to wear had changed because he eats and dresses according to what sounds good to him, rather than how it tastes or looks.

For Harbisson, and other thinkers, all knowledge is derived through our senses, so if we can extend our senses or create a completely new one, we can expand our knowledge. He imagines a time when apps will no longer be created for our smartphones but for our bodies or minds, which will serve as platforms to extend our senses in previously unimaginable ways, through an ecosystem of body hack applications. A brain scan showed that Harbisson's neural activity appeared to be behaving differently since it created sounds when looking at images, even as he slept.[45] In 2013, he had his antennae upgraded to include Wi-Fi and Bluetooth, so his friends could send colors directly to his headgear. However, the connectivity enhancements raised some serious concerns associated with connecting our minds directly to the Internet. As was discussed extensively in Daniel Wagner's book, *Virtual Terror: 21st Century Cyberwarfare*[46], anything that is connected to the Internet is vulnerable to hacking.

Many in the scientific community thought that our species stopped evolving a long-time ago but recent advances in genetics have demonstrated that the human genome is more malleable than previously thought. For example, 12,500 years ago residents of the Andes adapted to the high altitude through natural selection; today, their descendants exhibit a genetically determined trait wherein their hemoglobin can bind to more oxygen than the average person.[47] It may have taken many generations for this genetic trait to appear but sometimes athletes desire these traits and do not want to wait thousands of years to acquire them. Lance Armstrong was an American cyclist who won the Tour de France a record seven consecutive times but had a spectacular fall from grace when he was banned from the sport due to allegations of doping, in 2012. Armstrong had wanted to acquire an increased capacity to process oxygen and perform at high altitudes, so he manipulated his blood through transfusions and hormones. His story demonstrates that, regardless of the ethical implications, if technology exists that can hack our biology, some people will use it to push the boundaries of personal performance and ethics.

Clearly, technology is becoming more closely integrated with our physical bodies as the time it takes for our intentions to manifest themselves as actions is becoming shorter. At some point in the future, the distinction between our intentions and actions will become indistinguishable. Even before we become conscious of it, technology may carry out a desired action in real-time through brain-computer interface devices (BCI). A lot of research has already been done to show how BCI devices can improve the lives of disabled people. Jan Scheuermann was a quadriplegic woman who, at the age of 52, agreed to undergo surgery to implant two electrode arrays in her brain as part of the Brain Interface Project, to allow her to control a robotic arm through her thoughts.[48] The implants were specifically placed in the regions of the brain which activate when a person moves his or her hand and right arm. Scheuermann had not been paralyzed for her entire life, which made her a good candidate for the study, given she knew how it felt to be able to move. She had succumbed to spinocerebellar degeneration in 1996.

During the beginning of the exercises, doctors informed Scheuermann that their AI system would periodically intervene to help her pick up and move objects on a table. Eventually, they turned off the AI assistant but did not tell Scheuermann. She was able to control the robotic arm quite skillfully on her own. It is also possible for scientists to record the neural activity of a paralyzed person moving objects with her thoughts. Such data can be used so that a robot can learn to move around more fluidly, based on human dexterity. Perhaps the next breakthrough in teaching robots to move around will not come from observing humans but directly from the thoughts of paralyzed people who are unable to move.

Elon Musk, who is revolutionizing the auto industry and space exploration with his startups companies Tesla and SpaceX co-founded another company–Neuralink–to build BCI. The ambitions of Neuralink are quite profound and unsettling because the goal is to directly connect our brains to a computer or, more specifically, AI, via a neural lace implant. In theory, a small polymer mesh can be injected into the blood and spread out in the brain to monitor neural activity and potentially send and receive information.

There are many obstacles to this sort of technology becoming viable in the near-term. One of the main challenges is that we do not know enough about the complex interactions between neurons in the brain. Another major obstacle is that the brain has around 80 billion neurons and current state-of-the-art technology only has around 100 electrodes that can record the activity of a single neuron. When technology does become commercially viable (and it will at some point in the future), the possibilities could range from instant access to the Internet simply by thinking to telepathic communication to reliving past experiences to learning complex tasks in near real-time.

The risks associated with BCI cannot be understated, however. Once again, fundamental questions are raised, such as, do we really want to risk exposing our minds to the Internet and AI? Could a hacker download our thoughts, or worse, insert thoughts into our minds? Connecting our minds to the Internet could expose us to things we may not have even considered before, such as remotely induced schizophrenia, as we struggle to discern what is real versus fake. Could it merely be the realm of science fiction that one day our minds could literally be hacked by virtue of having been connected to the Internet? There are also serious socio-economic implications to consider, such as who will be the first people to gain access to BCI technology and whether ordinary people can compete in the job market with enhanced cyborgs.

Humans have evolved to survive on Earth with a drop of consciousness to monitor for potential threats. BCI could diminish the intimate contact we have with the real world by artificially enhancing our mental faculties. It is not clear how this may affect us long-term, of course, but one result could conceivably be voluntary confinement in controlled buildings, due to our increased capacity to process digital information but our decreased capacity to adapt in dynamic environments. The rise in "selfie"-related deaths, from driving, by falling off cliffs, and from wild animal maulings may just indicate that some people have become desensitized to danger in the real world and are willing to risk their lives to enhance their digital personas. Could we modify ourselves so extensively that we are unable to survive in dangerous natural environments, and simply visiting a national park would be unthinkable for the city dwelling cyborgs of the future?

Master Manipulator and Mind Reader

There is already a DARPA-funded research project underway to manipulate emotions of the human brain through AI, which is in its preliminary stages and is testing how brain implants leverage specialized AI algorithms to detect patterns of mood disorders. The idea is to stimulate the brain with electrical impulses periodically (as opposed to constantly), to enhance a subject's mental state. This type of therapy—called deep brain stimulation—is already being used for Parkinson's disease, but researchers believe it could help with other conditions, such as chronic depression.[49]

Researchers are not only creating AI-driven systems that can intervene with brain activity in real-time through electric stimulation, they are also building systems that can read our minds. AI researchers in China, Japan, and the US have published researched showing that functional magnetic resonance imaging (fMRI) machines that monitor brain activity can be used with deep neural networks to reconstruct the images people are looking at, or, in some cases, what they are thinking about. In a 2017 paper, researchers from Japan took DL algorithms a step further to not only reconstruct images from brain activity but to enable the AI to recognize new images that it was not trained to see.[50]

This is one of the unique strengths of DL using multi-layered neural networks: basic concepts can be established in lower layers and more complex abstractions can be made in higher layers. The fact that Japanese researchers could reconstruct images—even basic shapes—from brain activity is somewhat remarkable because it implies there is a distinct structure to our thoughts about images that can modeled.

Researchers from Carnegie Mellon took a different approach to mind reading and, instead of trying to reconstruct an image based on brain imaging, they used ML algorithms to predict what types of complex thoughts a subject was contemplating, based on the brain systems that were activated. The 2017 study[51] was funded by the US government's Intelligence Advanced Research Projects Activity (IARPA). The study suggests that the mind has building blocks for building complex concepts that are not word-based.

A 2016 study by Carnegie Mellon[52] also supported the idea that words are not the most fundamental components of thought, because sentences read in English and Portuguese triggered the same neural activation patterns in the brain, which suggested that AI can predict what a person is thinking regardless of their native language or culture, since humans share common brain activity patterns.

In the 2017 study, it was shown that the content category of a sentence mattered (such as focusing on people, places, actions, and feelings), but so did the context of how the sentence was represented. This interaction between the content category and context activated different parts of the brain, which was recorded by fMRI machines for 239 sentences that were used to train the AI. The AI was tasked with predicting the content of the last sentence based on how the subject's brain regions were activated, with an accurate rate of 87%.

In another 2017 report,[53] scientists from Stanford University described how they implanted one or two small electrodes into the surface of the brain of three paralyzed participants. The electrodes were implanted near the motor cortex to record signals that were transferred to a computer and converted by algorithms to generate point-and-click commands, guiding a cursor along a digital keyboard on a screen. The study was part of a multi-institutional collaborative effort known as BrainGate which is focusing on restoring communication, mobility, and independence for people who suffer from a disease or injury that impedes their ability to move around or control their environment. One of the long-term goals of the research is to create implants which can seamlessly integrate with the IoT of a person's home, so they can control the temperature, turn on the radio, and even eat independently by controlling digital devices with their thoughts.

All the BCI devices that scientists have been researching have been intended to address serious illnesses and restore some level of autonomy back to the patient. It is inevitable that BCIs will make their way to the consumer market where they serve no essential medical purpose but function as a convenience technology. The secretive research unit at Facebook—Building 8—is already working on a BCI that can type for a person by monitoring neural activity. As of 2017, Facebook had a team of 60 scientists working on the project, which would start with an implantable device in a medical setting like other BCI studies.[54] Ultimately, the goal would be to create a non-invasive neural cap—as opposed to relying on brain implants because those do not scale and pose more liability risks–that could monitor and translate thoughts to text.

The head of the Building 8 research group at the time, along with the former head of DARPA, stressed that the project was not about invading a person's thoughts but simply decoding a person's chosen words that they already decided to *share* with their network by decoding neural activity from their speech center. The 2018 Facebook-Cambridge Analytica data scandal raised questions about whether we should want to give Facebook (or social media firms like it) access–no matter how limited–to any of our thoughts. Soon enough, we will not need to make that decision; it will presumably be made for us, by AI.

Conclusion

The race for AI supremacy in science is real and well underway. AI will revolutionize science by drastically increasing the amount of work a research team, or even a single scientist, can conduct in a short period of time. The scientists who use AI in the future will not be researchers any longer but rather managers. One scientist may "manage" multiple AI systems that will be trained to handle specific tasks, such as combing through text in scientific publications and predicting what specific parameters of an experiment are most likely to succeed. While universities will continue to lead most domains of cutting edge research, there will be an increasing number of academics who will join well-funded startups to commercialize the long line of scientific breakthroughs.

Healthcare is one of the sectors of an economy that can benefit the greatest number of people from the race to achieve AI supremacy. The industry is in desperate need of disruption as healthcare costs continue to soar, antibiotic resistant bacteria continues to threaten to create widespread epidemics, and the shortage of doctors to treat a growing and aging population become ever more acute. AI has already had some major successes, such as more accurately diagnosing a cancer patient and providing the miraculous gift of a moving robotic limb controlled by the thoughts to a paraplegic who has not moved any part of her body below the neck for years.

The large banks on Wall Street have already realized they are in a war for talent and are at increasing levels of competition with the leading technology companies. The major hospital systems are also coming to realize that in many countries they are not only competing with other hospitals, but with the leading technology companies. This realization will manifest itself in many ways, but one way is that the major hospital systems will not be as eager to hand over medical data to AI firms; rather, they will be building their own AI systems. Some have already begun to do so, because the power these organizations hold is not necessarily by providing healthcare through their limited number of doctors, but through the data that shows how they provide effective healthcare.

That said, some smaller hospital systems may still be willing to give up their data to AI firms in exchange for more accurate medical diagnoses because they will not be able to fund their own AI innovation labs. As AI becomes more mainstream in healthcare, we could see an increased number of hospital systems buying smaller hospitals to disrupt data flows to the major technology firms playing in the healthcare space. However, the AI firms have a backup plan and they will probably offer high-powered AI systems to hospitals in developing nations in exchange for a continual stream of medical data.

AI is truly driving historic changes in medicine. Humanity can reap the rewards or we can be the architects of our own demise. Genetic sequencing and editing are erecting a whole new paradigm of possibilities for disease treatment and prevention. We are essentially at the beginning of an age where we can control and manipulate the most basic building blocks of life with the utmost precision. However, this level of control on something so profound and with so many unknowns—such as our own genetic code—can be mishandled and used for a variety of potentially illicit purposes.

As AI begins to invade all aspects of healthcare, there are some very serious risks that need to be considered, such as the threat of cyberattacks, being held for ransom, or invasions of privacy. There are, in addition, some perplexing ethical questions that we must ask our leaders not only as citizens of nations, but as members of the human race. In many ways, the risk is not that too much AI will exist in the healthcare arena, but that too little AI will be deployed and that we will fail to maximize the potential benefits to be gained from it.

Daniel Wagner and Keith Furst

6. *An Evolutionary Communications Landscape*

Infrastructure

Data is the fuel that powers AI and communication networks are the pipelines that move data to where it needs to go to be processed. Without communication networks, AI's potential usefulness and power would be greatly diminished. As companies and countries race to attempt to achieve AI supremacy, the speed of data transfer is influencing the competitive landscape. Average Internet speeds between the top 10 countries remain within a reasonable speed range of approximately 10 megabits per second (Mbps), but that gap is likely to widen.

In 2017, South Korea had the world's fastest Internet speed, with an average speed of 28.6 Mbps, and was the only country to exceed the 25 Mbps threshold.[1] Despite this, the South Korean government did not get complacent. In 2018, it announced plans[2] to fund a communications infrastructure that supports Internet speeds of 10 gigabits per second (Gbps – 10,000 times faster than 1 Mbps, and roughly 350 times faster than its 2017 speed), which was planned to be deployed to 50% of the country by 2020. If South Korea achieves that objective, it will influence the competitive AI landscape and prompt other countries to want to achieve the same.

Some companies in the AI space will certainly be incentivized to migrate to countries that can provide that kind of Internet speed. While the higher bandwidth and faster processing speed will naturally come with a higher operating cost attached, the flip side is that consumers will produce more data, and companies (and the government, of course) can collect more data more rapidly. Given the speed with which the South Korean government intended to implement its plan, there was no way that the regulatory environment could move at a similar pace, which raised a host of potential issues, ranging from the rule of law and governance considerations to the ability to adapt to changes in the operating environment.

There is another major change on the horizon as we move from fourth generation (4G) to fifth generation (5G) mobile technology. Over the past decade, 4G mobile technology has become the de facto standard for many consumers across the globe. The increased capacity to transfer large amounts of information quickly gave rise to entirely new industries, such as live-streaming, social media applications (apps), and highly accurate GPS location data to support disruptive ride sharing apps such as Uber and Lyft. The 4G network can support connection speeds up to 1 Gbps; the 5G network will increase that to about 10 Gbps.

The emergence of 5G and the IoT is expected to increase data traffic by ten-fold in the next decade. This massive increase in data traffic from the exploding number of connected devices will fundamentally change the dynamics of the mobile network and make it even more unpredictable.[3] The increased data speeds will undoubtedly make it difficult for Internet Service Providers (ISPs) to respond to congestion, equipment failures, and random traffic spikes quickly by leveraging existing tools without impacting network performance. The industry is exploring how ML could dynamically respond to network-related issues and resolve them autonomously.

In 2017, the International Telecommunication Union (ITU) established the Focus Group on Machine Learning for Future Networks, including 5G (FG-ML5G), to draft technical reports and specifications related to the 5G rollout. The FG-ML5G published a terms of reference document which highlighted some of the ITU's main priorities, such as keeping personal data confidential and establishing standard interfaces, processes, and data formats.

The objectives of the FG-ML5G are to:

- Study, review, and survey existing technologies, platforms, guidelines, and standards for ML in future networks;

- Recognize and highlight various perspectives for the future of networks and computing systems involving ML;

- Identify aspects enabling safe and trusted use of ML frameworks;

- Spur the adoption of ML in future networks, including architecture, interfaces, use cases, protocols, algorithms, data formats, interoperability, performance, evaluation, security, and protection of personal information;

- Review and study how to train, adapt, compress, and exchange ML algorithms in future networks, and how multiple algorithms interact with each other;

- Identify possible requirements of ML applied to future networks (taking into account a variety of fixed and mobile communication stacks) and promote the development of new ML methods that will be able to meet these requirements;

- Ascertain possible requirements on network functionality, interfaces, and capabilities to use ML;

- Categorize challenges in the standardization activities for ML in communications;

- Produce a gap analysis of ML in order to identify the relevant scope of ITU recommendations for these topics and develop a roadmap for ML; and

- Establish liaisons and relationships with other organizations that could contribute to the standardization activities for ML.[4]

The FG-ML5G objectives are very important because it is one of the first instances in which an international organization has tried to establish standards for the interaction of two or more ML algorithms, with potential impact for the majority of the world's population.

Today, it is very common to standardize interfaces when two or more systems communicate, but one of the FG-ML5G goals is that two or more ML algorithms must be trained so they can interact, in a predictable way, with one another, in an unpredictable environment. The ML algorithms that will power 5G are expected to collaborate for a common goal, which is the exact opposite of how ML algorithms tend to interact with one another today—in a zero-sum game.

It is not clear whether these 5G ML algorithms will have equal access to information from which to base their decisions, or whether some will become smarter than others (presumably, some will). This could create scenarios where the smarter (or dumber) ML algorithms will expect other algorithms to act in a certain way. If they do not, it could create unanticipated responses that cascade through a network. As discussed in Chapter 4, algorithms appear to have contributed to the 2010 flash crash in the US stock market. Will we see similar volatility, due to ML algorithms gone wild, in the mobile technology networks of the future? It is hard to imagine that we would not.

The increased speed of the 5G network will greatly accelerate the number of potential applications for AI because large amounts of real-time data will be transported all over the world at speeds which are exponentially greater than is currently possible. Importantly, *we cannot get to the next level of 5G mobile technology without ML, and the next level of 5G mobile technology will deliver more accurate and powerful applications of ML, so they are dependent upon one another*. In other words, as AI and ML manage the complexity of the high-speed networks of the future, there will be even more opportunities for other AI and ML applications to emerge.

The notion of faster Internet speeds accelerating the race for AI supremacy could be enhanced by significant future breakthroughs in technology. Researchers from University College of London described in 2016[5] how they set a record for the fastest ever data rate transfer for digital information. The team achieved a rate of 1.125 terabits per second (Tbps), which is roughly 50,000 times faster than the average UK broadband connection speed of 24 Mbps. To put this into perspective, at this speed, someone could have downloaded the entire Game of Thrones video series (50 episodes as of 2015) in a single second.

As we begin to train AI with massive amounts of live streaming data, companies with smart operating systems will cut production times from months or weeks to days or even hours. These systems are likely to be so complex that it would take a team of human researchers months to test all of the possible scenarios to ascertain how accurate an AI-driven system's predictions are. Even then, they would be only scratching the surface. *We are rapidly advancing to a point where resolving AI-driven issues will require more AI, and we are barreling towards a future in which AI could become both the question and the answer.*

Prioritization and Efficiency

While advances in Internet connection speeds will give rise to new uses for AI in global communication systems, major advances continue to be made with existing infrastructure. Smartphone, email, and social media usage are far outpacing the growth of the global population. A 2017 study[6] estimated that the total number of emails sent and received *each day* will reach *320 billion* by the end of 2021. Some email providers have been using AI and ML to help address spam for years, but new use cases have emerged.

For example, Gmail extended its spam feature to automatically organize emails in different subcategories, such as Social, Promotions, and Updates, in essence pre-categorizing messages by presumed level of importance. In 2017, Google estimated[7] that 50% to 70% of the messages Gmail received were spam and claimed that its AI had blocked 99.9% of those malicious emails. Google and LinkedIn are taking electronic communication management beyond message classification and have already rolled out a *smart reply* feature, driven by ML algorithms, in their email and messaging platforms. The smart reply AI is trying to solve a similar classification issue as the spam and email categorization algorithms. However, the smart reply feature takes the problem a step further with NLP by not only understanding what the message is about, but how to respond, in a few words or symbols, intelligently.

In a 2016 research paper,[8] Google described how a study of several million email-reply pairs showed that approximately 25% of the responses contained 20 tokens (words) or less. This means that a good portion of responses to emails are rather brief, which led researchers to consider whether ML could help provide short and sensible responses. Google's approach was to use DL (a neural network machine translation model) called sequence-to-sequence learning. Google also used a semi-supervised learning approach wherein labels were given to a small cluster of emails to act as an anchor (or seed) for the algorithm. The researchers wanted to arrive at three possible replies that were of high-quality, yet diverse enough to capture the user's sentiment. Gmail-enabled users can activate, or deactivate, the smart reply feature through the Gmail settings tab.

LinkedIn also offers smart reply functionality. For existing business relationships, a request for a meeting can usually be simply answered with a smart reply of either yes or no. However, some of its smart reply options are framed so as to appear more *human* since, sometimes, even a thumbs up emoji or use of Internet slang (such as laugh out loud) is a preferred option. For recruiters and sales representatives contacting people outside of their network, the smart reply option can be very helpful. LinkedIn has had a dramatic impact on the recruiting industry because its members' skills, experience, and work history are all listed on the platform. This enables recruiters to search for talent via LinkedIn and contact them directly, as opposed to combing through various job boards where the most qualified candidates may not even post their resumes. LinkedIn's smart reply feature will offer multiple options to the person being contacted to permit a reply to recruiters with a click of a button.

In 2016, Microsoft bought LinkedIn for $26 billion[9] or $196 per share. The closing share price on the day of the purchase was priced at a 50% premium. A significant driver for the *seemingly* absurdly overpriced acquisition cost was to get ahead of—and begin to merge—leading technology trends such as AI, cloud computing, mobile, and social media. Another factor was that many professionals have LinkedIn accounts, which would provide Microsoft with a large repository of data regarding companies, skills, and hiring trends. It is hard to put a value on this data because there is no telling what AI could be used to predict and what type of restrictions might be imposed on the use of such data in the future.

While much of the data on LinkedIn is publicly available, Microsoft could gain additional insights in the M&A space by detecting when smaller tech companies get into trouble, indicated by, for example, high-profile executives leaving, or lower ranked employees exiting the company in droves. Mass or strategic employee departures could be a sign of turmoil inside a company. Such intelligence could be used as leverage in a buyout. Job postings, course content enrollments, and member listings are other types of data that Microsoft may use to deploy AI to analyze, in order to gain unique insights into their customer base, competitors, and overall market dynamics. Microsoft could, of course, also integrate LinkedIn with its other flagship applications or borrow some of the core technology, such as smart reply AI, to improve the efficiency of it businesses by offering single click responses to its large corporate customer base. This raises privacy, legal, and ethical concerns, which will only become more prominent with time as Microsoft develops new applications for acquiring and making use of more and more data.

The smart reply feature has already covered new ground through Google, which announced in 2018[10] that its new "smart compose" Gmail feature sought to compose emails in real-time. As a user is writing an email, the feature will prompt him or her to accept Gmail's suggestion by pressing the tab button. AI will then finish a sentence. While some may see this as presumptuous, intrusive, pompous, or silly on Google's part, many others will, no doubt, embrace Gmail's suggestions and be glad they do not need to finish a given sentence themselves. The smart compose feature can finish a word, suggest several words to complete a sentence, suggest a meeting time, or even autofill an address. The feature is good at generating generic greetings and general filler conversation, which is how many business emails begin.[11]

However, it appears that AI still has a lot to learn and sometimes suggests peculiar responses to messages it is not able to comprehend. Do most people really want AI to read our minds and compose emails for us? Where do we draw a line, or will it be AI that draws the line for us regarding when human touch is required? Google is drawing from a massive repository of emails in order to train its AI. Could doing so exacerbate poor email impulses, such as promoting poor use of a given language or making us even lazier than we already are? Will companies attempt to create policies regarding employee use of smart replies for internal versus external communication, or will they simply cede governance-related issues to organizations such as Google? On one hand, the smart compose feature could be useful for employees conducting repetitive tasks such as responding to customer email inquiries, increasing employee output, and possibly fostering a better experience through rapid feedback. On the other hand, it implies a further erosion of our control over basic tasks and our desire to think for ourselves.

The Future of Customer Service

AI has already embedded itself in call center interactions by guiding conversations that occur between company representatives and customers. Some people fear that the rise of AI-driven chat bots will be the last nail in the coffin for the customer service employee. However, a 2017 survey[12] highlighted that more than two-thirds of all customer service interactions occur with agents via voice or chat, and over two-thirds of customers still prefer agent assistance over self-service. While self-service through AI-powered chat bots could help resolve simple issues, they are not yet robust enough to handle many of the more complex inquiries that can arise. Yet, AI is becoming an integral part of how companies remotely interact with customers. Agents will continue to use AI to enhance their ability to provide exceptional experience. Some will even offer customers the option of leveraging other AI-attended channels for issues they could otherwise resolve on their own.

More companies are also using AI to drive proactive customer interactions. For example, there are approximately 12 million elevators in use globally that carry around 1 billion people daily. In 2015, ThyssenKrupp launched a predictive maintenance service[13] to collect data through IoT sensors that were integrated with its customer relationship management (CRM) systems. One of Thyssen's challenges was that general rules cannot be written for elevator failures because each elevator has its own unique usage patterns, and although specific maintenance routines are supposed to be followed, they are not necessarily followed exactly the same way for each elevator. ML algorithms are therefore vital to predict failures based on the unique usage of each elevator, when compared to a repository of utilization and failure patterns. Thyssen increased field operation efficiencies and decreased costs by managing resources based on predicted needs, as opposed to a random influx of service requests. In the process, the company ended up providing a superior customer experience. In this sense, the future of customer service, and the standard upon which companies may be judged may become their ability to fix an issue before it happens.

AI is not only involved when predicting if or when customers may experience product or service failures but also in identifying their successes. Some companies are using CRM systems that are integrated with AI to identify instances when they can engage with their customers in a positive way. For example, relationship managers enter the data of their customers in a CRM system which can then be monitored for mentions of the customer in open data sources. Alerts from the AI-enhanced CRM system can then be generated for customers who recently received an award in their local community, for example, which allows the relationship manager to contact and congratulate him or her. This sort of AI monitoring must be handled with care, however. On one hand, the customer could be flattered that the relationship manager reached out. On the other hand, some customers may feel put off because, if they do not have such a strong relationship with the manager in the first place, they could feel that their privacy is being infringed upon, or that the manager is overstepping her bounds.

How can companies craft a sensible policy about when and how to use AI to monitor customers to improve their relationship? The use of AI in communications could get more convoluted as time passes. Some organizations may need to offer an AI "op-out" option. If AI privacy becomes a serious public concern, they may choose to use another type of AI to predict how sensitive a person would be to AI monitoring, based on his or her public presence in cyberspace. This again highlights the virtuous cycle of AI. It could turn out to be the case that the companies that offer some sort of AI monitoring protections to their customers will end up having a higher degree of loyalty and satisfaction, as opposed to organizations who use AI to predict everything about their customers indiscriminately.

Negative public sentiment towards AI could grow if companies use it haphazardly. In 2012, a *New York Times Magazine* story[14] described how Target sent pregnancy-related product coupons to a teenage girl because the company predicted she was pregnant, based on a basket of goods she had purchased, even before her own father knew. *It will be the companies that learn to strike the right balance between using AI to provide good customer experience and added value through accurate predictions while maintaining policies to ensure their consumers' right to privacy, that come out on top.* Such awareness and acknowledgement of the potential misuse of AI will reduce the risk of an algorithm breaching sensitive topics thoughtlessly or providing recommendations to minors without obtaining the prior consent of their parents or legal guardians.

Bricks and Mortars

One consequence of the digital revolution is that an email address is required for many online transactions. But the manner in which businesses and customers communicate with one another in brick and mortar stores is also becoming a race to see how quickly transactions can be completed so that the next customer can be served. The trend of consumers providing data to confirm their identity is gradually shifting to consumers providing their identity in order to confirm their data. In other words, today, a customer may enter a coffee shop and identify all his past transactions by providing a bar code on his smartphone, an email address, or a phone number. AI is shifting the focus from data the customer knows or possesses to where the customer is and what he may be likely to order by leveraging biometric identification.

As an example, CaliBurger, in Pasadena, California has rolled out an AI-powered kiosk[15] that lets customers log on to their loyalty accounts and pay for their order by using facial recognition software. The system still requires the customer to enter the card verification code (CVC) code from their debit or credit card in order to approve the transaction. The goal is to remove the need to verify the CVC code or similar data (such as an email address) as facial recognition algorithms and their reliability improve. As we will see in Chapters 11 and 14, some businesses have already achieved this on a larger scale. It would appear to be a question of time until it is used more widely.

Since so many of us are willingly providing restaurants and other establishments with our biometric information, what else might they do with it? Restaurants could certainly use biometric information to predict what will taste best to an individual based on personal preferences and other profile demographics. Analytical Flavor Systems is a New York-based startup that has created the Gastrograph[16] that uses ML to discover patterns in taste preferences based on the assumption that each of us experiences flavor differently. The tool is available as a smartphone app that includes a 24-spoke wheel divided into categories of sensory experience, such as rich, spices, or dairy. The user rates the intensity of each sensory experience on a scale from one to five and selects an overall score for the food of between one and seven. The company must address how to collect enough detailed data on various foods to make accurate predictions for a diverse group of people and taste preferences.

This type of AI could change the communication dynamic of consumers. If the Gastrograph were to become well-known among foodies, then when a customer asks a waiter for a recommendation for a main course or a bottle of wine, a common server response might be to ask a question about what the person liked. The restaurant experience of the future could evolve to include customers' food profile barcodes, which would then be scanned to make unique recommendations for food, wine, and other drinks. If a restaurant patron scored high on rich flavors but low on spices, AI might recommend to the server a dish with a complementary glass of wine that uniquely matched his palette.

AI is also changing the operational efficiency of physical retail stores to encourage more people to shop in them. In 2017, the world's largest retailer, Walmart, announced[17] that it would be using robots to scan shelves so that store associates could restock items faster. The robots created by Bossa Nova Robotics were fitted with multiple cameras and used ML algorithms for computer vision and other cognitive tasks. The robots could also identify when prices were missing from shelves or an item had been placed in the wrong location. As the robot scanned the shelves, it would upload the information to a server which would process crucial information to associates for them to act upon.

Walmart said the robots were 50% more productive than their human counterparts and could scan shelves more accurately and roughly three times faster. As noted in Chapter 3, Amazon has been using Kiva robots in its warehouses to handle selecting and packing items, which saves the company an estimated 20% in operating costs. The difference with Walmart's robots is that they are actively scanning the shelves while customers are in the store. The communication dynamic in these stores is changing because people are not telling machines what to do; rather, machines are telling people what to do.

Amazon has become masterful at collecting data on its users, making recommendations for future purchases, and advertising its suite of AI-oriented products, such as Alexa. It remains unclear just how much data Amazon is collecting, how long it intends to keep it, or what it might be used for in the future, but Amazon could potentially build a digital profile of our buying habits online and merge them with insights into our physical bodies. For example, Amazon could identify purchasing patterns among people of similar weight, height, gender, body type, age, clothing preferences, or ethnicity. It may make real-time purchasing recommendations based on previous buying habits or taking into account what others in your household may have purchased. If Amazon could discover insights into what people would want based on who they are (not what they clicked online), how else might the company use such data?

What seems clear is that communication in some brick and mortar stores will evolve from being less about sounds, words, and body gestures between people and be more about constant surveillance of our actions while in the establishment —what we say to the people at our table, whether we appear to be in a good or bad mood, and whether we are more or less inclined to purchase something expensive, merging that information with our online behavioral profile to create a more holistic view of who we are, what we value, and what might want at any given point in time.

The way consumers engage with businesses will also change based on the relentless march of AI into every nook and cranny of consumer interaction. In 2018, Google showcased[18] a video of its AI, which it calls Duplex, interacting with a hair salon employee and scheduling an appointment on behalf of a real person. Duplex made an appointment at a hair salon on behalf of its human boss and spoke with a restaurant employee who appeared to be a non-native speaker of English. Duplex was somehow able to navigate through the ambiguity of the conversation to determine that no reservations were required. Duplex, and AI applications like it, will clearly get smarter over time. Not all businesses will be pleased to deal with AI, but some could see the rise of AI as an opportunity to ride with the presiding tide of the times. Will businesses begin displaying AI-friendly stickers or flags on their front doors to symbolize their openness to engaging with AI? That seems to be just a question of time.

Master Negotiator and Debater

We are on the brink of having AI transact with other human beings on our behalf, and the purveyors of AI are busily crafting ways in which this will occur more easily and naturally. Perhaps one day, in the not too distant future, two or more AI-powered virtual assistants will deal directly with one another to negotiate on behalf of their owners. This reinforces the notion that ML algorithms ought to be developed based on common standards so as to be able to interact with one another in predictable ways.

In 2018, IBM showcased[19] its AI system—Project Debater—to engage in the first ever live debate with humans. The 2016 Israeli national debate champion, Noa Ovadia, and Project Debater both prepared arguments for and against the statement "we should subsidize space exploration". Both then delivered an opening statement, rebuttal, and summary. The AI argued that space exploration provides benefits to humanity by helping advance scientific discoveries and inspires young people to think beyond themselves. Ovadia made the opposing argument, that there were better uses for government funds than space exploration, such as focusing on Earth-based projects. The AI responded with a rebuttal asserting that the potential technological and economic benefits of space exploration outweigh other government spending initiatives. A short poll after the closing summaries showed that the audience felt the AI enriched their understanding of the issue more than their human counterpart.

Project Debater does not have a body. It cannot sweat or look sick. So, is AI its own medium of communication? Could AI potentially come to represent the embodiment of purely rational thought? It may be beneficial to businesses, in certain circumstances, to keep AI in a box and make it look like a computer, as opposed to anthropomorphizing it into a human-like body. On the surface, it may appear that AI cannot have biases, but as we noted in Chapter 1, this is simply not the case. The quality of an AI system's arguments and predictions could be linked to many factors, such as the quality of the data it was trained with and whether that data contains any hidden or historical biases. Regardless, such debates could play an important role in challenging our own cognitive biases as individuals, members of families, organizations, companies, communities, and governments. If we were to use an AI debater system to enhance our own understanding and challenge our existing ideas and biases of the world, it would undoubtedly promote our own cognitive development and growth.

Deception

Humans have a comparative advantage over machines when it comes to intuition, gut feeling, and the ability to "read" peoples' emotional state. AI researchers are already on the path toward infusing some of those capabilities in machines and AI has already made profound progress in reading the likely path of an individual's actions. Now, researchers are busy trying training AI to detect deception. In 2017, researchers from the University of Maryland described how they developed an AI system–Deception Analysis and Reasoning Engine (DARE)–that autonomously detects deception in courtroom trial videos[20]. DARE was trained to identify and classify micro-expressions such as lips protruding, eyebrow frowns, and vocal pattern indicators of lying. Researchers trained the AI based on video data wherein actors were instructed to tell the truth or lie in certain circumstances. The DARE system outperformed humans when identifying lying.

While the results of the DARE system's performance were encouraging, the data set it used should be questioned.

How can we really know if lying in a courtroom is representative of AI's true capabilities or due to flaws in a controlled setting? An actor could have an elevated heartbeat and sweat during a camera interview, while a diagnosed psychopath could have minor physiological reactions to a barrage of questions from a prosecutor. The US Department of Homeland Security and authorities in Canada and the EU are testing a system called the Automated Virtual Agent for Truth Assessments in Real-Time (AVATAR) to help determine deception from individuals trying to cross borders.

The AVATAR system uses ML algorithms to detect changes in the eyes, voice, posture, and other body gestures, to flag potential risks that could require a person to be interviewed by a human investigator. The AVATAR system reportedly detects deception with up to 75% accuracy rate, compared to its human counterparts, which perform at a 54% to 60% accuracy rate. The AVATAR system could be useful to law enforcement authorities by focusing their limited resources on the individuals who pose the greatest risks. Some critics have urged caution, stating that there is no reliable link between physical stimuli and what a person is thinking.[21]

According to a 2017 study,[22] US businesses lose about $50 billion each year to employee theft. Converus launched an AI lie detection system called EyeDetect[23] to help combat it. EyeDetect is a less expensive and less biased alternative to polygraphs. The test is administered via a tablet in the form of true-or-false questions as an infrared camera tracks eye movements and pupil dilation. The system then scores the subject on a scale of 0 to 100 for deceptiveness. As of 2017, EyeDetect was being used in more than 30 countries as part of job interviews and corporate investigations. Financial institutions in Latin America are using it to screen possible bank tellers to ensure they can be trusted.

Just as companies are developing AI that can detect if we are lying, as described in Chapter 5, others are using AI to develop drugs and, possibly, other means to hide our lying. The EyeDetect system appears to be highly dependent on the normal functioning of eye movements and pupil dilation. It is not clear, however, whether AI can be fooled if the person used an anti-dilation eye drop solution before taking the EyeDetect test. For example, dapiprazole[24] (otherwise known as Rev-Eyes) was a drug that was used after eye examinations or surgery to reduce the size of the pupil by blocking smooth muscle contraction of the iris. Rev-eyes is no longer available in the US, but if an eye test that focuses on pupil dilation were to become the gold standard for deception detection then it, or drugs that cause similar effects, could reappear on the black market. As can be expected with any range of AI technologies, software developers may be expected to seek to develop machines that will do the exact opposite of what other AI and ML were designed to do.

One method of stealing money from employers is vendor fraud, where an employee creates phony invoices for a shell company and siphons money to it for work that never occurred. Employers could require employees under suspicion to take an AI lie detection test, or even schedule them periodically. If a similar approach were taken for politicians and government employees, the impact it could have on global corruption could be profound. Politicians who voluntarily take such tests could greatly enhance their chances of getting elected or reelected. Similarly, politicians who refuse to take such tests could potentially imperil their chances of getting elected or reelected. In the future, it will not be not hard to imagine politicians' or business leaders' "honesty scores" being displayed every time they appear in the media or give a speech.

Yet, how can we validate when a lie is truly a lie, or, for that matter, when any result from AI is necessarily 100% accurate? Bugs inherent in or introduced to AI systems could remain undetected for years as they produce false results. A central issue associated with blindly accepting the predictions of AI is that we really do not know enough about the human brain and how it, or any one of its underlying functions, actually works. Is there truly an irrefutable link between physiological stimuli and what a person is actually thinking? Deception detection systems operate on the fringes of certainty and while businesses could potentially reduce fraud costs and governments could combat corruption by utilizing AI designed for that purpose, such technology obviously needs to be handled with extreme care. What safeguards are being built into existing AI systems and those being developed? Are AI's creators willing to reveal such safeguards? Many AI developers are, and will remain, silent on that.

The Workplace of the Future

AI is, of course, doing more than assessing a person's integrity—it is also monitoring their behavior in the workplace. AI is already giving more control and leverage back to employers, but is this necessarily a good thing? Amazon has patented a wristband[25] that tracks the hand movements of warehouse workers and uses vibrations to nudge them in the direction of performing more efficiently. Some may view this type of technology as invasive because it is essentially attempting to use AI to rewire the neural pathways of their employees to become more efficient. It is one thing to have an employee go through a training program to learn the most efficient way to pick up boxes but quite another to use wristband vibrations to tell them what to do in real-time. Is Amazon encroaching on the sanctity of their employees' own minds as they meddle with their nervous systems? This is a slippery slope. Could anything stop companies from using AI to attempt to read employees' thoughts or program them to think a certain way once such technology becomes available?

Some high-tech startups are offering AI to monitor employee activity in the office and on company networks. Humanyze is a startup that uses smart badges to not only track employees' movements around the office but to attempt to predict how well they will interact with colleagues.[26] There are many privacy concerns associated with collecting this type of data, but there are also some potential benefits, such as identifying managers who invite equally qualified women to meetings less frequently than their male co-workers. Clearly, there could be many benefits into integrating AI in the workplace if it is done wisely and fairly.

In some organizations, a person's direct manager has a major influence on the evaluation of employee performance, which inevitably includes their own biases. If a company uses AI to conduct employee performance analysis, it can potentially level the playing field by conducting evaluations on all employees, including upper management. This could create a more open conversation between employees at different levels of the corporate hierarchy, which could make the company a better place in which to work. The flip side, of course, is that the AI may have a different kind of subjectivity built into it, which could defeat the purpose.

Employers already monitor their employees' online activity. Imagine the impact introducing an AI-driven system into the equation could have. Would employees be more inclined to over-adapt and engage in Internet searches that are specifically intended to encourage AI to believe they are doing something, or believe something, that they are not? If AI were to alert an employer that a given employee is doing something outside his or her norm, would that be an indication that the employee is about to leave the company or might be engaged in illicit behavior? There are certainly instances where employee monitoring could be justified, such as when certain employees have access to confidential information or funds. Of course, employees can manipulate existing monitoring systems in the same way, but AI will presumably not be able to put a given behavior into context and would not, for example, have been privy to a private lunch an employee may have had with her manager, revealing an intended course of action.

Do we really need AI to manage employees at all, and just how effective can AI possibly be in adding value to that process? There is not necessarily a correlation between corporate conformity and the value proposition of an individual employee. Some companies may view employee monitoring as something that actually *interferes* with getting the job done. Will the companies that choose to deploy AI to monitor employees be filled with people who do not take risks and simply conform to corporate culture? How beneficial could that be toward common long-term corporate objectives, such as encouraging entrepreneurship and risk taking?

Some of the people that can have the most transformative impact on a company are the same ones who underperform in their evaluations. For example, Steve Jobs was fired from Apple in the mid-1980s, only to return years later to make the company the global force it is today. Employers need to strike a balance between the insights AI surveillance can bring to the workplace versus the conformist behavior it may encourage, and the visionaries it may inadvertently push away. Transparency about how a company is using AI to evaluate employee behavior may be an important ingredient in getting this right.

Extended Reality

As we have seen, AI is fundamentally reconfiguring how people communicate with each other and with companies. For most people, AI remains an application that exists purely outside the realm of our physical bodies, but there are some exceptions, such as individuals who have had devices installed in their heads to prevent epileptic seizures, as described in Chapter 5. Another manifestation of the merger between human and machine is extended reality (ER),[27] wherein real-and-virtual elements are combined with human-machine interactions generated by computer technology and wearables. It sounds incredibly futuristic, and even hard to imagine, but it is becoming a reality.

ER consists of three parts: virtual reality (VR), augmented reality (AR), and mixed reality (MR). VR is fully immersive technology and creates the impression that users are in another environment, based on an intense barrage of information to a person's sensory systems. The VR experience is usually delivered through a headset where visual and auditory information is received. AR overlays digital information onto the real world, which can be experienced via a headset, a pair of glasses, or a smartphone. Pokémon GO is a well-known example of AR, as well as the dangers associated with this type of technology. For example, a 2017 study[28] estimated that Pokémon GO caused 100,000 traffic accidents. MR allows a user to interact with both physical and digital elements at the same time; in other words, a user can be in both the physical and digital world simultaneously.

Several technology companies have developed smart eyewear that can display digital information in their field of vision. Some of the eyewear presents a digital image or information in a person's field of vision, based on voice commands. Other versions of the technology allow a user to control computer applications by using their vision as a focus mechanism and their hands to interact with the programs. As the technology develops, the computer will no longer be a screen we stare at but something that we immerse ourselves in as it overlays our physical world.

Today, most virtual assistants are activated by voice command and respond with their *own* voice, but AR can be delivered by smart eyewear to allow virtual assistants to respond to us through our own field of vision. There are several reasons why smart eyewear technology has not been widely adopted. One has been the price, since many of these devices are expensive. Another subtler reason could be that people do not want to wear technology on their face. If smart eyewear were to be made to look like eyeglasses instead of a bulky and uncomfortable contraption, more people would probably be willing to buy it. One reason smart watches became so popular is that they were manufactured to appear like ordinary watches.

This was the approach that Intel had taken[29] with its Vaunt Smart Glasses project, which was discontinued. Instead of creating elaborate 3D objects that a user could interact with, its plan was to develop smart eyeglasses that could provide a stream of contextual information on demand. For example, imagine activating the eyewear through your virtual assistant by saying, "Alexa, how do you make mojitos?" The idea was that the instructions to make the drink would appear in one's field of vision. It was not supposed to be obvious to anyone looking at the person wearing the glasses that they were seeing digital information at a particular point in time. Another use case for smart eyewear could be simply to look at a restaurant in one's field of vision and obtain contextual information about a Yelp review or Zagat rating.

This is hardly a new idea. In the film *Terminator 2*, Arnold Schwarzenegger was able to focus his field of vision on an object to obtain a live stream of contextual information about it. This sort of technology requires AI and ML to be embedded into it in order to process language, identify objects with computer vision, and learn what information is helpful to the user. Vision is arguably the dominant sense for *most* humans, so it is natural that AI will eventually make its way into our field of vision. As with other AI applications, there are risks associated with data privacy, and precautions must be put in place to help ensure that someone is not watching a movie through eyeglasses while they are driving a car.

The broader trend of AR and computing will move us to an interactive experience that engages our entire bodies. While, today, we can log on to a company's website to interact with a chatbot to get our questions answered, in the future, we may be able to put on a pair of smart glasses and the digital projection of a human will appear in our field of vision offering to help us resolve our issue. It merges the idea of remote service with personal attention, as the representation of a physical entity could provide us with the impression of personalized service more than a phone call, chatbot, or email can do.

AI will expand the range of possibilities for ER in global communications as use cases are enhanced beyond the constraints of existing software tools. VR is another area where AI is having an impact. Many of today's VR applications involve video games, but there are commercial applications such as flight simulators. Another emerging trend is that several large technology firms are directly or indirectly involved in preserving historical sites by leveraging photography, AI, and VR.

For example, the Syrian Civil War destroyed or damaged all the country's six United Nations Educational, Scientific and Cultural Organization World Heritage sites. Iconem is a firm that creates 3D models of historical sites that are threatened by time, war, and nature. The company used drones (so as to avoid landmines) to take 50,000 photos of the ancient city of Palmyra, which had a long history of different Greek, Roman, and Islamic cultural influences. Iconem used Microsoft AI to stitch thousands of photos together to create a 3D digital model of the historical site.[30] As VR continues to develop, it will allow future generations to *experience* history through technology, as opposed to reading about it or watching it passively on television.

As AI and ER are continually integrated into commercial applications, its possibilities are seemingly endless, as the distance between consumers, products, and services slowly evaporates. People will be able to visualize how a 3D digital projection of a sofa looks in their living room, vacationers will be able to view the inside of a specific hotel room before booking it, and pathologists will be able to diagnose disease as groups of living cells become the size of a room which a doctor can walk through and examine in detail.

Yet, for all the potential benefits of VR and AI, there are many drawbacks to this type of technology. The Japanese game maker Illusion Software has become infamous[31] for its erotic games that involve inappropriate interactions between the game player and a young virtual girl. While the girl is supposed to be 18 years old in the game, her age is left to the imagination of the players. The game appears to cross ethical boundaries on many levels regarding what many cultures would consider immoral, but AI allows individuals to transcend such boundaries. The realm of fantasy takes on an entirely new dimension with the help of AI.

What are the implications of creating VR games that cater to peoples' innermost and darkest secrets and desires? Will AI enable businesses and governments to similarly transcend existing boundaries of legality, ethics, and governance? As the Dark Web has spread terrorist propaganda, child pornography, and murder fantasies, will VR experience a similar nefarious evolution? That seems inevitable. As people begin to use VR to immerse themselves in anti-social behavior, will doing so make users more likely to commit such crimes in the real world? Will they even be able to discern fantasy from reality? AI will elicit the best and worst in human nature. Individuals, businesses, and governments must ultimately decide for themselves where the boundary lines are to be drawn.

Conclusion

The race for AI supremacy will continue to accelerate in an exponential manner, having a profound impact on most sectors and industries. The degree to which its impact is felt on any given sector will in large part depend on where and how R&D dollars are spent and the extent to which it is embraced and adopted by individual businesses. There is no conceivable way that AI's continued growth will be equal or similarly effective by sector or business. As we have seen throughout this chapter, growth will be unpredictable and fraught with possibilities as well as potential perils.

In the communications space, faster Internet connections and the explosion in the growth of smart phones will spur rapid growth in the number and type of AI applications. Email providers are providing a real service by filtering out spam and categorizing email, and the smart reply features are helpful for people who are increasingly pressed for time, but such developments also appear to be little more than another way for people to take the path of least resistance and turn off their thinking caps. As we willingly give away the sovereignty of our own minds to AI, it will clearly reduce our collective ability to formulate our own thoughts. Relying too much on AI for our cognitive tasks will inevitably cause some of our brains to become sluggish, tired, and uncreative.

AI SUPREMACY

Our interaction with companies will continue to evolve as we interact with AI in the form of chatbots and virtual assistants. Our voice patterns, tone, and word usage are already being tracked and consumed by machines. Companies will continue to use AI to personally craft messages, products, and services to cater to our needs based on our digital profiles in cyberspace and based on our physical attributes and quantifiable demographics. AI to AI (AI2AI) communication is starting to emerge, enabling two or more algorithms to interact in a predictable way, but whether this is achievable in a consistent manner, and under what circumstances, remains to be seen. As is the case with so many other aspects of AI, AI2AI interactions must ultimately be governed by universal standards, which simply do not yet exist.

The future of people interacting with hyper-cognitive stores is perhaps one of the most promising—and insidious—uses of AI to date. It seems unlikely that Amazon would create its cashier-less store (Amazon Go) simply to provide its members with a faster and easier shopping experience. To achieve that, the company could have simply focused on grocery delivery instead, and while that may also be in its future, its management chose to develop an AI store that tracks our every movement. To be fair, other large technology companies are working on similar technologies. The goal is to take the digital profile they already have on their customers and merge it with our physical profiles, which they can only get if we choose to walk within the walls of a smart store or use another physical device that transmits our biometric information. The cashier-less store may well be the bait to get us to walk in, but we are very much guinea pigs for AI, rather than the other way around. If these types of AI-driven stores clearly state what type of data they are collecting, for what purpose, and how the consumer can benefit, it would be a step in the right direction.

ER is perhaps one of the most transformative developments of the AI era and could well define how we interact with technology in the future, but at what cost? There are clearly benefits to people and businesses in having the ability to immerse in a digital environment. As AR becomes more commonplace in computing it could allow employees to get up from their desks, improve their health, and interact with business applications with their entire bodies all around the office. We know all too well the addictive nature of smartphones and the difficulty many people have in simply turning them off.

AI could end up decreasing our ability to communicate with our own minds. Anyone who has experienced a lucid dream or out-of-body experience (OBE) knows that VR will never come close. OBEs are the original VR experiences and can be felt without the use of drugs. As we fall asleep there is a point in time where our minds begin to create the reality being experienced, as opposed to being based on our bodies receiving input from the outside world. This is just one of the existential risks AI may bring to humanity. The risk is not only that we lose our ability to communicate with others but that we also lose our ability to communicate with our own minds as we are lured deeper into a seductive digital fantasy land.

Regardless, AI is having a profound impact on the way in which we can communicate and will do so in the future. It is up to us to craft boundaries and safeguards to ensure that more good comes from the process than potential harm, and that humans remain in the driver seat over the course of time. We must not allow our desire for never ending progress and greater convenience and efficiency to take precedence over the pace at which AI proceeds and the manner in which it functions. The risk is that we lose sight of what is really important as the race for AI supremacy takes on a life of its own in the future.

7. *The Future of Learning*

Overhaul

The education sector has been more resistant to the relentless assault of technology and automation than other parts of countries' social infrastructure, a natural byproduct of its size and general unwieldiness. But as AI begins to rearrange markets, the distribution of goods and services, and how societies function, it will put additional pressure on education systems to produce graduates who can function and adapt in the AI-driven economies of the future. This raises many issues, such as how educational institutions can anticipate and account for the market turbulence that lie ahead of us, where the money will come from to fund public school adaptation to the new normal, and how graduates will cope with rapidly changing employment conditions.

In 2016, the on-time graduation rate for US high school students rose to 84%,[1] which was the highest since 2011. It appears that, at least in some cases, the increase in the on-time graduation rate is attributable to the lowering the high school graduation standards rather than an improvement in the caliber of American students. For example, in 2016, the Ohio State Board of Education reduced minimum proficiency standards[2] on high school math tests because student performance was lower than expected. Some studies have suggested that fundamental deficiencies exist in American adults' reading abilities, and in the basic science, technology, education, and math skills of America's youth. Can AI help improve adult reading ability, enable students to perform better on standardized exams, and improve overall learning outcomes?

The democratization of knowledge via the Internet means that anyone can have access to some of the best college courses. In 2011, two Stanford professors offered a massively open online course (MOOC) that popularized the online course concept. This resulted in a wave of MOOC initiatives being launched by prestigious universities and private education companies. Despite such open access, some students do not perform as well academically in MOOCs, when compared with traditional classrooms, and MOOC dropout rates have been as high as 90%.[3] Online education could greatly benefit from the deployment of AI and ML to create smarter teaching platforms that encourage higher learning rates and lower dropout rates.

In 2018, Google announced a program to invest in startups to boost its digital assistant ecosystem. One of the first companies it invested in was Edwin, an AI-powered English tutor, to prepare students for the Test of English as a Foreign Language (TOEFL).[4] The ability to learn a new language appears to be highly correlated to the degree to which a person's practice of the language is consistent, frequent, and repetitive. This is precisely why digital assistants are laying the foundation for an AI tutor.

A student can interact with a digital assistant, which can potentially process his or her native language and provide real-time translation, guidance, and pronunciation for the new language. Edwin allows students to prepare for the TOEFL test on Facebook messenger at lower cost than a traditional tutor. The technology is new and will improve over time, but it offers a wide range of benefits, such as a safe place to practice. AI-powered language learning platforms could get people to speak at a basic proficiency level much more quickly than conventional methods. These AI-powered language tutors will not eliminate the benefits that come with learning a new language in a classroom setting by interacting with real people, but they could provide the foundation to explore other learning alternatives after successfully completing AI-powered language training.

In 2018, online learning platform Quizlet[5] began producing ML algorithms that used anonymized student study habit data to enhance future lesson creation. It was one of many companies moving toward creation of autonomous AI tutors that will be able to dynamically adapt students' learning queues in real-time to provide instant feedback and guidance. The creation of more powerful and user-friendly AI tutors will not, of course, guarantee that participants will successfully complete the courses they start, but doing so may prove to be an impetus for a grander vision for the global educational landscape of the future among policymakers.

The Collective Mind

While developed nations can do the most to improve the cognitive abilities of students, it is the developing world that stands to benefit the most from the future spread of AI in education. In 2014, a United Nations Educational, Scientific and Cultural Organization (UNESCO) report[6] noted that about 9% of children ages 5 to 11 were not in school in the developing world. This number increased to 16% for children ages 12 to 14. We should presume that the actual numbers are much higher, for more than 70% of children not in school live in Southern Asia and Sub-Saharan Africa, and girls generally have fewer opportunities than boys.

Another major issue looming for developed and developing countries alike is that most are facing a massive teacher shortage. In another 2016 UNESCO report,[7] it was noted that the world needs to add approximately 20 million primary and secondary school teachers, and to replace nearly 49 million teachers expected to retire by 2029. It is unrealistic to think that enough high-quality teachers can be trained quickly enough, and distributed widely enough, to satisfy growing global demand, especially in the developing world. Teacher pay in the US is so low in some school districts that administrators are recruiting qualified teachers from overseas.[8]

The ubiquity of smartphones provides an opportunity to introduce AI in classrooms throughout the world, which can, at a minimum, assist students to prepare for standardized tests. Smartphone-based apps such as Socratic already leverage AI to tutor students by allowing users to take photos of math problems with their phones and provide real-time explanations about how to solve them. The ease of accessing a comprehensive explanation from a user-friendly app can make the difference between a student giving up or feeling empowered to persist and solve more problems.

In this sense, AI is already having a positive impact on learning outcomes. The mass repository of knowledge hidden in homework problems can be unleashed by AI that can extract, process, and contextualize data to make accurate predictions about what concepts need to be taught to students. Socratic is nearly a one-stop shop app for math problems, giving us a glimpse of the power of AI and ML in education achieved by mining the collective intelligence of the world's nations.

Tiger Moms

In New York City, there are nine specialized high schools that serve around 18,000 students per year. Their alumni tend to get accepted into elite universities, and some of them go on to have prestigious careers. The single metric used to determine whether a student will be offered a place in any of eight of the nine schools is their score on the Specialized High Schools Admissions Test. Asians overwhelmingly dominate these tests. In 2018, the student body in Stuyvesant High School—arguably the most prestigious of all of the New York City specialized high schools—was 75% Asian. Overall, Asians made up 62% of the student body at all the New York City specialized high schools.[9]

Contrast that with the fact that, in 2016, roughly two-thirds of all New York City students were black and Latino. That led Mayor Bill de Blasio, to propose a radical idea:[10] get rid of the test entirely. This, of course, proved to be highly controversial. While the underrepresentation of black and Latino students in these specialized high schools is unfortunate, Asians are also a minority in New York. Many felt that his plan sought to penalize Asians at the expense of the majority of other students, and his plan did not address the need to improve the academic performance of students throughout the city's public school system.

There are many reasons why Asians outperformed other ethnic groups on standardized tests, but one factor may be that Asian parents are willing (and/or financially able) to spend more on tutoring services than parents of other ethnicities. Asian parents may also place a higher value on the importance of education than other parents. A 2017 study highlighted that households in Asia spend 15% of their income on supplemental education services, compared with 2% in the

US, which supports the notion that culture is a factor in attitudes towards education spending.[11]

A 2015 ProPublica study[12] found that Asians were nearly twice as likely to pay a higher price from Princeton Review for test preparation services. One of the study's conclusions was that a *Tiger Mom tax* was inferred from the density of ethnic populations in specific geographic locations based on zip codes and associated pricing from nearby Princeton Review locations. It appears that Princeton Review used a smart pricing algorithm, as many companies do, which dynamically adjusted pricing based on market demand and other factors. Princeton Review said that the price differences were *incidental*.

Local communities would clearly benefit from governments which support AI-driven education projects. Perhaps New York City should hold competitions to benefit from ideas among local AI startups, of which there are many in what is commonly referred to as *Silicon Alley*. On the other hand, introducing AI into the New York City schools could actually *increase* the Asian demographic among these high-profile schools, who will undoubtedly leverage the same tools as other students and may still outperform everyone else.

While discussion of ethnicity and the extent to which that is correlated to academic outcomes is taboo, if AI is to assume a meaningful role in public school systems, we need to take a hard and honest look at such these issues and not be afraid to talk about them, even if doing so may prove to be politically incorrect. We are not going to maximize the potential value of AI and ML in general, and in the education arena specifically, by tip-toeing around sensitive issues.

Education and Innovation

In 2011, former President Barack Obama made the following statement in his State of the Union address: "If we want to win the future–if we want innovation to produce jobs in America and not overseas–then we also have to win the race to educate our kids."[13] As AI continues to march into every corner of America's economy and society, Obama's words have become more relevant than ever before. As countries grapple with the many ethical and moral questions that emerge with the rise of AI, they must find a way to embed it into their respective education systems in order to have any chance of getting ahead and staying ahead of other nations economically, politically, and socially.

In 2016, the White House held a Symposium on the Future of Education, R&D, and Digital Learning, where approximately 100 researchers came together to examine the progress made through supporting educational innovation by the outgoing administration and via the Advanced Research Projects Agency for Education (ARPA-Ed).[14] The ARPA-Ed was not a new idea, as it had been proposed in Obama's 2012 budget, but it failed to gain wide support. The goals of ARPA-Ed stemmed from the need to out-innovate other countries (such as China) which have clear national objectives, in an effort to retain US leadership in education. The ARPA-Ed agency would have, in theory, functioned in a manner similar to the DARPA, which, as discussed throughout the book, has a long history of innovation success. The focus was to be on transforming education and learning to prepare students for the demanding and fast-paced nature of the 21st century.

A 2014 study[15] highlighted how the DARPA and Navy supported the development of a digital tutor that was used to train new Navy recruits in IT administrative skills. The study found that the group of sailors that were trained by the digital tutor outperformed their peers, who had received traditional classroom training, by a wide margin. This reinforced the idea that technology-guided education can lead to better student outcomes, which is desperately needed in large, segregated school systems. If digital tutors were introduced in schools, even on a pilot basis, teachers would, in some cases, take on the role of manager of a group's learning, as opposed to acting as the tutors themselves.

One of the bottlenecks typically encountered when trying to implement self-directed and active student learning is that one teacher is not able to create and manage personalized learning plans for a room full of students. The first set of applications of AI in the classroom could help free up teachers' time to focus on managing the classroom, as opposed to teaching. Not surprisingly, China is already exploring how to leverage AI to reduce administrative tasks and increase free time for teachers. In 2018, 25% (or 60,000) of Chinese schools (involving more than 120 million people) participated in a program to evaluate student essays with AI.[16] The AI-powered essay grading system was designed to assess a student's writing ability, grammar, word usage, and flow. Teachers provided feedback, concluding that the AI used was not perfect and there were incidences in which the algorithms gave a low mark for what should otherwise have been perceived as a brilliant essay. Despite such glitches, a Chinese government document highlighted that the AI-powered essay grader gave the same score as a human 92% of the time.

This is one of the many reasons why China is forging ahead in its race for AI supremacy at startling speed. Most countries are barely scratching the surface, if at all, with respect to introducing AI into their education systems. While many are debating such topics as privacy concerns, China is taking an aggressive approach, which, as is noted elsewhere in the book, is putting them light years ahead of other countries. The sheer number of people who participated in China's AI-powered essay grading program is phenomenal, and the amount of data it provided to the government is staggering.

The program was, in many ways, an experiment, but it is this willingness of the Chinese government to test things and its ability to execute initiatives with precision, based on centralized coordination, that gives it a distinct advantage over other countries in the race. Chinese teachers will be some of the first able to provide more personalized support to individual students as AI-powered grading programs, and other programs, are leveraged to remove the drudgery from reviewing student essays, among other administrative tasks. If the US Department of Education were to back AI initiatives as China has done, it could begin to use student clicks, feedback, and answers to become more competitive globally.

Machine Learning for Learning

Student data will be one of the main sources of fuel that powers AI for education, but given the US government's hands-off approach to education, it will probably not take ownership of the process. One of the basic questions that may arise when discussing education and AI is where and how student data should be collected, processed, and analyzed. Technology companies could be the natural place for this data to be stored and used to train an AI.

Many of Google's products—such as Gmail, Drive, and Classroom—have already been integrated into schools across the globe. Some schools allow students to submit assignments through Google products and they can even interact with other students, in real-time, using Google Docs for group projects. Given Google's deep expertise in ML and its existing market penetration in the education sector, it is not hard to imagine that Google will be one of the first companies to leverage student data to build comprehensive AI-powered education apps that could begin to transform education for the better.

The Google for Education privacy policy[17] states that it does not own school or student data, but collects technical information about logins, IP addresses, and device information. Technical data collection could be more relevant for detecting inappropriate access to or hacking of student and administrator accounts, as opposed to building AI-powered education apps. It is not clear whether Google is doing anything with student data to try and build AI for education applications but, given its history of creating new products to utilize its data, we should expect that, if it is not already in the process of doing so, it soon will be. The company may be waiting for a nod from governments before it becomes aggressive using student data. This nod from governments will undoubtedly eventually come, but other things should be considered in the process. For example, should policymakers enact legislation that will ensure that AI for educational purposes is developed as a public good and will not be used as another lever to pull in order to increase profits on demand?

Perhaps policymakers could sway, or compel, companies like Google to cover the costs of building AI for education, but not profit from it directly, since many schools are already paying for some of their other services. The risk here is that once AI becomes an essential component in an education system, a single company will be in complete control it. Nations and schools could in theory be held hostage to a company's will and expose them to exorbitant price increases and/or cessation of service. This is another argument in favor of the open access model, which will be discussed later in this chapter.

As more and more students interact with AI-powered education apps, ML will incorporate the data that is collected and rate their feedback. As students begin to learn a new concept, the AI will *know, or predict*, when the student has reached an acceptable level of knowledge and mastery before allowing him or her to move on to the next concept. This is important because in the traditional classroom model the pace of learning generally remains static. Some students may be able to learn various subjects or concepts faster than others. However, when AI-powered education apps infused with ML algorithms are introduced into schools, students can follow the same roadmap, but get there at their own pace. Some kids may be faster than others on nearly all topics, while others might be stronger addressing specific problems. AI can enable them to excel at what they do best.

Immersive Learning

The other major trend that ARPA-Ed identified was that video games provide a template for creating interactive and engaging learning environments. The many applications of VR, AR, and MR make these mediums highly viable candidates for interactive learning in the future. Google created a program called Tilt Brush that allows users to paint with a handheld controller in VR in a 3D space. The VR experience is quite engaging and could prompt students to become excited about going to an art class, for example, or find other ways to use the application to express their creativity. Some would argue that this sort of technology should not be released in schools because we do not know whether the results will be positive or not. However, if ARPA-Ed had its way, it may not be about proving whether a particular method works with absolute certainty, but rather, introducing the technology on a limited basis to track the results. Once, the results demonstrate positive student outcomes, the program could be rolled out to more schools.

Introducing more technology and AI into the classroom still takes money, which is why AI-powered education apps that can be delivered through a computer or smartphone may offer the best bang for the buck, given that they can be easily accessed by many students simultaneously. Doing so on a mass basis provides interactive feedback that makes AI smarter. Immersive learning is not limited to having students step into virtual worlds but could also be defined as a hands-on experience with physical things. Other technology, such as

3D printers and VR headsets, are more cost prohibitive, but schools should still allocate funds to purchase these tools on a limited basis and create classes around their use to drive interest in technology and innovation.

Another potential for VR technology is to enhance the learning experience by explaining abstract concepts and making them interactive and fun. For example, some may recall learning Newton's Second Law, which states that the only force on a body in freefall is the force of gravity. The body accelerates at 9.8 meters per second, which could be just another equation for a student, but if teachers take students into VR and show them the concept of Newton's Second law by slowing down an object in free fall and visually representing gravity by warping space and time, it provides a completely different experience.

In other words, many of the phenomena that humans have studied over centuries, and some of the most interesting concepts in science, lose their glory and importance by being reduced to a multiple-choice question on a test. By deploying VR, AR, and MR, students are limited only by their own imagination, or lack thereof. The many forms of alternative reality learning that will emerge will allow students to step into the imagination of great minds in order to learn complex and abstract concepts in an immersive experience, as opposed to being exposed to a limited stream of language that lacks the ability to enthrall.

It immersive learning's nature is to lend itself to divergent thinking, sparking a student's own creativity and inquisitiveness. This ability to influence interpretation of key concepts could improve learning outcomes, increase self-confidence, and allow for greater economic opportunities. Immersive learning must still be managed, for there is a risk that some students will become so tantalized by the immersive nature of ER that they could refuse to learn in any other way. Convincing students to use immersive technology to learn about the real world, while simultaneously finding ways to pull them out of these seductive environments to appreciate the real world with all of its flaws and a lack of control, is another challenge that will need to be addressed.

Open-Access

If the goal of introducing AI into the classroom is to improve learning outcomes, then should policymakers consciously decide whether to allow public access to student data, or limit it, for purposes of building AI-powered apps for education? Would open access to large volumes of student data such as homework assignments increase the risk of cheating? Should schools actively participate in sharing large volumes of homework assignments with these types of AI-driven learning platforms? Should open access even be considered? Students could still upload homework assignments on their own, of course, but most of these apps require parental permission for students under a certain age; if schools agree to sharing and publishing homework assignments or other types of student data, it could rapidly increase the repository of data available.

The hard sciences that involve solving equations, such as mathematics, chemistry, and physics, are conducive to sharing, because most of the problems can be modified. However, not all subjects are conducive to an open-access model, given that they may be more difficult to change on demand. This points to the need for a thorough assessment to be conducted to determine what types of homework problems would be the most potentially beneficial and should be shared publicly without AI, versus those for which AI-based learning platforms should be developed, while minimizing the risks and consequences of moving towards an open-access education model.

In 2016, one teacher decided to take matters into his own hands and automated his exam creation process. He created a program in the R language to dynamically generate random values for different variables in his exam questions, solve the generated equations, and save the answers for each version of the exam.[18] Each student took a unique version of the test, to reduce the risk of cheating as well as the amount of time required to create and maintain exam content. This could be one of the missing pieces in moving towards an open-access publishing approach to homework problems. If teachers can leverage applications to dynamically create exams and homework assignments for the hard sciences that involve solving equations, a school's administration should have no problem with publishing the data because it would not increase the risk of cheating.

Education policymakers need to become aware that publishing student homework assignments in a coordinated and standardized fashion will allow more effective AI-driven learning platforms to emerge, which will improve learning outcomes for the students they serve. Doing so may be a hard pill for education systems to swallow, especially in primary and lower schools, because teachers have generally been regarded as an omniscient force in the classroom. If schools were to openly publish homework assignments, it would make teachers more vulnerable, while at the same time subjecting them to being challenged. Assuming schools around the world do not ban smartphone use—as France did in 2018,[19] a student could simply snap a photo of a teacher's work and question any inconsistencies or potential errors.

While AI can extract the insights hidden away in homework assignments, emerging trends in AI will present new challenges for students who may be inclined to cheat. This problem already exists, and has existed for years, since the answers to some exam questions can be easily accessed by a Google search. Other questions are more technical in nature and could be harder to find. Given the range of new wearable technology, it is conceivable that some students could in the future live-stream the questions in their field of vision through smart eyewear.

The recipient could search for the answer on the Internet or leverage an AI-based learning platform to determine the answer and send the solution back to his or her accomplice using the same eyewear (or possibly communicate the answer vocally through a small earpiece). While most smart eyewear is still bulky and expensive today, it will become less obvious and cheaper over time. Eventually such eyewear will be indistinguishable from a normal pair of glasses. How can teachers prevent cheating when the technology people may use to communicate with other people and AI will become nearly invisible?

In 2017, a headmistress in the UK[20] sent a letter to parents stating that she invested in signal blocking technology and restricted Wi-Fi access so as to hamper students' relentless obsession with their phones. Her plans were derailed when the North Yorkshire County Council's IT support team and the UK Office of Communications informed her that it was a criminal offense to use signal blocking technology under the

Wireless Telegraphy Act. The issue was that signal blocking technology could interfere with other people's mobile services (which was, of course, the point of the exercise) but, also, the emergency services and air traffic control. Perhaps the UK authorities will take the opportunity to follow the French lead and ban cell phone use in schools.

Revamping the Recruitment Process

In 2017, Dutch-British consumer goods giant, Unilever,[21] inserted AI into its recruitment process with great success. The company employs more than 170,000 people globally and its management was looking for a way to revolutionize its hiring process to reduce costs and bias. Unilever revamped its recruitment process by leveraging AI to screen candidates in order to get the most qualified people in front of recruiters the most efficiently.

First, candidates learn about the job opportunities online and submit their LinkedIn profile so that no resume is required. LinkedIn profiles are digital and have structured formats, so ML algorithms can learn how to predict—based on a company's objectives and evaluation metrics—who the best candidates are for the job. Second, the candidates play neuroscience-based games that evaluate cognitive abilities. Finally, the candidate conducts a video interview on a smartphone, tablet, or computer and responds to a series of predefined questions which are recorded. The video interview is then fed into an AI platform that analyzes a candidate's facial expressions to assess his or her mood and personality traits. If a person scores well on all phases of the AI screening process, they will be invited to an in-person interview.

In a short period of time, Unilever demonstrated quantifiable success utilizing AI by increasing the number of colleges it recruits from, from 840 to 2,600[22] for entry-level jobs. This was a significant achievement because many large corporations have a limited number of recruiters and send them to the best schools in search of the brightest minds. AI will not solve the diversity problems that plague many corporations across the globe today, but it will modify them. While corporations will be able to recruit from a wider network of colleges, they will inadvertently redistribute the second round of job interviews to people who perform well on standardized tests.

Anyone who has worked in a large corporation knows that a person's intelligence quotient is not always the most accurate predictor of their effectiveness within the organization. Sometimes it is soft skills, a person's emotional intelligence, and the ability to navigate corporate politics, manage executive expectations, and remediate conflicts of interests among a diverse group of stakeholders that will make an individual indispensable. It is not clear that AI will be able to assess a person's emotional intelligence as well as hiring managers. In the future, highly effective leaders may never be hired because they do not make it past the first round of an AI screening process.

In a 2017 Deloitte report,[23] 33% of survey respondents confirmed that they already used some form of AI in their hiring process. As companies embed AI in all aspects of the recruitment practice, it could result in greater oversight of hiring managers and their decisions. Nepotism could be reduced and replaced with a more balanced form of a hiring meritocracy. For example, hiring managers may need to explain why they recommended a certain job candidate who happened to have attended the same university as the manager over someone who did not, although the selected candidate scored significantly lower than other qualified candidates in the AI-driven screening assessment.

As companies begin to leverage AI in the interview process, it will create more competition among candidates. Services will undoubtedly be offered to potential job candidates to increase their AI screening scores, and people will practice in order to get better at presenting themselves in a way that results in higher scores through AI. However, AI screening can quickly become another form of standardized testing. Some candidates will inevitably come to fully comprehend the AI screening process and solve complex problems quickly, scoring higher in certain parts of the screening process, just as some individuals "crack" more conventional testing systems. This implies that the testing process must undergo regular revision to prevent candidates from gaming the system.

Overemphasizing quantitative metrics can lead to situations where certain demographics of the population will excel at screening assessments and progress through the interview process more often than their counterparts. Clearly, manual overrides need to be inherent in the AI hiring process, but such human intervention could create opportunities for hiring managers to game the system to their own liking, as was the case before AI systems became available. On one hand, there is a risk to over relying on AI to make decisions about human lives. On the other hand, once people and organizations can call into question the conclusions of AI, its insight into the people it was monitoring could become meaningless.

Migration

One area of governing where quantitative metrics do not stand much of a chance of becoming irrelevant, where the stakes are extremely high, and where policymakers have a lot to learn, is in the area of immigration. One of the biggest problems countries face when trying to integrate immigrants is ensuring that they learn the local language. Can AI actually teach immigrants to learn a new language and behave in a manner that helps ensure that they assimilate quickly in an unfamiliar society, even though some of their beliefs and behaviors are so deeply ingrained in their psyche that they cannot be easily transformed?

In 2018, the polarization among EU member states was abundantly clear when Hungary and Poland refused[24] to accept the EU's refugee quotas. This was driven by an effort to relocate 160,000 migrants out of a total of 2 million that had arrived in Europe since 2015. Hungary had erected a border fence and created a program to train and arm border hunters[25] to assist the police and military in capturing and detaining migrants trying to enter the country illegally. Given the depths the Hungarian government had already gone to in an attempt to control the flow of illegal immigrant across its borders, it is not hard to imagine that the government would opt to invest in AI-based surveillance and sensor technology to deploy their limited resources more effectively as needed.

Some EU countries are using language skills as a metric to determine eligibility for social benefits. In 2018, it was reported[26] that Austria would tie the level of welfare benefits to be received to a person's German language proficiency. Supporters of the plan argued that language is a key component of integration and the program provides an incentive for newly arrived migrants to push themselves and learn the language. Critics maintained that the standards for German fluency were set too high and could lead to humanitarian-related issues, such as a rise in child poverty.

In 2018, researchers from Stanford[27] created an ML algorithm that analyzed historical data regarding refugee resettlement in the US and Switzerland. The study found that refugees' level of financial self-sufficiency depended on three main factors: education level, language proficiency, and geographic location. The researchers found that refugees with certain educational backgrounds and skills did better in some locations than in others. They projected that a refugee's chance of finding a job could be increased by between 40% and 70% by using an ML algorithm to match a person with a specific geographic location that could most benefit from his or her skills.

It seems clear that AI will continue to play an increasingly central role in teaching people new languages, evaluating their knowledge and skill sets, and assessing their current language abilities on behalf of schools, governments, and corporations. Perhaps immigration appointments in the future will include an AI-powered language assessment and–based on the policy of the country conducting the interview–an applicant may not even be given a chance to enter the country without possessing strong language skills, among other factors.

Conclusion

The invention of the Internet allowed all human knowledge to be accessible but not necessarily understandable. The power of AI and ML is that it will make sense of human knowledge in ways that were never previously deemed possible. Citizens participate in the economy by working and they contribute to the development of AI by clicking, searching, and buying online. It is the digital actions of a country's citizens that create the fuel to train AI to be used for a variety of applications, including for educational purposes. In most countries, people engage in vigorous debate about the exchange between businesses and workers, and whether that relationship is perceived as being reasonably equitable. This will become a more central issue in national political discourse as the inextricable link between AI and economic growth becomes clearer to more people.

The future of learning in the classroom involves the convergence of AI, high student engagement, and personalized learning. Perhaps, this is a necessary step in the evolution of an educational system in which teachers are no longer restricted to the role of an instructor imparting knowledge to students as they sit captive and listen. Rather, some teachers are likely become more like guides or managers, that still lecture and impart knowledge but also allow students to play a more active and hands-on role in the learning process.

The global economy is continually being upended by technology in unforeseen ways. It is the obligation of policymakers to ensure that educational systems teach students how to think critically, solve problems, and promote innovation and risk taking. If our education systems continue to promote the illusion that we have all the answers, then they will be doing humanity a disservice by not preparing our children for a future that needs to be disrupted in order for all of us to survive.

AI and immersive learning environments can spark creativity and a passion for learning that is desperately needed as the future approaches us faster than ever before. One of the key pivots in the future will be to use AI to drive self-learning, group learning, and experimentation. As students become more inquisitive, they should be allowed to mold the learning process to suit their interests. If it takes a village to raise a child, in the ML era it will take an entire classroom, increasingly utilizing AI, to teach a student.

Daniel Wagner and Keith Furst

8. *Smarter Government*

Strategy and Inspiration

Most democratic forms of government seek to achieve similar objectives, such as establishing institutions to ensure law and order, maintaining a military capable of defending citizens, developing infrastructure, and encouraging mechanisms to promote free trade and entrepreneurship. AI is positioned to reconfigure how governments achieve their objectives, the policies they need to adopt to allow the benefits of AI to flourish, and the manner in which they manage the risks and negative consequences of AI-inspired growth.

Since 2017, more than 20 countries have announced national strategies[1] to work through the political, economic, social, ethical, and legal implications of adopting and promoting AI. There are recurring themes in many of these countries' strategies, such as ensuring job growth and developing an ethical framework for AI to operate within. However, each country's AI strategy is unique, naturally reflects its own economic strengths, and is often intentionally focused on enhancing those economic sectors for which it has a comparative advantage.

It was not surprising that Canada became the first country to announce a national AI strategy in 2017, given its vibrant AI history, and that it is the home of some notable figures in the AI industry, such as Geoffrey Hinton (known as the Godfather of Deep Learning), a professor at the University of Toronto. Canada's AI strategy is unique when compared with other nations because it focuses primarily on scientific research in the belief that education is the foundation upon which economic growth is built. Canada has chosen to pursue a collaborative AI model. In 2018, Canadian Prime Minister Trudeau and French President Macron announced an AI collaboration,[2] consistent with both national AI ambitions and strategies.

Canada has a vibrant AI startup scene, as is illustrated by the MaRS' 1.5 million-square-foot facility in Toronto, which houses hundreds of startups across a wide range of sectors. Montreal is also pursuing a non-predatory model[3] of knowledge sharing, in the belief that scientific progress should benefit everyone. Canada has drawn a wide array of students from across the world to its universities. In 2018, international students compromised nearly 25%[4] of the total student body of the University of Toronto, which has more than doubled in size since 2007. That said, Canada could invest more in specialized high schools or charter schools that focus on AI to prepare native-born Canadians to compete with the world's best and brightest minds. The Canadian government could also develop a more purposeful plan to increase the likelihood that the world-class AI researchers it produces will remain in the country to contribute to its future AI and economic development.

Other countries are prioritizing their AI strategy to address the unique challenges of their homeland. Japan had a population of about 127 million people in 2017, which is expected to shrink by a third over the next five decades. The low birth rate, combined with increasing life-expectancy, means that there are roughly twice as many jobs[5] as there are applicants. One way to address the problem would obviously be to allow more foreign workers into the country, but Japan has historically severely restricted the number of immigrants it allows to migrate. Prime Minister Abe views the country's unique challenges as an opportunity to increase the use of AI and robotics in manufacturing and job markets.

Japan is the global leader in the production of industrial robots, having shipped 115,000 units (or 52%) of the global supply in 2016[6]. Its expertise in robotics is allowing it to tackle some of its most enduring social issues. In a 2016 survey,[7] 80% of 1,200 Japanese respondents, aged 40 or older, were open to the idea of having robots care for them in old age to ease the burden on their families. Some have argued[8] that Japan's openness to robots can also be traced to the country's main religion of Shintoism, which posits that both animate and inanimate objects have a spirit. There are already a wide variety of robots available to the elderly, such as smart walkers, animals, video phones, objects with sensors, and lifts. In a 2017 Japanese government study,[9] robots were found to be effective at improving the autonomy of 34% of seniors in assisted living facilities. It will take time for a completely autonomous robot caregiver to emerge, but this is undoubtedly one of Japan's long-term goals in the AI arena.

Governments, of course, play a crucial role in promoting AI development by setting tax policy to encourage investment, altering regulatory frameworks to reduce operating costs for startups, and partnering with the private sector to help define AI standards and other benchmarks. Another challenge that governments must address is how to spark innovation in the private sector to generate commercially viable products. The DARPA has provided an excellent blueprint for what government supported innovation could look like. The agency's consistency in producing groundbreaking new technologies demonstrates that its strategy works.

The DARPA's many achievements include the integral roles it has played in the development of the Internet, stealth aircraft, miniaturized GPS technologies, UAVs, flat-screen displays, and brain-computer interfaces. It ascribes the following key factors as being central to its success in innovation:[10]

- Limited tenure and the urgency it promotes;
- A sense of mission;
- Trust and autonomy; and
- Risk-taking and tolerance of failure.

One of the DARPAs signature strategies is to host challenges for teams to compete. In 2004, the first DARPA Grand Challenge was held for American autonomous vehicles, but no winner was announced because no vehicle actually finished the race. However, in 2005, DARPA hosted its second Grand Challenge and the participating vehicles' performance improved remarkably. A team from Stanford University, led by Sebastian Thrun, was announced as the winner. Thrun is the CEO of Kitty Hawk, which is an ambitious startup working on developing flying vehicles which could be fully operational by 2023.[11] Thrun may just have committed to such an ambitious, long-term project as a result of having participated in and won the DARPA Grand Challenge. Countries that follow the DARPA innovation principles and competition model may similarly lead to innovation in unexpected ways by inspiring a creative, risk-taking culture.

Streamlining the Legislative Process

While the adoption of the right types of laws and regulations with the right mix of leniency and simplicity may similarly encourage innovation and risk taking, the process of law making itself often acts as an impediment to progress in a host of sectors. Many national lawmakers have little real understanding of the laws they are passing or any real appreciation of their implications on individuals, businesses, or societies at large. In the US, for example, many lawmakers do not even necessarily read the bills they vote on, relying instead on feedback from specialized staff members to review and explain legislation, and/or the independent analyses produced by non-governmental organizations (NGOs), think tanks, or lobbyists.[12] AI is likely to be another interpreter of legislation in the future, though not necessarily to create legal summaries directly for lawmakers. It is more likely that NGOs, think tanks, lobbyists, and staff will use AI to help them analyze the impacts of legislation more accurately.

Another emerging trend is the ability to create readable reports from structured and unstructured data. Primer, an AI company, has developed SML and UML algorithms that can mine through vast amounts of Internet data and summarize a topic into key concepts and trends.[13] For example, a user can enter information into an AI platform about a topic she may be interested in, such as coal mining in Indonesia, and a report will be produced by algorithms behind the scenes. The technology is already being used by some big banks and members of the intelligence community to summarize topics of interest on-demand, which allows analysts to focus on other things while staying abreast of key trends and risks.

If an AI system can create human readable reports from unstructured Internet data, then it can also decipher legislation. It will take time to train AI how to process legislative language effectively, but as ML algorithms become ubiquitous, more easily deployable, and more affordable to run, it is likely that someone will develop AI to make legislation more transparent.

AI can transform the legislative process by moving it from lawyers manually reading and writing bills to modeling them. Perhaps one analyst may read and write bills as another leverages AI, NLP, and data visualization to model their impact within existing complex legislative frameworks.

Corruption and the Election Process

In a 2018 Gallup poll,[14] only 5% of those surveyed had a high degree of confidence in the US Congress. In many countries, simply trying to understand what an elected official or candidate running for public office believes or has historically voted on can be a daunting task. AI can decipher with ease how lawmakers have voted, their public stances in the media, sources of campaign funding, possible conflicts of interest, and other publicly available information. Doing so could make the legislative process easier to understand and might also encourage more people to run for office and participate in the political process by virtue of it being made more transparent and making elected representatives more accountable.

The relentless parade of corruption scandals across the globe amplifies the need to introduce more transparency and accountability across the board. The International Monetary Fund estimated[15] that the annual global cost of bribery as of 2016 was between $1.5 and $2 trillion, or approximately 2% of global GDP. Some have criticized the study and questioned the reliability of the data and the extrapolation techniques used, but anyone who has traveled throughout the world, follows the news, and has read Transparency International's Corruption Perceptions Index understands the gravity of the problem.

As has already been discussed, AI is being used to detect a person's honesty based on their eye movement and pupil dilation. In the future, some politicians may choose to use their willingness to undergo an AI-based lie detection test as a platform from which to build their campaigns. Similarly, if an NGO developed AI to monitor the public data of lawmakers for any potential concerns about conflicts of interest, it might reduce the level of corruption because of the fear of getting caught.

The voters of the future may be able to visit a website to browse a politician they are thinking of voting for and get a lot of useful information from AI-driven platforms that could score politicians for honesty, graft risk, historical voting patterns on key issues, publicly available data on personal assets, and hidden links with companies and other high-risk individuals. While some politicians may not want to give up their personal information voluntarily, others who may be an underdog in a political race may think they have nothing to lose by submitting themselves to an intrusive AI cavity search. Also, the use of AI could help promote their campaign through viral media posts and reduce the need for wealthy campaign donors. In other words, AI could alter the power dynamic among the political elites, level the playing field, and give more power back to where it belongs—with voters!

If AI were to provide us with a wealth of information about potential political candidates, would people actually use it? Also, would a candidate's honesty or propensity for corruption impact the outcome of elections in the future, given the low bar that has been set on the "acceptability" of corruption, its widespread practice, its institutionalization in the halls of government, and the relative political apathy of some voting populations? The caliber of the AI used in the political process of the future will, of course, only be as good as the data used to calibrate it and the degree to which voters use and benefit from it.

Evaluating the effectiveness of processes and laws boils down to the values societies, lawmakers, and the voting public attribute to them. Instead of being deliberately transparent and neutral, AI systems could be trained to promote radically different value systems. Could a truly objective political AI that is immune to partisanship even be created, and would it stand a chance of being accepted and effective if the noting public remained partisan in nature? These are among the many basic questions that have yet to be addressed.

Smart Cities

A key function of government is to ensure that cities operate efficiently as centers of economic activity that drive a country's growth. As briefly referenced in Chapter 2, Singapore's National Research Foundation launched a program called AI Singapore[16], in 2017 that will use AI and ML technologies to address major social issues, such as creating healthcare schemes for an aging population and managing vehicle traffic congestion more effectively during peak hours. The program will be funded with approximately $107 million over 5 years.

One of the issues AI Singapore has proposed to address with AI and ML is the Land and Liveability National Innovation Challenge (NIC). Approximately 90% of the world's countries have more land area than Singapore but, in 2017, Singapore's GDP was greater than roughly 80% of the world's countries[17]. The government of Singapore views its limited land capacity as a precious resource it must manage carefully. It thinks that a key part of the answer to this problem is AI and smart urban planning to ensure the prosperity of the country going forward. The objectives of the NIC that can potentially be addressed with AI R&D are:

- Creating useable underground spaces in caverns and areas with poor soil quality;
- Devising new land reclamation methods that require less sand;
- Developing alternatives to land reclamation;
- Intensifying low-density land uses for industry and infrastructure;
- Achieving a good quality living environment;
- Reducing space demands for public infrastructure such as transportation;
- Co-locating urban infrastructure, utilities, and services with other spaces; and
- Reducing the unproductive use of land.[18]

By teaching ML algorithms to look for specific patterns in the data, a government can track how a city's green areas are maintained, whether they are shrinking or expanding, and whether certain building development projects are not in accordance with zoning laws. AI and ML are also being used to analyze satellite data to help improve urban planning. In 2018, several of the World Bank Group's research teams collaborated[19] to merge the abundance of satellite images with computer vision capabilities, driven from ML algorithms, to offer stakeholders in Ho Chi Min City, Vietnam, the ability to track the urban development process over a set period of time. The research teams used SML and trained the algorithms with survey data and other classification labels based on information from comparable cities.

Networks of nanosatellites can capture high-resolution images of the Earth for a range of applications, such as tracking commercial vessels, predicting crop yields, monitoring livestock movements, and even surveilling the structural changes of cities. In 2017, Bank of America Merrill Lynch estimated that the space industry will be worth $2.7 trillion by 2045.[20] The explosion in the number of space-enabled AI will lower the price of high-quality Earth images over time. Once this data is available at scale, many governments across the globe can procure it and use AI to assist with urban planning and resource management.

City sustainability is more relevant today than it has ever been. In the UN's 2018 World Urbanization Prospects report,[21] it was estimated that 2.5 billion people will be added to the world's population by 2050, and that 90% of this growth will come from Asia and Africa. Only three countries (China, India, and Nigeria) will account for 35% of the population growth in urban areas. As the population increases, cities will grow up and out, as population density dramatically increases. Regional ecosystems will come under even greater strain and governments will be forced to think longer-term about ensuring basic services such as water, energy, and transportation at greater scale.

According to a 2017 Deloitte report,[22] there are more than 1,000 smart city pilot projects underway and in excess of 500 of them are in China, making it the world leader by a wide margin. China has many challenges to address with its smart city projects, such as improving healthcare for its citizens, reducing pollution, regulating waste management, and managing traffic congestion. Smart cities face some common implementation challenges, such as lack of proper planning, funding challenges, inadequately integrated data siloes, and information security gaps. A major advantage China has over many other countries when it comes to smart city planning is that the government can access nearly any data it wants, any time, and in virtually any way, by law. This should result in making China's smart cities smarter than in other countries because obtaining access to critically important data will not be blocked by stringent privacy regulations—something a lot of other governments can only dream about.

Transportation

As countries continue to roll out smart city initiatives, an important but often overlooked contributory factor will be the successful construction and operation of smarter transportation infrastructure. AI will be part of what powers smart transportation such as parking, self-driving cars, and drone delivery services. Barcelona piloted a smart parking system[23] with sensors installed in asphalt to detect whether a car was present in a parking spot. If the space was available, the system would direct a driver to the spot, in a small way reducing emissions and increasing the city's tax revenues.

In 2018, Drive.ai announced[24] its collaboration with the city of Frisco, Texas, to pilot a self-driving car program that will shuttle about 10,000 people to a geographically defined area where autonomous vehicles can operate. This means that self-driving cars need to operate in a controlled environment until they are ready to share the road with human drivers. Quarantining self-driving cars is likely to remain the case for several more years until self-driving technology becomes more advanced. For it to really take off, cities will need to create the infrastructure for them to operate within, such as dedicated lanes on roads, sensors embedded into streets, and traffic lights with ML algorithms to dynamically adjust traffic signal patterns. The AI-powered roads in the smart cities of the future could adjust the timing of traffic lights in real-time based on congestion, construction projects, and accidents. Some of the many promises of AI in smart cities may not get implemented in certain countries—even if the technology is sufficiently mature—because of an aversion to risk on the part of governments and their fear of lawsuits in some of the more litigiously-advanced countries.

Another aspect of transportation that AI will transform is package delivery. According to a 2015 Accenture report[25], the parcel delivery business is expected to grow to $343 billion globally by 2020. One of the consequences of the digital revolution is that choice and the availability of information has tipped power away from companies and back toward consumers. Consumers now have access to massive amounts of information such as customer reviews, search engines, and delivery options, which enables them to compare similar products and select the best option for their needs. This has led to pricing wars among companies in highly saturated industries including parcel delivery. Robotics has already entered the delivery business via pilot programs. Last mile delivery models include postal mail-runs, courier delivery to lockers, crowd sourcing delivery to homes, and crowd sourcing delivery to lockers.

In 2017, Estonia became the first country in the EU to allow Starship Technologies delivery robots[26] onto sidewalks to deliver lunches. The robots are not very tall and, given many people's obsession with walking and looking at their smartphones, it is only a matter of time before they start tripping over these robots. Eventually, someone will get hurt and prompt an inevitable call for stricter regulations, which could include proposals for designated robot paths (similar to bike lanes on streets) that allow robots to arrive at their destination with lower risk to pedestrians. Legislators may also restrict ground delivery robots to areas where small children and elderly people are less likely to roam, such as in business districts.

The risk to pedestrians falling over land delivery robots is real because of an increasing lack of awareness to our surroundings, the result of the increasingly high volume of digital media consumption and our insistence on making smart phones a priority over our own safety. We increasingly text while driving and surf while walking—so much so that, in 2014, the Chinese city of Chongqing created a "phone lane" for texting pedestrians.[27] If we permit it, AI will continue to decrease our capacity to think for ourselves and perform basic tasks.

In 2017, the US Department of Transportation (DOT) announced the Integration Pilot Program, designed to allow state, local, and tribal governments collaborate with the private sector to safely integrate Unmanned Aircraft Systems (UAS) with the national airspace infrastructure. The DOT announced 10 awardees that will work with some large tech companies from the private sector to create pilot programs that focus on specific applications and benefits for a specific sector or type of consumer. Some of the issues that need to be worked through to help develop an appropriate regulatory framework include nighttime flight operations, flying over people and beyond the line of sight, package delivery, autonomous collision avoidance technology, data latency and reliability, and cybersecurity and privacy concerns.[28]

There are a host of other applications being tested to make government smarter, such as using AI to spray insect populations, inspect pipelines, make time-sensitive deliveries for medical equipment, and enhance border protection. To prevent the collision of drones with people, planes, and other infrastructure, a common communication platform will need to be created and perfected. Drones could have built in AI capabilities to avoid and detect other drones and objects, but it will be difficult to anticipate and account for every type of scenario, including extreme weather or unforeseen behavior. However, if a central command were built to track all drones in real-time, information could be exchanged between the command center and drones to react to obstacles in their environments. Given how far behind the curve governments are in regulating drones, we are long way away from being able to accomplish such a feat.

Electricity Grids

The increase in demand for energy and the push towards more renewable energy sources is driving AI toward managing the energy grids of the future. By 2040, US electricity demand is expected to increase by 11%, but the US electric grid is notoriously inefficient and wastes most of the energy it produces. A 2017 flow chart, produced by the Lawrence Livermore National Laboratory and the US Department of Energy (DOE),[29] estimated that the US wasted approximately 67% of the energy it produced, with 25% of that amount coming directly from the electricity grid. The primary culprit is heat waste through the utilization of inefficient technology. The inefficiency of the electric grid drives up the price of energy as more resources must be consumed to fund the inefficient distribution of energy. For example, during times of peak demand, utility providers must pull energy from the regional system and pay fees for doing so, which increases costs to the end consumer.

However, the US is moving forward with a strategy to use AI and ML to create an autonomous grid that can dynamically predict and interact with user demand. In 2015, a Vermont utility provider, Green Mountain Power, was the first provider to offer Tesla Powerwall batteries to its consumers, as well as a package with the Nest thermostat.[30] Nest uses AI and ML to automatically adjust temperatures based on the patterns of people in the home, personal preferences, and changes in temperature outside the home. Some users have reported as much as a 40% reduction in their energy bill from smart home devices. Independent studies have shown a savings of between 7% and 17% for gas heating and electric cooling.[31] In 2014, Google bought Nest, for $3.2 billion, as part of its push into the smart home market and its broader strategy of acquiring more data on people outside of Internet searches.

As consumers continue to cognify their homes with smart devices, this data can be used or resold to help utility providers understand more about their consumers' behavior and consumption patterns. However, the utility providers also need to understand more about the behavior and operations of their own electric grids. One of the first steps in applying AI and ML into electricity grids is to gather large data sets from satellite images, utility operations, and other forms of data, to model its behavior and potential vulnerabilities.

Part of what makes infrastructure generally vulnerable to cyberattack is the fact that so much of it is dependent upon software and so many of its systems are interconnected. Much of the world's critical infrastructure utilizes supervisory control and data acquisition (SCADA) systems, which automatically monitor and adjust switching, manufacturing, and other process control activities based on digitized feedback data gathered by sensors. These tend to be specialized, older computer systems that control physical pieces of equipment that do everything from route trains along their tracks to distribute power throughout a country. SCADA systems have increasingly become connected to the Internet but were not designed with cybersecurity in mind.

This is a big problem that has been widely written about but becomes more of a threat each year. Virtual terrorists could do tremendous damage if they wanted to, ranging from taking control of water treatment facilities to shutting down power generation plants to causing havoc with air traffic control systems—and *all of these systems are extremely vulnerable to attack*. Despite the fact that cyberattacks occur with greater frequency and intensity around the world, many either go unreported or are under-reported, leaving the public with a false sense of security about the threat they pose and the lives and property they impact. While governments, businesses, and individuals are all being targeted on an exponential basis, *infrastructure is becoming a target of choice among individual and state-sponsored cyberattackers, who recognize the value of disrupting what were previously thought of as impenetrable security systems. This has served to demonstrate just how vulnerable cities, states, and countries have become.*[32]

The locations and period of peak or weak energy usage can be identified in electric grids, and strategies to recover from failure are being developed. The Grid Resilience and Intelligence Project (GRID),[33] developed by the DOE, is one such effort, one of seven projects seeking to address grid resiliency and security. As smart devices provide feedback to grids the risk of a cyberattack increases, since that information can be hacked. In the future, grids may become heavily reliant on feedback from users. Malicious hackers can potentially manipulate these devices to send bad information back to the grid as part of a broader cyberattack strategy.

One of the GRID projects is the Visualization and Analytics of Distribution Systems (VADER),[34] designed to address renewable energy integration into existing electric grids. AI and ML will be at the center of new grid operating paradigms that will dynamically apply intelligent controls to voltage regulators and capacitor banks on a daily basis, as opposed to waiting for manual data collection in the field. The VADER project is part of a broader effort to monitor resources at the edge of the grid or beyond the substation.[35] As renewable energy assumes a more prominent role in the grids of the future, tools such as VADER will become increasingly important in monitoring and enhancing grid performance.

The application of AI to the electric grid is similar to what is being proposed for managing 5G and other advanced communication networks of the future with ML. Ultimately, both systems are too complex, fast, and dynamic to manage manually. ML algorithms can autonomously route resources to the appropriate places based on current and forecast demand. They will eventually reduce consumption in grid resources and reduce energy costs, which will spur greater demand for renewable energy equipment. The cognification of the grid should also allow less energy to be wasted, which should reduce consumer energy costs and benefit the environment by decreasing the need to burn as much coal to generate the same amount of electricity.

Algorithms have been known to behave unexpectedly. The designers of the AI-managed grid of the future will have to think about how to contain algorithmic contagion and switch to a manual or fixed distribution mode of control for impacted parts of the grid. Stock markets have circuit breakers that halt trading of individual stocks if there is a significant drop in the percentage of its price. Smart grids may also need to have algorithmic breaker capability, because ML is dynamic and there is no way to know how a system will react in all circumstances. We need to devise a way to manage it when it happens.

Managing Complexity

Governments are attempting to address the complexity of running a country by embedding intelligence into their cities, but the rapid speed of technological change has clearly outpaced how fast many of them can react. A 2017 Deloitte report[36] estimated that automation, including AI, could save the US government approximately 97 million federal man hours, resulting in a potential annual savings of $3.3 billion. If automation were to be widely adopted, the man hour savings could rise to as much as an estimated 1.2 billion hours, resulting in a potential annual savings of $41 billion.

Some of the functions that governments perform are complex. The US government-funded healthcare program known as Medicaid involves 51 agencies, tens of millions of beneficiaries, tens of thousands of providers, and hundreds of managed care plans. The US Office of Management and Budget has designated 19 programs as high-priority[37] to identity, monitor, and report on improper payments. The agency defines an improper payment as any payment that should not have been made, was made for an incorrect amount, or had insufficient documentation that resulted in an inability to determine its legitimacy.

In 2017, Medicaid had the highest dollar amount of improper payments, totaling $36.7 billion (or 10%) of the $364 billion the US federal government spent on the program. It is important to note that improper payments are not the same as fraud and only $3.8 billion of the $36.7 billion total was attributed to a known monetary loss to the government.[38] In 2017, the GAO evaluated the Centers for Medicare and Medicaid Services' (CMS) fraud management capabilities[39] and discovered, unsurprisingly, that some were inadequate. Its report found that the CMS did have an entity committed to managing fraud risks, but it lacked some other key components that are ordinarily part of an effective anti-fraud strategy. For example, regularly scheduled employee training was not being conducted and the CMS has not implemented a risk-based approach to fraud monitoring.

This suggests that the CMS is only scratching the surface of the actual fraud being detected. Most fraud detection systems have a scoring mechanism to assign a level of risk to an event or series of events so that an organization's limited resources can focus on high-priority potential fraud. If the CMS has not conducted a risk assessment of Medicare and Medicaid and does not apply a risk-based approach to fraud monitoring, then its anti-fraud capabilities are bound to be rudimentary. The CMS anti-fraud framework is in desperate need of an overhaul. The National Healthcare Anti-Fraud Association has estimated[40] that tens of billions of dollars are lost to healthcare fraud in the US each year. In 2017, the DOJ arrested[41] more than 400 individuals suspected of being involved in $1.3 billion of fraud losses. The fraud involved doctors, nurses, and other medical professionals prescribing unnecessary medication that was never purchased or distributed to beneficiaries.

AI and ML could be used to detect healthcare fraud more accurately and identify suspicious activity in historical data. One of the main data sources that contain red flags for healthcare fraud are claims records, which number in the hundreds of millions. Each claim record could have up to 300 data points, and there are 67,000 diagnosis codes and 87,000 procedure codes that categorize the nature of an illness and the services provided. These codes can be entered in error by poorly trained clerical staff or maliciously manipulated to increase the amount of a claim. It is a notoriously difficult problem to address.

The massive repository of data makes monitoring healthcare fraud an ideal problem for AI and ML to solve. As AI platforms begin to succeed at identifying more fraud, the US government and private health insurance companies will be motivated to feed AI's insatiable appetite for more data. Publicly available information about health care providers could be ingested to identify scenarios that could contribute to a person's propensity to commit fraud. For example, if a healthcare professional had a lien placed on his house for not paying local taxes, it could put him in a higher-risk fraud category, signaling the need for closer monitoring.

The CMS could label or identify transactional patterns and other behavior that are indicative of healthcare fraud and load it into an anti-fraud ML platform. Once ML algorithms are adequately trained, they can predict the probability of fraud for future transactions using a risk-based approach and allow analysts to investigate and determine whether the activity was fraudulent. The CMS analysts who investigate potential healthcare fraud events would provide feedback that would be sent back to the ML algorithms so that parameter weights and other factors could be updated automatically, to more accurately predict the likelihood of fraud going forward. The CMS could also load a large repository of historical data into an AI platform and use UML to identify clusters of similar behavior. CMS analysts could then investigate pockets of abnormal activity to identify historical incidents of fraud that could potentially be prosecuted or used as labels for the SML to make better predictions in the future.

Apart from governments providing healthcare services to their citizens, they must also take measures to protect their people from getting sick. In the US, foodborne illnesses affect approximately 48 million people annually, with about 120,000 hospitalized and 3,000 people dying from those illnesses every year. Researchers from the University of Rochester developed an AI program (called nEmesis) that mines through Twitter feeds to identify tweets referencing food poisoning and link them to restaurants using geotagging to identify hot spots. The researchers estimated that the health violations prompted changes in food preparation practices by some restaurants and led to roughly 9,000 fewer food poisoning incidents and more than 550 fewer hospitalizations in the Las Vegas area for the period.[42]

Pipelines

Accurately identifying the source of water pipe leaks has historically been challenging because many parts of pipeline networks may not be easily accessible. Acoustic devices called geophones have been used to identify leaks in pipeline networks, but this requires someone to listen and interpret feedback regarding whether a leak exists and where it is located. A Brazilian startup, located in Sao Paulo, has leveraged AI and ML to create a portable leak detection tool to make tracing leaks easier. The company's AI system, called Fluid, has a hand-held device that is placed against parts of the pipeline (such as under a manhole covers) to record vibrations in the network for approximately 15 seconds.

An initial version of the system is providing 800 data samples per day from 10 water companies.[43] The idea is that the company will use the information it has accumulated as training data so that the ML algorithms can improve over time. The ML algorithms must first predict whether the pipe vibrations suggest that a leak is present and then predict where the leak is located. The company believes that, armed with enough data, its AI can be trained to use recordings from two different points in the network to more accurately and quickly identify the location of the water leak. This type of AI system can help governments manage their pipeline infrastructure by deploying limited resources with precision, as opposed to randomly checking points in the network using antiquated detection techniques or not fixing the leaks at all.

Researchers from the Massachusetts Institute of Technology developed a small rubber robot device called PipeGuard that is placed inside pipeline networks and gathers data on water pressure. The data can be fed into ML algorithms to identify variations in water pressure and predict what sections of a network are most susceptible to a pipe burst. Following successful tests in Mexico in 2017, the researchers plan to make a more flexible, collapsible version of their robot that can quickly adapt itself to pipes of different diameters. Under the streets of Boston, for example, are a mix of 6-, 8- and 12-inch pipes—many of which were installed so long ago that the city has no accurate maps of their locations. The robot would expand like an umbrella to adapt to each pipe.

The value of the robot is not only to reduce water losses, but to make water services safer and more reliable. The ability of the system to detect much smaller leaks could enable early detection and repair well before serious pipe breaks occur. Not only could the system find leaks in virtually any kind of water pipe, it could also be used for other kinds of pipe distribution systems, such as for natural gas. Such pipes, which are often old, have produced serious gas buildups and result in explosions in some cities, but leaks are hard to detect until they become large enough for people to smell. The team hopes that the robot will eventually be able to repair leaks on the spot.[44]

Water leaks are usually unintentional, but leaks in oil pipelines are often deliberate. Mexico's state-run oil company Pemex loses in excess of $1.5 billion per year due to fuel theft, which was roughly 3% of its 2016 revenue. Pemex is important to the economic stability of Mexico because it provides the country with approximately one-fifth of its income. Some 2,274 illegal taps in Pemex's oil pipeline network were recorded in the first two months of 2018, an increase of 38% over the same period in 2017, and a 352% increase over the same period in 2014.[45] The fuel theft has gotten so out of control that the Mexican military must escort Pemex employees to oil refineries in heavily armed vehicles. The Mexican government could use AI systems to identify leaks sooner to plug leaks faster. AI-enabled drones and satellite images could be another way to obtain more data for AI to process and identify abnormal activity around illegal tapping hot spots.

Public Safety

Governments have access to a lot of information, but not all types of data are accessible or searchable on a computer. Perhaps one of the most blatant examples of information that is not digitized in the US is gun ownership records. The US Bureau of Alcohol, Tobacco, and Firearms' Explosives National Tracing Center (ENTC) is the only crime gun tracing facility in the US. In 2017, the NTC received 408,000[46] requests for traces of guns suspected of being used in crimes.

Routine traces are usually completed in an average of 9 days. The reason it takes the agents so long to trace a gun is that the information is not stored in a searchable database with multiple attributes. Nearly all modern databases allow a search to occur based on information with multiple parameters, but agents are forced to find information about gun dealers and are restricted in the type of information they may seek. Agents are then forced to manually review sometimes thousands of images before the gun they are looking for may be identified.[47] Law enforcement could potentially feed the gun ownership records of individuals suspected of being involved in crimes into an AI system to gain insights and uncover hidden links among larger criminal networks.

Since gun ownership is an excellent example of a component of law enforcement that is highly politicized, a number of questions arise, such as, should governments be allowed to use AI to analyze gun ownership records and merge that with other data to prevent crime, or would doing so be perceived as giving governments too much power? Does it make sense to have AI scan someone's emotional expression before buying a gun? Where do we draw the line between public safety, privacy, and AI? Is it even possible to use AI to predict whether a person is more likely to use a gun inappropriately? Do legally purchased guns used in crimes depend more on personal circumstances which could change over time, and if so, of what potential benefit is the use of AI in trying to predict what will happen next?

There is a thin line between governments ensuring public safety and monitoring all law-abiding citizens indiscriminately; AI can blur the line even further, but if the net result is that crimes are prevented or criminals are apprehended, does not the argument in favor of the greater good trump any argument about infringement of privacy rights? For instance, in 2017, China had more than 170 million closed circuit television cameras across the country,[48] many of which had been enabled with AI facial recognition software to identify suspects in real-time. In 2018, China's AI facial recognition system detected a fugitive through security camera footage at a concert. The suspect was wanted since 2015, after allegedly fleeing following a payment dispute over a commercial shipment. When he went through the stadium's security checkpoint, a warning was issued informing police that he was a wanted person.[49] He was later apprehended. Although their privacy may have been violated, all of the attendees were subjected to the same screening procedure, even though they did not know it. Do we have the luxury of quibbling about privacy rights in the AI era when so many capabilities that were previously little more than fantasy have become reality and give law enforcement capabilities they did not previously have?

Not all countries have laws and regulations in place to force private businesses to share live data feeds with law enforcement. Even if such regulations exist in other countries, the limits regarding how each government uses such information is inherently different in each case. Privacy is being thrown out the window in some countries in favor of the greater public good—or rather, what is best for a given government. For example, China's Social Credit System aims to "score" 1.4 billion[50] of its citizens by 2020, based on how they behave (this is explored in more detail in Chapter 11).

The massive repository of online genetic information—growing each year as more people upload their DNA—gives AI programmers an opportunity to find hidden links among relatives. Genetic information can also be enriched with labels that AI can be trained on. For example, if DNA information can be uploaded with associations to other family members, it can be processed by ML algorithms. Genealogists will not have to scour through genetic profiles, looking for indications of hidden links, because, when enough data and accurate labels exist, the ML algorithms can program themselves to predict who is related to the DNA profile being investigated.

Could AI and ML be part of the answer to solving nearly any crime where DNA can be recovered at the scene? In theory, yes. Based on the technological developments in facial recognition, genetic sequencing, and social media monitoring, it is more than likely that emerging AI programs can help solve more crimes faster. It is no longer a question of law enforcement not having the tools to complete an investigation. It is now up to lawmakers to enact legislation that will allow these agencies to do their jobs the best way possible.

Digital Government

The future of effective governments is to be able to offer services entirely online, and to exchange information digitally between agencies, reducing costs and improving efficiency. Estonia is considered by some to be the global leader in digital government. It was the first country to implement groundbreaking innovations such as enabling online elections, permitting the use of delivery robots, and offering e-residency.

Estonia has digitized nearly its entire government, as 99% of its state services are online, earning it the title of "the most advanced digital society in the world"[51] by *Wired* magazine. However, some critics have pointed out that not all of its digital services are as secure as they may appear to be. A team of international e-voting experts analyzed Estonia's system and concluded that it used an inadequate security architecture. Experts ran tests in their own lab and found multiple ways to attack the system, which could alter future election results and would be undetectable.[52]

In 2007, Estonia suffered a series of cyberattacks[53]– presumably courtesy of the Russian government in retaliation for relocating a Bronze Soldier statue from World War II–which prompted the small nation to adopt even more digital technology. Estonia had the advantage of being a young nation when it gained its independence in 1991 and was not burdened by legacy IT systems. Its feeble digital infrastructure, a remnant of the Soviet era, was a weakness at the time, but became a strength that allowed the government to build something new from the ground up. The country's population of just over 1 million people is much smaller than most nations in the world, which makes it easier to get consensus from citizen stakeholders and implement policies with greater ease.

The digital infrastructure that Estonia built is based on public key infrastructure and eID systems which leverage encryption technology and require two-factor identification for access. Estonia uses a system called X-Road to provide a wide range of services to citizens and power its e-government. X-Road allows data to be shared among the different government agencies on a need to know basis, with the proper permissions, and maintains audit trails of what data has been accessed and by whom.

According to the Estonian government, the three basic mandates[54] that an interoperability system must achieve in order to offer a wide array of government services to a country's citizens are to ensure that data is securely exchanged only with authorized users, enforce data integrity by preventing the alteration of data as it is in transit, and protect data confidentiality by preventing unauthorized users from viewing it as it is in transit. The cost savings and efficiency of having this system are profound because the data is decentralized and not duplicated across various participating institutions but, rather, simply accessed and used on demand.

Estonia provides a blueprint for other countries to emulate. Its model is to only exchange information between users and agencies that have the right to do so, while auditing who accesses what information. It puts citizens in charge of their own data and what services they may apply for. This type of data exchange model is desperately needed in many areas of government, such as fighting money laundering, fraud, and other financial crimes. The various agencies within a government could potentially request the data they need to build specific AI to predict something they may be interested in. The exchange of information can be tracked and audited. Individuals suspected of being involved in a crime would have their privacy rights revoked, so there could be circumstances wherein a person is not alerted when their information is accessed, or why. As a paradigm for digital government, Estonia stands tall above the rest.

Conclusion

According to a 2017 PWC study,[55] AI has the potential to create up to $15.7 trillion in GDP gains, with the top three potential GDP beneficiaries of AI growth being China (26.1%), Europe (21.4%), and the US (14.5%) by 2030. Part of the frantic pace with which some countries are setting national AI strategies has as much to do with future GDP growth as with a preoccupation not to lose what growth they have already achieved. Most governments with AI capabilities have already begun to deploy them in earnest, or are in the process of doing so, as noted in Chapter 2.

AI is positioned to reconfigure economies in profound ways. There could be periods of time in which market upheavals create situations where more citizens need to be supported by their government. If large swathes of workers are displaced in the race for AI supremacy, then reducing wasteful government spending will become essential to maintaining a functioning nation. One way for governments to achieve this is to use more AI with their operations and slowly migrate to a digital and interoperability model of offering government services online and exchanging information between agencies. Only the most enlightened of governments are likely to end up doing so with urgency, however. Most are likely to do enough to appear to be in the game, but not enough to position them to actually win the race.

The governments that develop comprehensive AI strategies, and that can account for the interactions between a diverse group of adaptive agents (people, companies, and other governments) in a rapidly changing ecosystem, will have the best chance to succeed in the race for AI supremacy over the long haul, as they prepare for the inevitable disruptions the transition to an AI-dominated world will cause. Government AI strategy clearly cannot be developed in a vacuum. Overemphasizing one area of potential benefit at the expense of another could have long-term repercussions that negatively impact a country's ability to compete successfully in the race.

AI will be an important part of planning and managing cities, sectors, and populations in the future. In the end, governments that will develop policies determining what data is collected, how it is used, who can see it, and when it gets deleted (if at all). The world is in desperate need of global AI best practices and standards to ensure that there is a carefully crafted balance between the needs and objectives of governments and the desires and requirements of their people. It remains to be seen whether civil liberties are either properly taken into consideration or are tossed by the side of the road in the process. There is no single "correct" approach, for what is appropriate for a given government in a given circumstance at a given point in time is, and will remain, a moving target.

There is no guarantee that personal freedoms and privacy will even be central ingredients of national AI policies. If they are, in some countries the average citizen is unlikely to know anything about it, or to necessarily care whether their civil liberties are even part of the mix. They may simply be informed of the latest version of the evolving new AI normal, in the knowledge that "normal" today is very likely not what it will be one year, five years, or ten years later. So much about AI is, and will remain, entirely within the scope of control of governments and businesses, and outside the control of the average citizen.

Daniel Wagner and Keith Furst

9. The AI/Cyber Nexus

Computers and software are not naturally self-aware, emotional, or intelligent the way human beings are. They are, rather, tools that carry out functionalities encoded in them, which are inherited from the intelligence of their human programmers.[1] If programmers teach a machine to learn in the wrong way, it may completely defeat the purpose of the learning exercise. Software can be designed to learn like the human brain, but it can be rather difficult for machines to be taught correctly.

For example, in attempting to teach a machine to learn the difference between dogs and wolves, one machine was shown photographs of wolves on snow and dogs on grass. It learned to differentiate the two animals by looking at snow and grass. In the absence of being taught to ignore the background, a dog on snow and the wolf on grass would be completely misunderstood. In the same vein, an army wanted to teach a machine to distinguish between camouflaged tanks and a forest. Photos of the tanks were taken on cloudy days, while photos of the forest were taken on sunny days. The machine distinguished between clouds and sun rather tanks and forest.[2]

We know that algorithms already play increasingly active roles in a growing range of businesses and governments. Computer-aided diagnostics help physicians interpret ultrasound and magnetic resonance imaging results, and do the same to identify crime suspects through facial recognition and fingerprint-specific software. AI e-discovery software can sift through millions of pretrial documents and rank them for potential evidentiary value much faster than a human would ever be able to do, and for a fraction of the cost. There is a lot of potential value to be gained by using AI for a seemingly endless array of purposes, but, by the same token,

AI's existence and widespread use has also opened the door for virtual terrorists to hack, steal, malign, and sow fear.

What Say You, Mr. Watson?

In 2016, it is estimated that 4.2 billion records were exposed globally in more than 4,000 known security breaches (many more remained unreported). That is a lot of data to keep track of. Before AI became widely used, most businesses, governments, and security firms had the ability to analyze 10% of the data at their disposal. IBM has applied AI to security in the form of its Watson cognitive computing platform. The company has taught Watson to read through vast quantities of security research, with some 60,000 security-related blog posts published every month and 10,000 reports being produced each year. The company has scores of customers using Watson as part of their security intelligence and analytics platform.

The program becomes smarter with every piece of information it consumes and is able to provide detailed insights on specific threats and how such campaigns evolve. IBM believes that Watson performs 60 times faster than a human investigator and can reduce the time spent on complex analysis of an incident from an hour to less than a minute. While ML and AI speed up the laborious process of sorting through data, quantum computing—which makes direct use of quantum-mechanical phenomena such as superposition and entanglement to perform data operations[3] —will eventually be able to look at every data permutation simultaneously. Whereas common digital computing requires that data be encoded into *binary digits* (bits), each of which is always in one of two definite states (zeros or ones), quantum computation uses *quantum bits*, which can be in superposition (i.e. sub-atomic level) states.[4] That is a fancy of way of saying that digital computers represent data as ones or zeros but, to a quantum computer, these can be both zeros and ones, or neither at the same time.[5]

The most critical cryptographic protocols used to secure the Internet and financial transactions are all susceptible to attack by quantum computers. While cryptocurrencies are in general relatively resistant to the enhanced speed of quantum computers, the elliptic curve signature scheme used by Bitcoin is vulnerable and could by some estimates be "broken" in the coming decade. This raises the broader question about potential vulnerability of blockchain schemes, which its adherents believe are impenetrable.[6] Of course, there is no such thing as a truly impenetrable computer or software system. Every time a super sensitive (and well protected) government or military system is hacked, we are reminded of this.

Software advances are proceeding at such a lightning fast pace that many of the things that we considered to be science fiction a decade ago appear to be within reach today. Some scientists who work in the AI field believe that *AI robots will be living and working with us within the next decade*. Autonomous animated avatars are already being created, such as "Baby X", a virtual infant that learns through experience and can "feel" emotions. Adult avatars can be plugged into existing platforms like Watson, essentially putting a face on a chatbot. Within a decade, humans may well be interacting with lifelike, emotionally responsive AI robots, very similar to the premise of the HBO series "Westworld" and the film *I, Robot*. However, before that becomes a reality robotics will need to catch up to AI technology,[7] and that will still take at least a decade or two.

The Rise of Homo Deus

Israeli Historian Yuval Noah Harari has chronicled how Homo sapiens ("wise man") is a temporary creature that will soon to be replaced by Homo Deus ("god man"), believing that, within a century or two, Homo sapiens, as we have known ourselves to be for thousands of years, will disappear. That is because we will use technology to upgrade ourselves into something which is far more different from us than we are different from Neanderthals. Breakthroughs in biotechnology, including gene-editing methods such as CRISPR, hint at the power we will soon have to change our genes, our bodies, and perhaps even our brains.

AI may soon allow us to build brain-computer interfaces that will blur the line between man and machine. We are already moving in that direction, as we depend on our smartphones for a staggering number of the decisions we make each day. Harari believes that in 2050 our smartphones will be indistinguishable from us, literally being embedded in our bodies via biometric sensors that will monitor our heart rate, blood pressure, and brain activity 24 hours a day. By constantly analyzing the data generated, our smartphones may come to know our desires, likes, and dislikes even better than we think we do.

Amazon already proclaims to know our taste in books and music better than we do. Harari believes that people are already delegating more responsibility to AI by, for example, merging their smartphones with their social media accounts. He contends that smart devices are intelligent entities that constantly study us, adapt to our unique personality, and actively shape our worldview as well as our innermost desires.[8]

In 2017, the police force in Dubai enrolled its first robotic officer, with plans for robots to account for up to a quarter of the city's police force by 2030. Wearing a police cap and moving on wheels, the robot features a computer touch-screen on its chest that can be used to report a crime or inquire about speeding tickets. The robots are mostly to be deployed at tourist spots, equipped with a camera that transmits live images to the police force's operations center and identifies suspects wanted by police with its facial recognition software.[9] A Silicon Valley-based firm has taken this a step further by creating entirely autonomous vehicles that guide themselves through complex environments. The vehicles are armed with 16 microphones, speakers, a camera, 360-degree video capability, two-way audio, thermal imaging, real-time human interaction, and an interactive link with IBM Watson to provide 24/7 real-time monitoring capabilities.[10]

Since law enforcement and intelligence agencies use AI to enhance their capabilities, there is no reason to believe that virtual terrorists (who operate via the Internet) would not want to do the same thing. Imagine a world where Watson is hijacked and its cognitive capabilities are used to hack into government computers, engage in murder for hire, or distribute child pornography. It may seem far-fetched, but no more so than in 1987, when the film *RoboCop* envisioned a superhuman cyborg policing the streets of Detroit. Thirty years later, it is nearly a reality.

As Stephen Hawking noted, in 2014, "Whereas the short-term impact of AI depends on who controls it, the long-term impact depends on whether it can be controlled at all." He believed we were on the cusp of the kinds of AI that were previously exclusively the domain of science fiction films. In 2014, Google purchased Deep Mind Technologies (an AI company) and, in 2016, it picked up Robotics maker Boston Dynamics. The amount of money that Google and other commercial companies will continue to pour into robotics and AI could create a world in which smart robots could indeed roam our streets, *I, Robot*-style.[11] The question is not how soon it will come but when, and how swiftly, that new reality will become the new normal.

Smart Bodies and Augmented Realities

Apple created a paradigm shift with the launch of the iPhone in 2007. Ten years later, it was estimated that 44% of the world's population owned a smartphone[12]–they had become as natural a part of our lives as ordinary cell phones were in 2006. We quickly adapt to change. Yet, as rapidly as the technology (and our comfort levels) have evolved, by 2030 it may not be hard to imagine that the suite of apps and services that today revolve around the typical smartphone will have migrated to other, even more convenient and capable smart devices, giving us "smart bodies", with devices and sensors residing on our wrists, in our ears, on our faces, and perhaps elsewhere on our bodies.

Apple is diving deep into augmented reality–in a self-imposed race against time with the other large social media and Internet companies–to be able to overlay digital images onto a person's view of the world. Interfacing the digital world with our smart homes, smart cars, and even smart cities will constitute a new way to interact with computers. Apple hopes to make the Siri of the future capable of doing nearly everything an iPhone does currently through its touch interface.

Apple and other firms in the digital space are likely to release smart eyeglasses that are almost unnoticeable and have a tiny display in the lenses that is neatly concealed in the frame. They would be easily operated with a few fairly inconspicuous touch gestures, eye movements, or voice commands[13] and could convincingly supplement our visual and auditory reality, delivering information at what is deemed to be an appropriate time and place. Instead of using your phone to get away from the world, in the future you would use it to engage with the world in a much deeper and more meaningful way. We are already catapulting our way toward this brave new augmented reality. A variety of our smart devices (and associated apps) are involved in many of our daily tasks, ranging from our health and home security to our home appliances and banking. Armed with the predictive power of AI, Apple and firms like it seek to become the "default interface" to our lives.

Imagine having our daily lives structured and planned out in advance, with calendars automatically being filled using existing AI scheduling software, directing us from one task to another, and suggesting the fastest form of transportation to get us to the desired destination (ridesharing, mass transit, or a flying car), and booking it for us in advance. Taking it a step further, AI may one day recommend the ideal place for you to have your afternoon coffee because it has selected a person who is looking for a new boyfriend or girlfriend, is also having coffee there, and matches what you have stated you are looking for in a mate.

Apple and its competitors will by then have crafted a world of tradeoffs: *If you want your life enhanced by AI you will have to submit to constant surveillance—by your devices and by the tech giants themselves.* These companies are betting that hundreds of millions of their customers will happily comply, whether to say they have the latest in technology or simply for the convenience of not having to think about it all. These companies will, of course, swear they will encrypt your data and keep it confidential, but then, these are the same companies whose ToS require you to hand them the keys to your life in the process.[14]

By creating augmented reality, these companies may also be (wittingly or unwittingly) creating entire societies that cannot think for themselves—just as so many students in today's world cannot read, write, or sign their own names on a piece of paper—because of their overreliance on and fixation with smart devices. *While a future rich with possibility and "vision", a future dominated by augmented reality also portends further erosion of what passes for "privacy" in this era.*

Former Amnesty International Secretary General Salil Shetty believed checks and balances must be in place, especially because no one truly understands the implications of many of the algorithms that are being written. Shetty was particularly concerned about military use of AI in weapons (so-called "killer robots") and other potential applications.[15] AI technology is already being used in the US for "predictive policing", which has rights groups calling for enhanced regulation. One concern is that algorithms based on historic trends could reinforce existing biases against people of certain ethnicities.

A computer program called COMPAS, a risk assessment algorithm used by the state of Wisconsin, produced a risk assessment score that judges may consider in court to determine the length of prison sentences, which some consider a violation of the right be sentenced based on the unique characteristics of a person and case. In 2017, the Wisconsin Supreme Court rejected a formal challenge to its use, and the US Supreme Court declined to hear the case, meaning that a majority of justices effectively condoned the algorithm's use. The risk of relying on COMPAS is that judges could abuse unchecked power to sentence based not only on directly relevant factors, such as the seriousness of a defendant's offense, but also on variables that are morally and constitutionally problematic—such as gender or race.

This is precisely why some US states are abdicating taking responsibility for sentencing to a computer. Some states, such as Colorado, even require its use, believing that computers will be less biased than even the most well-intentioned humans. Of course, the bias risk inherent to a computer does not necessarily *eliminate* bias—it can *compound* it. Algorithms like COMPAS merely mimic the data that is downloaded into them, replicating its subjectivity and biases. Lawyers and judges have legal training, are bound by ethical oaths, and must account for their decisions and reasoning in published opinions. Programmers lack each of these safeguards. And computers may be intelligent, but they are not wise. They cannot un-learn what programmers taught them; they are dependent on humans for that, too.[16] There are so many unknown unknowns with augmented reality that it is difficult to determine where to even begin to examine implications, pass laws, and anticipate future ramifications.

Autonomous Malware

While the debate about AI and augmented reality rages, virtual terrorists are getting smarter and thinking of new ways to benefit from both, creating methods to operate autonomously in this brave new world. Malware is being designed with adaptive, success-based learning to improve the success and efficacy of cyberattacks. *This coming generation of malware will be situation-aware, meaning that it will understand the environment it is in and make calculated decisions about what to do next, behaving like a human attacker: performing reconnaissance, identifying targets, choosing methods of attack, and intelligently evading detection.*

This next generation of malware uses code that is a precursor to AI, replacing traditional "if not this, then that" code logic with more complex decision-making trees. Autonomous malware operates much like branch prediction technology, designed to guess which branch of a decision tree a transaction will take before it is executed. A branch predictor keeps track of whether or not a branch is taken; when it encounters a conditional action it has seen before, it makes a prediction so that, over time, the software becomes more efficient.

Autonomous malware is guided by the collection and analysis of 'offensive intelligence', such as types of devices deployed in a network to segment traffic flow, applications being used, transaction details, or time of day transactions occur. The longer a threat can persist inside a host, the more adept it becomes at operating independently, blending into its environment, selecting tools based on the platform it is targeting and, eventually, taking counter-measures based on the security tools in place.

Cross-platform autonomous malware designed to operate on and between a variety of mobile devices is also being developed. Such cross-platform "transformers" include a variety of exploit and payload tools that can operate across different environments. This evolving variant of autonomous malware includes a learning component that gathers offensive intelligence about where it has been deployed (including the platform on which it has been loaded), then selects, assembles, and executes an attack against its target. Transformer malware is being used to target cross-platform applications with the goal of infecting and spreading across multiple platforms, thereby expanding the threat surface and making detection and resolution more difficult. Once a vulnerable target has been identified, these tools can also cause code failure and then exploit that vulnerability to inject code, collect data, and persist undetected.

Autonomous malware can have a devastating effect on our connected devices and, as a result, our ability to perform daily tasks we usually take for granted. Fighting against it will require highly integrated and intelligent security technologies that can see across platforms, correlate threat intelligence, and automatically synchronize a coordinated response.[17]

AI and Cybersecurity

A 2017 study[18] found nearly 50 cybersecurity vulnerabilities in robot ecosystem components, many of which were common in home, business, and industrial robots, as well as the control software used by other robots tested. Although the number of robots tested was not a large sample, the fact that dozens of vulnerabilities were uncovered, in such a broad spectrum of robots, is concerning. Most of the robots evaluated were using insecure forms of communications, with mobile and software applications connected to the Internet, Bluetooth, and Wi-Fi without properly securing the communication channel. Although some had weak encryption, most of the others sent data to vendor services or the cloud without *any* protection.

Most of the robots exposed one or more services that can be remotely accessed by computer software, mobile applications, or Internet services. These services included complex and critical functions, such as programming and receiving external commands, as well as simpler functions such as returning basic robot information. Some robot services did not even require a username and password, allowing anyone to remotely access those services. In some cases, where services used authentication, it was possible to bypass the process, allowing access without using a correct password.

Robots can store sensitive information, including encryption keys, user social media, e-mail accounts and vendor service credentials, and send that information to and from mobile applications, Internet services, and computer software–meaning that channel encryption is mandatory to avoid data compromises. Robots also receive remote software updates, so proper encryption is necessary to ensure that these updates are trusted and have not been modified to include malicious software.

Most of the robots that were tested did not require sufficient authorization to protect their functionality, including critical functions such as installation of applications in the robots themselves or updating of their operating system software. *This enables cyber attackers to install software without permission and gain full control over the robots.* If hackers were to gain control of sex robots, for example, they could be programmed to kill someone during sex with a knife or via asphyxiation.[19]

Most of the robots were also either not using encryption or improperly using it, exposing sensitive data to potential attackers. Some of the robots with insecure features could not be easily disabled, having features with default passwords that were either difficult to change or could not be changed at all. Others were easy to hack because default passwords are usually publicly known or could be easily obtained since many robot models share the same default passwords.

Many robots use open source frameworks and libraries. One of the most popular is the Robot Operating System (ROS) 29, used in several robots from multiple vendors. The ROS suffers from many known cybersecurity problems, such as cleartext communication, authentication issues, and weak authorization schemes, all of which make robots insecure. In the robotics community, it is common to share software frameworks, libraries, and operating systems for robot development and programming.

Not all of the robots tested were vulnerable to every one of the cybersecurity issues identified, but each robot had many of the issues, ultimately because many researchers and enthusiasts in the robotics space use the same (or very similar) tools, software, and design practices globally. For example, it is common for robots originally intended as research projects to become commercial products with no additional cybersecurity protections—with the security posture of the final product remaining the same as the research or prototype robot. This results in poor cybersecurity defenses, since research and prototype robots are often designed and built with few or no protections.

Vulnerability to Cyberattacks

Each robot naturally has different features, and the more features, the more advanced and smarter a robot typically is, but these features can also make robots more vulnerable and attractive to hackers. Certain features are common, intended to improve accessibility, usability, interconnection, and reusability (such as real-time remote control with mobile applications). Unfortunately, many of these features make robots more vulnerable from a cybersecurity perspective, with both critical- and high-risk cybersecurity issues present.

The following list provides an overview of some possible threats for common robot features:

- Network connectivity: Sensitive robot services are vulnerable to attack from home, corporate, or industrial networks or the Internet.
- External services interaction: Robot owner social networks, application stores, and cloud systems could be exposed, providing access to private user information, usernames, and passwords.
- Remote control applications: Mobile applications or microcomputer boards can be used to send malicious commands. Mobile phones could be an entry point for launching attacks; if a user's phone is hacked, the robot can be hacked also. An attacker could use a hacked robot to launch attacks against the owner's phone.
- Modular extensibility: When a robot allows installation of applications, it can also allow installation of custom malware. Malicious software could cause the robot to execute unwanted actions. Ransomware could also take robots hostage, making them unusable and allowing hackers to extort money to make them usable again.
- Safety features: Human safety protections and collision avoidance/detection mechanisms can be disabled by hacking the robot's control services.
- Firmware: When a robot's firmware integrity is not verified, it is possible to replace the robot's core

software and change its behavior in a malicious way by installing malware or ransomware.

- <u>Autonomous robots</u>: A hacked autonomous robot can move around as long as its battery continues to provide power, which allows hackers to control an "insider threat", steal information, or cause harm to nearby objects or people for a set period of time.
- <u>Network advertisement</u>: It is common for robots to advertise their presence on a network using known discovery protocols. Attackers can leverage this to identify a robot in a corporate or industrial network with thousands of computers, and interact with its network services.
- <u>Fast installation/deployment</u>: Since many vendors do not highlight the importance of changing the administrator's password in their documentation, a user may not change it during fast deployment, meaning it can be hacked easily.
- <u>Backups</u>: Configuration files and other information may be backed up on the robot vendor's cloud or the administrator's computer. An unencrypted backup could result in a compromised robot and a leak of sensitive data if obtained by an attacker.
- <u>Connection ports</u>: Physical connectivity ports lacking restriction or protection could allow anyone to connect external devices to the robots and either extract information or insert malware.
- <u>Microphones and cameras</u>: Once a robot has been hacked, microphones and cameras can be used for cyberespionage and surveillance, enabling an attacker to listen to conversations, identify people through face recognition, and record videos.

A hacked robot has many of the same vulnerabilities as computers and can suffer the same consequences. A hacked robot operating inside a home might spy on a family via the robot's microphones and cameras. An attacker could also use a robot's mobility to cause physical damage in or to the house. Compromised robots could even hurt family members and pets with sudden, unexpected movements, since hacked robots can bypass safety protections that limit their movements. Hacked robots could also start fires in a kitchen by tampering with electricity, or they could potentially poison family members and pets by mixing toxic substances with food or drinks, or utilize sharp objects to cause harm.

Robots that are integrated with smart home automation could unlock and open doors and deactivate home alarms. Robots that are not integrated into a smart home could still interact with voice assistants, such as Alexa or Siri, which integrate with home automation and alarm systems. If the robot can talk, or allow an attacker to talk through its speaker, it could tell voice-activated assistants to unlock doors and disable home security. Once robot firmware is hacked, it may be difficult or even impossible to reset the robot to factory defaults and recover data. If the core robot software is hacked, it would be impossible to recover the robot.

Many business robots are customer-facing, so a hacked robot could easily impact one of their most valuable assets. A hacked robot could be made to deliver incorrect orders, share confidential information, or go offline, all of which would negatively impact a business. An attacker could also instruct a robot to physically hurt customers or company employees, and if robots have access to customer information such as personal and credit card data, that could also be put at risk. Robots could cause destruction by physically damaging business assets. The potential financial impact for a business is similar to that of a home, but on a larger scale, since business robots are generally much more expensive and can perform more complex tasks. A company with several hacked robots could suffer losses, not only to the computers themselves, but in terms of potential liability and other forms of financial loss. Compromised robots could be used to hack other computers or IoT devices on the same network, becoming a platform to attack and/or exploit vulnerabilities in other network devices.

Improving Robot Cybersecurity

Building robots is not an easy task, nor is securing them from cyberattacks, but taking some basic precautions can go a long way toward preventing attacks. It starts with robot manufacturers, who need to do a better job of encrypting their products and advising purchasers that default passwords must be changed upon first usage. Consumers need to be smarter about taking a common-sense approach to cybersecurity. What follows are some basic recommendations for doing that:

- Software: Vendors should implement Secure Software Development Life Cycle processes from day one.
- Encryption: Vendors should properly encrypt robot communications and software updates.
- Education: Vendors should invest in cybersecurity education for everyone in their organization, with training not only for engineers and developers but also for executives and everyone else involved in product decisions.
- Vulnerability Disclosure: Vendors should have a clear communication channel for reporting cybersecurity issues and clearly identify an individual or team to be responsible for handling reports appropriately.
- Security Audits: Vendors should ensure that a complete security assessment is performed on all of the robot's ecosystem components prior to going into production.
- Authentication and Authorization: Vendors should make certain that only authorized users have access to robot services and functionality.
- Factory Restore: Vendors should provide methods for restoring a robot to its factory default state, and ensure that its default configuration is secure.
- Secure the Supply Chain: Vendors should make sure that all of their technology providers implement best practices in cybersecurity.

Research from the University of Washington conducted in 2009 found that household robots did not protect users' security and privacy. In the time that has passed since then, robot cybersecurity has actually gotten worse. Since robot adoption is not yet mainstream, there is still time to improve the technology and make it more secure.[20] That will take sustained effort on the part of both robot manufacturers and users.

It would also undoubtedly help if governments around the world would *pass and implement legislation requiring unique passwords for both drones and robots, when they are produced*. Governments and regulators should also update existing laws and regulations on an ongoing basis, rather than once every 5 or 10 years, as is so often the case. As quickly as the drone and robot landscape is evolving, existing regulations quickly become obsolete, and governments fall further and further behind the curve. With stakes as high as they are for governments, businesses, and individuals, we need to build a better mousetrap.

Maximizing AI Benefits

If machines can be taught to think like humans, they can also be tricked like humans. AI is itself becoming a target, with hackers attempting to exploit ML by gradually teaching a security system that unusual behavior is normal (known as "behavioral drift"). AI can also be used by hackers to fake human voices and create video images that could let criminals into a network. There are hosts of other ways that virtual terrorists could exploit AI for their nefarious purposes.

In order for the cybersecurity industry to get the most out of AI, it should be able to recognize what machines do best and what people do best. It should also gravitate toward finding the sweet spot between SML and UML learning, so that businesses and governments can fully benefit from their existing knowledge of current threat vectors and combine it with the ability to detect new attacks and uncover new vulnerabilities. Advances in AI can provide new tools for threat hunters, helping them protect new devices and networks even before a threat is classified by a human researcher. ML techniques such as unsupervised learning and continuous retraining can keep us a step ahead of virtual terrorists.[21]

IBM has taught Watson to scour through cybersecurity incident reports written in natural language in a bid to help companies and cybersecurity researchers better detect and ward off breaches. Called Watson for Cybersecurity, the project comes with an app called IBM QRadar Advisor with Watson, which enables cybersecurity professionals to consult Watson's vast cybersecurity knowledge. The app draws information from multiple resources, including cybersecurity blogs, websites, and research papers, considerably shortening the amount of time required to find relevant clues. IBM has also introduced a Watson-powered chatbot that can advise customers of its global X-Force Command Center on cybersecurity issues.

As part of a new research project, code-named Havyn, IBM will attempt to develop a voice-powered security assistant leveraging Watson's conversation technology to respond to verbal commands and natural language requests from security analysts. Havyn will provide real-time information about newly detected threats as well as past incidents and give guidance about the steps required to restore the security of affected networks. Watson makes concealment efforts more difficult by quickly analyzing multiple streams of data and comparing them with the latest security attack intelligence to provide a more complete picture of the threat. Watson also generates reports on these threats in minutes, which greatly speeds the time between detecting a potential event and a security team's ability to respond.[22]

AI and ML have also enabled the creation of an intelligence banknote neutralization system (IBNS) that can automatically translate business requirements and apply them to the entire cybersecurity infrastructure. IBNS provides pervasive visibility across an entire distributed network and enables integrated security solutions to automatically adapt to changing network configurations, shifting business needs with a synchronized response to threats. IBNS is not a product or a market; it is a piece of networking software that helps to plan, design and operate networks to improve their availability and agility. Another way to describe it is the lifecycle management software for networking infrastructure.

The system takes a higher-level business policy as input from end users, converts it to the necessary network configuration, then generates and validates the resulting design and configuration for correctness. It can configure appropriate network changes across existing network infrastructure and ingest real-time network status for systems under its administrative control. Since it is protocol-agnostic, the system continuously validates (in real-time) that the original business intent is being met and can take corrective actions (such as blocking traffic, modifying network capacity or issuing notifications) when the desired intent is not met.[23] IBNS can also dynamically partition network segments, isolate affected devices and remove malware.

AI and ML are becoming significant allies in cybersecurity. ML is being bolstered by predictive applications to help safeguard networks. New security measures and countermeasures can also be provisioned or updated automatically as new devices, workloads, and services are deployed or moved from anywhere to anywhere in a network—from endpoints to the cloud. Tightly integrated and automated security enables a comprehensive threat response far greater than the sum of the individual security solutions protecting the network.

One of the biggest challenges of using AI and ML lies in the caliber of intelligence. Cyberthreat intelligence is highly prone to false positives due to the volatile nature of the IoT. Threats can change within seconds—a machine can be clean one second, infected the next, and brought back to clean again. Enhancing the quality of threat intelligence is critically important as IT teams pass more control to AI to do the work that humans otherwise would do. It is ultimately an exercise in trust and balancing operational control. As is the case with any relationship, as it matures, AI and ML applications for cybersecurity will become more and more effective.

Since a cybersecurity skills gap persists, products and services must be built with greater automation to correlate threat intelligence in order to determine the level of risk while automatically synchronizing coordinated responses to threats. In the future, AI in cybersecurity will constantly adapt to the growing attack surface, but for now we are connecting the dots, sharing data and applying that data to systems.

Humans make these complex decisions, which require "intelligent" correlation through human intelligence, but in the future, a mature AI system could be capable of making such complex decisions on its own. Total automation is not possible, so humans and machines must find a way to work together to maximize outcomes.[24] That said, the more we depend on adaptive AI, the more the issue of transparency and integrity become important. A system can only be as good as the data that is fed into it, along with the intentions of programmers. An important vulnerability is how an entire system may be corrupted by small or imperceptible flaws, biases, or malware that may be slowly fed into a system.

With that in mind, some cities have already begun to deploy AI technologies for public safety and security. By 2030, the typical North American city will rely heavily upon them, including cameras for surveillance, which can also detect anomalies pointing to a possible crime, and predictive policing applications. AI tools may also prove useful in helping police manage crime scenes for search and rescue events by helping commanders prioritize tasks and allocate resources, though these tools are not yet ready for automating such activities. Improvements in ML, in general, and transfer learning in particular (to facilitate enhanced learning in new scenarios, based on similarities with past scenarios) may promote the creation and adoption of such systems.

The cameras deployed almost everywhere in the world today tend to be more useful for helping solve crimes than preventing them. This is due to the generally low quality of event ID from videos and the lack of manpower to review massive video streams. As AI for this domain improves, it will better assist crime prevention and prosecution through greater accuracy of event classification and efficient automatic processing of video to detect anomalies. These improvements could lead to even more widespread surveillance. The New York Police Department's CompStat was the first tool pointing toward predictive policing, and many police departments now use it.

ML significantly enhances the ability to predict where and when crimes are more likely to happen, and who may commit them. Well-deployed AI prediction tools have the potential to actually remove or reduce human bias, rather than reinforce it, since AI techniques can be used to develop intelligent simulations for training law enforcement personnel. While international criminal organizations and terrorists from different countries are busy colluding, police forces from different countries still face difficulty in joining forces to fight them. The EU, through the Horizon 2020 program, supports such attempts in projects such as LawTrain. The next step will be to move from simulation to actual investigations, by providing tools that support such collaborations.

AI can assist in social network analysis to prevent those at risk from being radicalized by the IS or other violent groups. Law enforcement agencies are increasingly interested in trying to detect plans for disruptive events from social media, and also monitor activity in large gatherings of people to analyze security. Significant work on crowd simulations is being done to determine how crowds can be controlled. The TSA, US Coast Guard, and the many other security agencies that currently rely on AI will likely increase their reliance to enable significant improvements in efficiency.

AI techniques in vision, speech, and gait analysis can aid interviewers, interrogators, and security guards in detecting possible deception and criminal behavior. The TSA currently has an ambitious project to redo airport security nationwide. Called DARMS, the system is designed to improve airport security by relying on personal information to tailor security based on a person's risk categorization and the flights being taken. The future vision for this project is a tunnel that checks passengers' security as they walk through it. Developers of this technology need to be careful to avoid building in bias (such as about a person's risk level category) through use of datasets that reflect prior bias.

A recent multi-year study comparing privacy regulation in four European countries and the US yielded counter-intuitive results. Those countries, such as Spain and France, with strict and detailed regulations bred a "compliance mentality" within corporations, which had the effect of discouraging both innovation and robust privacy protections. Rather than taking responsibility for privacy protection internally and developing a professional staff to foster it in business processes, or engaging with privacy advocates or academics outside their walls, these companies viewed privacy as a compliance activity. Their focus is on avoiding fines or punishments, rather than to proactively design technology and adapt practices to protect privacy.

A vigorous and informed debate about how to best steer AI in ways that enrich our lives and society, while encouraging creativity, is an urgent need. Since future AI technologies and their effects cannot be foreseen with perfect clarity, policies will need to be continually re-evaluated in the context of observed societal challenges and evidence from deployed systems. Recent advances are largely due to the growth and analysis of large data sets enabled by the Internet, advances in sensory technologies, and applications of DL. While encouraging innovation, policies and processes should address ethical, privacy, and security implications, and should be made to ensure that the benefits of AI technologies will be spread broadly and fairly.[25]

A New Cyber Era

A new cyber era has begun, with AI and machines ready to fight battles, and sophisticated cyber attackers and criminal groups seizing any opportunity to take advantage of systemic vulnerabilities. The battlefield is corporate and government networks, and the prize is control of the organization— whether known or unknown to them. The stakes are extremely high. The target of this behind-the-scenes battle is not just stolen information or the ability to embarrass a rival but the ability to alter IT systems, with the ability to install kill switches that can be activated at will. These attackers are sophisticated, they use previously unknown code, and they silently breach boundary defenses without being seen or heard.

Conventional approaches to cybersecurity rely on being able to understand the nature of a threat in advance, but that approach is fundamentally flawed, since threats are constantly evolving, laws and policies are outdated, and the threat from insiders is growing. In the current cyber era, threats easily bypass legacy defense tools. New black hat machine intelligence need only enter an organization's IT systems a single time. Based on that point of entry, they listen, learn how to behave, blend in, and appear as authentic as the original devices, servers and users. These automated attackers can hide their malicious actions among ordinary daily system tasks, with at times devastating results.

Today's attacks can be so swift and severe that it is impossible for humans to react quickly enough to stop them. Based on advances in self-learning, however, it is possible for machines to rapidly uncover emerging threats and deploy appropriate, real-time responses against the most serious cyberthreats. Firewalls, endpoint security methods and other tools are routinely deployed in some organizations to enforce specific policies and provide protection against certain threats. These tools form an important part of an organization's cyber defense strategy, but they are quickly becoming obsolete in the new age of AI and ML-driven cyber threats.

AI SUPREMACY

We are experiencing the third great revolution in automation: the era of ML. Recent advances in science, engineering, and mathematics have made possible new classes of intelligent systems capable of making value judgements and carrying out higher value tasks, some of which require thought. As we wade through ever larger oceans of data, ML is helping us process and make sense of that information, while creating a pathway to the future. Yet, ML is difficult to develop and deliver, as it requires complex algorithms devised in a framework that permits information to be interpreted and results to be produced.

When applied correctly, machines can make logical, probability-based decisions and undertake thoughtful tasks. ML is already operating successfully in a broad range of commercial and industrial fields, such as payment processing, online video services, advertising, healthcare, and onboard computers in cars and airplanes. However, much of our existing ML is supervised, meaning that, in order for it to operate successfully, prior knowledge of the potential outcomes must be pre-programmed by a human. In industries where behaviors are well understood, this approach is more than adequate for assisting with product development or consumer safety. In an area as complex and obfuscated as cyber security, there cannot be complete knowledge of all forms of existing or emerging threats.

Traditional approaches to cyber security are based on identifying activities that resemble previously known attacks—the "known knowns". This is usually done with a signature-based approach, whereby a database of known malicious behaviors is created, new activities are compared to those in a database and any which match are flagged as threats. Other systems use methods based on SML, wherein a system is trained using a data set in which each entry has been labelled as belonging to one of a set of distinct classes.

In the information security context, the security system is trained using a database of previously seen behaviors, where each set of behaviors is known to be either malicious or benign and is labelled as such. New activities are then analyzed to see whether they more closely resemble those in the malicious class or those in the benign class. Any that are evaluated as being sufficiently likely to be malicious are again flagged as threats.

While SML has inherent benefits, it also has fundamental weaknesses because malicious behaviors that deviate sufficiently in character from those seen before will fail to be classified as such and will pass undetected. A large amount of human input is required to label the training data and any mislabeled data can seriously compromise the ability of the system to correctly classify new activities. It is also worth mentioning that, in a dynamic and evolving threat landscape that is constantly bouncing up against existing boundaries, SML can miss a lot of important operational variables in the process.

By contrast, UML presents a significant opportunity for the cyber security industry with the prospect of enhanced network visibility and improved detection levels resulting from enhanced computational analysis. UML can overcome the limitations of rules and signature-based approaches by learning what is considered normal within a network and not being dependent upon knowledge of previous attacks. It thrives on the scale, complexity and diversity of modern businesses—where every device, person, and operation is different—and it turns the innovation of cyber attackers against themselves by making any unusual activity visible.

Utilizing ML in cyber security technology is difficult, but when correctly implemented, it is powerful. Previously unidentified threats can be detected, even when their manifestations fail to trigger any rule set or signature. Instead, ML allows the system to analyze large sets of data and learn a pattern from its processes. ML can attribute human capabilities to machines, such as thought (using past information and insights to form judgements, real-time information processing, and self-improvement by adapting to the integration of new information. UML therefore allows computers to recognize evolving threats without prior warning.

As information networks continue to grow in scope and complexity, the opportunities for attackers to exploit gaps have naturally increased. Walls are no longer enough to protect the content of systems, rules cannot pre-emptively defend against all possible attack vectors, and signature-based detection methods fail repeatedly. Since cyberattacks are advanced, subtle, and varied, only automated responses based on ML can keep pace with them. ML technology is the fundamental ally in the defense of systems from the hackers and insider threats.[26] Used cleverly, it offers a real opportunity to gain an upper hand in the ongoing battle for supremacy in cybersecurity.

A Growing Protection Marketplace

AI can combat cybercrime by more swiftly detecting threats, adopting predictive measures, reducing response times, and identifying attacks that require immediate attention. Given its increasingly important role in identifying and responding to cyber threats, the AI component of the global cyber security marketplace is expected to reach $18 billion by 2023. ML is also gaining increasing traction in the space, and is projected to exceed $6 billion by 2023, largely attributable to better threat learning algorithms. Network security accounts for the largest portion of corporate spending on cyber security-related AI, driven primarily by growth in the adoption of wireless networks and their vulnerabilities.

Many organizations are doubling down by also investing heavily in AI capabilities for cognitive computing, improving cloud platform performance and resulting in a new generation of cloud computing technologies that will drive future industry growth. Cloud-based cyber security enhances intelligence gathering, threat modeling, and data collaboration, blocking attacks and minimizing the delay between detection and remediation. The result is more secure communication, which reduces the need to purchase, maintain, and hire support personnel. In sharing the cloud, there is as much the risk of inadvertent cross-fertilization of data as there is of an opportunity for all cloud users to learn from each other. The question is whether it is intended for or approved of in advance by users.

Hesitancy to rely too much on new forms of technology is naturally hindering industry growth. ML, in particular, is at risk of being tainted based on the subjectivity of the data that is input into the system, which is, of course, a risk with any information-fed system. The industry needs to find better ways of stripping subjectivity from the process, which will drive competitiveness and help drive prices down, while enhancing the effectiveness and efficiency of future ML applications in cyberspace.[27]

Integrating AI into Cybersecurity

Integrating AI into a mature cybersecurity ecosystem can provide some substantial benefits that may both enhance existing safeguards and help cybersecurity professionals protect against attacks in the future. New forms of malware and cyberattacks can be difficult to detect with conventional cybersecurity protocols, so more dynamic approaches are necessary. Since ML adapts to new threats and realities over time, responding effectively to new cybersecurity risks can become easier and somewhat more reliable. Cybersecurity solutions that rely on ML use data from previous cyberattacks to respond to new attacks with similar dynamics.

Another great benefit of AI systems in cybersecurity is that they make tech employees more efficient by making more time available to pursue other tasks. While AI will not make human cybersecurity experts obsolete, it will reduce the need to have as many people on staff and it increases their effectiveness. AI systems can categorize attacks based on threat level. As models become more effective at detecting threats, bad actors will devise methods to confuse them—a form of adversarial AI intended to poison the models.[28] While ML can be used to find anomalies in the form of malicious behavior, only a subset of "unusual" events may end up being worth pursuing; An increase in network traffic might be statistically significant, but may only rarely represent an attack.[29] The ability to distinguish between what is worth paying attention to and what is not becomes critically important.

SML is great at *classifying* data to determine whether it is good or bad and useful or not, but large amounts of data are required in order to do so. Supervised algorithms apply acquired knowledge to classify previously unknown data, having the biggest impact on cybersecurity, especially for malware ID and spam detection. The reason malware ID works so well is because of the availability of millions of *labeled samples* (training data), but a variety of other types of data—such as related to network traffic—do not produce good training data, so algorithms cannot be trained to identify the anomalies within them. Additional problem areas include the inability to deterministically label data, the challenges associated with cleaning data, and understanding the semantics of a data record.

By contrast, UML draws inferences from large data sets and is well suited for making data easier to *analyze* and understand. Unsupervised approaches are useful for data exploration, to reduce the number of dimensions or fields of data to be reviewed (known as dimensionality reduction), or by grouping records together (known as clustering). However, these algorithms are of limited use when it comes to identifying anomalies or attacks. The most common unsupervised learning method is cluster analysis, which is used for exploratory data analysis to find hidden patterns or groupings within data.

Other important building blocks include context, which is anything that helps better understand the types of entities present in data (such as devices, applications, or users). Context for devices include its owner, location, and role. Rather than looking at network traffic logs in isolation, context must be added in order to make sense of the data; for example, knowing which machines represent Domain Name System (DNS) servers on a network helps to understand which of them should be responding to DNS queries. A non-DNS server that is responding to DNS requests could be a sign of an attack.[30]

The Malicious Use of AI

While AI and ML have many broadly beneficial applications, human nature being what it is, there are a plethora of potential nefarious uses of both. Just as ammonium nitrate is both a fertilizer and potential explosive, AI is a dual-use area of technology. AI is dual-use in the same sense that human intelligence is, for it is not possible for AI researchers to avoid producing research and systems that can be directed towards harmful ends. Many tasks that would benefit from automation are themselves dual-use. For example, systems that examine software for vulnerabilities have both defensive and offensive potential applications; the difference between the capabilities of an autonomous drone used to deliver packages or explosives may not be all that great.

In addition, research that aims to increase our understanding of AI, its capabilities, and our degree of control over it, appears to be inherently dual-use in nature. AI systems are generally both efficient and scalable. For example, once it is developed and trained, a facial recognition system can be applied to many different camera feeds for much less than the cost of hiring human analysts to do equivalent work. Many AI systems can perform a given task better than any human can, as has been proven against top ranked players in games such as chess and Go.

AI systems can also increase anonymity and psychological distance in tasks that involve communicating with other people, observing or being observed by them, making decisions that respond to their behavior, or being physically present with them. By allowing such tasks to be automated, AI systems can allow actors who would otherwise be performing the tasks to retain their anonymity, resulting in a greater degree of psychological distance from the people they impact. For example, rather than using a handgun to carry out an assassination, someone who uses AI avoids both the need to be present at the scene and greatly increases the likelihood of never getting caught.

While attackers may find it costly to obtain or reproduce the hardware associated with AI systems—such as powerful computers or drones—since many new AI algorithms are reproduced in a matter of days or weeks, it can be much easier to quickly gain access to software and resultant scientific findings. In addition, AI research is characterized by a high degree of openness, with many published papers being accompanied by source code. AI systems also suffer from a number of novel enduring vulnerabilities, which include data poisoning attacks (introducing training data that causes a learning system to make mistakes), adversarial examples (inputs designed to be misclassified by ML systems), and the exploitation of flaws in the design of autonomous systems' goals.

Absent the development of adequate defenses, actors with malicious intent should be expected to expand existing threats, introduce new threats, or alter the typical character of threats. The diffusion of efficient AI systems can increase the number of actors who can afford to carry out particular attacks. Future attacks using AI technology should be expected to be more effective, finely targeted, difficult to attribute, and more likely to exploit vulnerabilities in AI systems. Increased use of AI should also be expected to expand the range of actors who are capable of carrying out attacks, the rate at which these actors can carry attacks out, and the set of plausible targets.

If the relevant AI systems are also scalable, then even actors who already possess the resources to carry out these attacks may gain the ability to carry them out at a much higher rate. One example of a threat that is likely to expand in these ways is that from spear phishing attacks, which use personalized messages as an attempt to extract information or initiate action from a target by using a superficially trustworthy façade to extract sensitive information or money from individuals. The most advanced spear phishing attacks require a significant amount of skilled labor, as the attacker must identify suitably high-value targets, research these targets' social and professional networks, and then generate messages that are plausible within this context. If some of the research and synthesis tasks can be automated, more actors may be able to engage in spear phishing. In doing so, attackers would no longer have the need to speak the same language as their target.

Attackers might also gain the ability to engage in mass spear phishing, becoming less discriminate in their choice of targets. If an actor knows that he or she is unlikely to be identified, they will presumably feel less empathy toward their target (if they have any empathy at all) and become even more willing to carry out the attack. The importance of psychological distance is easily illustrated by the fact that military drone operators who must observe their targets prior to killing them frequently develop post-traumatic stress from their work. Increases in psychological distance could, therefore, plausibly have a significant impact on potential attackers' psychologies.

AI is not the only force expanding the potential scale and scope of existing threats. Progress in robotics and the declining cost of hardware (including both computing power and robots) are contributing to the same phenomenon. Being unbounded by human capabilities implies that AI systems can enable actors to carry out attacks that would otherwise be infeasible. The proliferation of cheap hobbyist drones, which can easily be loaded with explosives, has made it possible for non-state actors to launch aerial attacks, which may use AI systems to complete tasks more successfully than any human could, or take advantage of vulnerabilities that AI systems have but humans do not. While most people are not capable of mimicking others' voices and creating audio files that resemble recordings of human speech, significant progress in developing speech synthesis systems that learn to imitate individuals' voices has vast potential negative implications.

Just as using fingerprints, retinal scans, and voice or facial recognition to unlock a smart phone can be convenient for users, there is a growing risk that biometric inputs can do more harm than good if they are stolen. The outputs of these systems could also become indistinguishable from genuine recordings, in the absence of specially designed authentication measures. Such systems would, in turn, open up new methods of spreading disinformation and impersonating others. Consider the implications in a political campaign, or a court case. Did the candidate, official, defendant, or witness actually say what they are accused of saying? How can they necessarily prove they did not when AI technology reaches the point of perfect mimicry?

AI systems could, in addition, be used to control aspects of the behavior of robots and malware that would not be feasible for humans to manually control. As an example, humans could not realistically be expected to monitor every drone in use at any given point in time, nor a virus designed to alter the behavior of a large array of air-gapped computers. The growing capability and widespread use of AI systems implies that the threat landscape will change through the expansion of some existing threats and the emergence of new threats that do not yet exist. The typical character of threats will likely shift in some distinct ways. Attacks supported and enabled by progress in AI could be particularly effective by being finely targeted, difficult to attribute, and exploitative of vulnerabilities in other AI systems.

Given AI efficiency and scalability, highly effective attacks will undoubtedly become more typical (at least absent substantial preventive measures), with attackers facing a trade-off between the frequency and scale of their attacks, and their effectiveness. For example, spear phishing is more effective than regular phishing, which does not involve tailoring messages to individuals, but is relatively expensive and cannot be carried out en masse. By improving the frequency and scalability of certain attacks, including spear phishing, AI systems can diminish the impact of such trade-offs. The expected increase in the effectiveness of attacks naturally follows from the potential of AI systems to exceed human capabilities. Attackers can be expected to conduct more effective attacks with greater frequency, and at a larger scale.

Efficiency and scalability—specifically in the context of identifying and analyzing potential targets—also suggests that more finely targeted attacks will become more prevalent, such as against high net worth individuals or with a focus on specific political groups. Drone swarms could be programmed by AI to deploy facial recognition technology to kill specific members of crowds. The increasing anonymity of AI systems also suggests that difficult-to-attribute attacks will become more typical, such as an attacker who uses an autonomous weapons system to carry out an attack, rather than carrying it out in person. We should also expect attacks that exploit the vulnerabilities of AI systems to become more typical.[31]

Changes in the Security Landscape

A central concern about the nexus between AI and cybersecurity is that AI might enable larger-scale and more numerous attacks to be conducted, given the amount of skill and resources available when compared to attacker capabilities without AI. There is clearly interest in conducting larger-scale attacks, two examples being how Russian attackers tailored messages carrying malware to more than 10,000 Twitter users in the US Defense Department, and how Russian bots were used in the 2016 US presidential election to influence voter thinking in social media. Community-targeted spam uses natural language generation techniques from AI to target an entire class of targets with common ways of writing. Armed with even more advanced natural language generation, one could envision more highly customized approaches spanning multiple communities at the same time.

The adaptability of AI systems may transform the strategic landscape of cybersecurity, though it is not yet clear how such adaptability will affect the offense versus defense balance. Many organizations currently adopt security systems called Endpoint Detection and Response (EDR) platforms, to counter more advanced threats. The EDR market already represents a $500 million industry in the cyber security arena. These tools are built upon a combination of heuristic and ML algorithms to provide capabilities such as next-generation anti-virus (NGAV) and Security Domains behavioral analytics.

Though these systems are effective against typical human-authored malware, research has already shown that AI systems may be able to learn to evade them. One research team created an ML model to automatically generate command and control domains that are indistinguishable from legitimate domains by human and machine observers. These domains are used by malware to "call home" and allow malicious actors to communicate with the host machines. They leveraged RL learning to create an intelligent agent capable of manipulating a malicious binary code with the objective of bypassing NGAV detection.

Other researchers have applied adversarial ML to craft malicious documents that can evade PDF malware classifiers. Attackers are likely to leverage the growing capabilities of RL so that attackers can benefit from the ability of AI to learn from experience in order to craft attacks that existing technical systems and IT professionals are not prepared to address. In doing so, AI can be programmed to teach itself how to perform nefarious (or other) tasks with greater efficiency and accuracy.

Hacking could be deliberately performed by an organization on itself, as an exercise in RL to determine which other hacks were and were not detected. Nothing would actually be stolen, but such exercises could lead to massive, sophisticated, and autonomous hacking programs in the future. Services such as Google's VirusTotal file analyzer allow users to upload variants to a central site and be judged by scores of different security tools. This feedback loop presents an opportunity to use AI to assist in crafting multiple variants of the same malicious code to determine which is most effective at evading security tools.

Large-scale AI attackers can accumulate and use large datasets to adjust their tactics, as well as alter the details of the attack for each target. This may outweigh any disadvantages they incur via the lack of skilled human attention to each target, and the ability of defenders like antivirus companies and IT departments to learn to recognize attack signatures. While the specific examples of AI applied to offensive cybersecurity mentioned above were developed by white hat researchers, it is reasonable to believe that similar efforts by cybercriminals and state actors are already occurring, as highly capable AI techniques and new applications become more widely distributed.

AI and cybersecurity are expected to evolve in tandem in the coming years, but it is clear that a proactive effort is needed to stay ahead of motivated and capable attackers. Educated consumers can identify telltale signs of certain attacks (such as poorly crafted phishing attempts) and practice better cyber security hygiene (such as using diverse and complex passwords and two-factor authentication), yet most end users of IT systems will remain vulnerable to even simple attacks (such as the exploitation of unpatched or otherwise poorly secured systems). This is concerning in light of the AI-cybersecurity nexus, especially if high-precision attacks can be scaled up to impact large numbers of victims.

Legal enforcement has been an ongoing challenge in the cyber arena, and AI will add to the challenge of policing cyberspace. Complicating the issue is the fact that AI is not explicitly addressed in most existing laws and norms. An important activity that cybersecurity researchers perform is the detection of vulnerabilities in code, allowing vendors to increase the security of their products. A number of approaches exist to incentivize such processes and make them easier, including:

- Payment of "bug bounties", in which participants are compensated for finding and responsibly disclosing vulnerabilities.
- "Fuzzing"—an automated method of vulnerability detection that involves applying many possible permutations of inputs to a program, which is often used internally by companies to discover vulnerabilities.
- Products that rely on ML to predict whether source code may contain a vulnerability.

Many large networks are constantly monitoring for anomalies, protecting those who use the networks if anomalies are correctly identified and acted upon. These systems benefit from economies of scale—it makes more sense to continue iterating a single spam filter for a large number of users than to have every user build their own (or have one installed) on their computer.

Similarly, cloud computing companies may enforce ToS that prevent their hardware from being used for malicious purposes, under the assumption that they can identify such behavior. Another example of a system-level defense is blacklisting of Internet Protocol addresses from which attacks are commonly launched, though skilled attackers can obfuscate the origin of their attacks. Centralization and the associated economies of scale may also facilitate the deployment of AI-based defenses against cybersecurity attacks by allowing the aggregation of large datasets and the concentration of labor and expertise for defense.

Centralization can raise the stakes if central systems are compromised, however, and attackers can, of course, learn how to evade system-level defenses. For example, they can purchase commercial antivirus software and analyze changes between updates of the protection protocol to see what is and is not being protected against. A necessary (though not sufficient) condition of successfully deterring and punishing attackers is the ability to attribute the source of an attack, which is a notoriously difficult problem. A compounding problem for those who would attribute an attack is that even if they have high-quality information, they may not want to reveal it, because doing so may compromise a source or method. Some entities may also choose not to punish certain actions, so as to avoid creating precedent and preserve the option to engage in such actions themselves.

Cybersecurity Defenses

Although a wide variety of cybersecurity defenses are available, there remains relatively little solid analysis of their relative effectiveness. Many of the existing approaches to cyber defense were produced before unique considerations that apply to AI were taken into consideration. There is a wide array of cybersecurity solutions, ranging from automatic patching of a vendor's own software, to threat detection, to incident response, to consulting services. Network and endpoint security products aim to prevent, detect, and respond to threats. Solutions include detection of software exploits, and the prevention or detection of attacker tools, techniques, and procedures. Core areas of defense include the endpoint security, internal network security, and cloud security.

Next generation antivirus solutions are leveraging supervised learning techniques to impact new malware variants. Behavioral tools monitor normal user or application behavior and detect deviations from normalcy in order to detect malicious behavior among the collected anomalies. AI has also been used to aid security professionals to more efficiently hunt for malicious actors within their enterprises by allowing interaction via natural language and automating queries for understanding potential threats.

That said, relatively little attention has been paid to making AI-based defenses robust against attackers that anticipate their use. Ironically, the use of ML for cyber defense can actually expand the attack surface due to this lack of attention and other vulnerabilities. As a result, some surveys of cybersecurity professionals indicate low confidence in current AI-based defense systems. Security Domains Regulation and technical research on defense have been slow to catch up with the global proliferation of weaponizable robots. While defenses against attacks via robots (especially aerial drones) are being developed, there are few obstacles at present to prevent a moderately talented attacker from taking advantage of the rapid proliferation of hardware, software, and skills to cause large amounts of physical harm through the direct use of AI or the subversion of AI-enabled systems.

There is a growing gap between attack capabilities and defense capabilities, since necessary defenses are capital intensive, and the hardware and software required to conduct attacks are increasingly widely distributed. Physical harm through the use of human piloted drones and land-based robots is already playing a major role in some intra- and inter-state conflicts. Unlike the digital world, where key nodes in a network such as Google can play a key role in defense, physical attacks can happen anywhere, and many people are located in regions with insufficient resources to deploy large-scale physical defenses. The resource and technological advantages available to large organizations to protect against attack will continue when such attacks become more heavily augmented by AI. However, it should be noted that some of the most worrying AI-enabled attacks may come from small groups and individuals who may operate outside what is considered the norm at any given point in time, which are difficult to anticipate or prevent, as with lone wolf terrorist attacks or mass shootings.

The Growing Use of Robotics

There has been an explosion in the number and variety of commercial applications for robots in recent years, with the number of industrial robots exponentially with and without AI components. Relatively primitive cleaning robots are in wide use and more sophisticated service robots appear to be on the horizon. Ambitious plans for drone-based delivery services are being proposed and tested, commercial opportunities for drones are continuously launched, and recreational uses are flourishing (such as drone racing and aerial photography). Driverless cars are robots, and they also are increasingly being used in uncontrolled environments outside of test facilities, though large-scale deployment of fully autonomous driverless vehicles must first tackle a plethora of technical and policy challenges.

A wide range of robots with autonomous features are already deployed within multiple national militaries—some with the ability to apply lethal force—and there is ongoing discussion of possible arms control measures for lethal autonomous weapon systems. Humanitarian, recreational, military, and commercial applications of robots are being explored around the world, and related supply chains are also global, with production and distribution dispersed across many countries. The diversity of robotics enables a wide range of applications, with drones already in use for competitive racing and photography.

Most current robotic systems are mostly not autonomous (humans play a significant role in directing their behavior), but a greater number of autonomous and semi-autonomous systems are being developed for purposes such as delivery and security in real world environments. The marketplace has advanced from relatively unstable and hard-to-fly drones a decade ago to drones that can stabilize themselves automatically, along with a steady increase in the autonomy of deployed systems. This has set the stage for a potentially disruptive application of AI with malicious intent that can be applied to existing and near-term robotic systems.

AI SUPREMACY

Non-automated drone attacks have been conducted by terrorist groups such as the Islamic State and Hamas, and the globalized nature of the robotics marketplace makes it difficult to prevent these groups from using drones and other forms of robotics. Greater degrees of autonomy enable a greater amount of damage to be done by a single person or small group, enabling large-scale attacks using robots to occur. The software being used to carry out such attacks is increasingly mature. For example, open source face detection algorithms, navigation and planning algorithms, and multi-agent swarming frameworks that can be leveraged for malicious purposes can be easily found.

Depending on their power source, some robots can operate for long durations, enabling them to carry out attacks or hold targets at risk over long periods of time. Robots are also capable of navigating different terrain than humans because of their varied perceptual capabilities (such as the ability to maneuver in the dark or in low-visibility fog) and physical capacities (like being undeterred by smoke or other toxic substances and not requiring oxygen underwater). As a result, a larger number of spaces that were previously considered relatively safe may become vulnerable to automated physical attacks.

The diffusion of robots to a large number of human occupied spaces makes them potentially vulnerable to remote manipulation for physical harm, as with a service robot hacked from afar to carry out an attack indoors. The Internet is, more generally, widely recognized to be highly insecure, representing an additional attack vector by which AI systems that control key operating aspects of an organization can be subverted, potentially causing more damage than would have been possible is such systems were under human control. In addition to traditional cybersecurity vulnerabilities, augmented Internet and robotic systems may be vulnerable to AI-specific vulnerabilities. There is also some evidence to suggest that people are unduly trusting of autonomous mobile robots, potentially creating additional sources of security vulnerabilities as such robots become more widely deployed.[32]

Building a Better Mousetrap

Enhancing cybersecurity so that it encompasses the emerging threats arising from AI and ML should be on the minds of lawmakers, IT specialists, and users alike. The tendency to be reactive rather than proactive is an unfortunate byproduct of being human, with all the potential ramifications that implies. Governments should, in general, work more closely with the private sector to investigate, prevent, and mitigate potential malicious uses of AI. This must include policymakers taking seriously their responsibility to avoid implementing measures that will interfere with or impede research progress, unless those measures are likely to bring commensurate benefits. Close collaboration with technical experts active in the space will help ensure that policy responses are appropriate, realistic, and a byproduct of the technical capabilities and constraints that frame any policy-oriented response.

Researchers and engineers in AI should take the dual-use nature of their work seriously, allowing considerations related to the misuse of ML technology to influence research priorities and norms, while proactively reaching out to relevant actors when harmful applications are anticipated. Given that AI is a dual-use technology, it is important that researchers consider it their responsibility to take whatever steps they can to promote beneficial uses of the technology and prevent harmful uses.

Nations should strive to achieve much more than they have to date, crafting and passing laws that are meaningful and have teeth. The European Union's (EU) General Data Protection Regulation (GDPR) is a good example of this, as is the work of the Partnership on AI, the 2016 White House series of workshops on AI, and the conferences that are regularly occurring around the globe to spur thought leadership and innovation in the space. Best practices should be identified in research areas with increasingly mature methods for addressing dual use concerns, and an even broader array of actors (including NGOs and other members of civil society) should be brought into the resolution mix.

Given the dual-use nature of AI, many of the malicious applications of AI may have legitimate uses (for instance, to battle bad actors with malicious forms of AI). The difference between legitimate and illegitimate uses of AI could be a means of ensuring that appropriate safeguards are put in place to manage malicious use. Surveillance tools can be used to catch terrorists as well as monitor ordinary citizens, and information content filters could be used to bury fake news as well as manipulate public opinion. That is why a transparent, public debate about AI's potential applications should be maintained, and why more people from a great variety of disciplines should become more educated on the subject.

It is clear that AI and ML will figure prominently in the cyber security landscape of the future, that malicious applications will proliferate, and that much more can and should be done to combat it. AI will remain intimately linked with digital security, physical security, and political security, creating an even more challenging dynamic that will require constant security management. In the cyber domain, AI can be used to augment attacks on and defenses of infrastructure and other critical aspects of society, implying that its future negative impacts are probably not being adequately contemplated.

Preparing for the potential malicious uses of AI is already an urgent task. As AI systems extend further into domains previously believed to be uniquely human (such as social interaction), more sophisticated attacks drawing on the social domain will occur. The many vulnerabilities that cyber attackers can identify, and the many platforms and methods from which they may choose to attack, are very difficult to defend against and may result in an explosion of network penetrations, personal data theft, and an epidemic of intelligent computer viruses. While AI is a looming threat in the cyber arena, one of the best ways to defend against automated hacking is also via AI, through automation of our cyber defense systems.

AI-based defense is not a panacea, however. More work needs to be done in understanding and achieving the right balance of transparency in AI, while developing improved technical measures for verifying the robustness of systems and ensuring that policy frameworks that were developed in a less AI-infused world adapt to the new world we are living in. Much has been published about problems which might arise accidentally as a result of highly sophisticated AI systems capable of operating at high levels across a very wide range of environments. Given that intelligence systems can be deployed for a range of goals, highly capable systems that require little expertise to develop or deploy may eventually be assigned new dangerous assignments as a result of being hacked.

Advanced AI systems may inflict unprecedented types and scales of damage in specific domains, requiring more acute forms of preparedness before more potent applications become realizable. Though the specific risks of malicious use across the digital, physical, and political domains are myriad, understanding their commonalities—including the role of AI in enabling larger-scale and more numerous attacks—is helpful in designing better prevention and mitigation efforts.[33]

The Future of AI and Cybersecurity

Many cybersecurity pundits have lost faith in traditional malware and antivirus software tools because they are only stopping an estimated 15 to 20% of malicious code. AI and cognitive computing offer a distinct advantage for protecting valuable data, since *AI gathers intelligence through learning as it anticipates the next attack on security, allowing for immediate and effective threat response*. Given that AI can be used for good or evil purposes, embedding controls on the use of this technology may be effective for a time, but they will eventually be penetrated. The insight and prediction capabilities provided by AI will change the current landscape of threats and begin to level the playing field.

Cyberspace is the ultimate 3D chessboard, with layers upon layers of moves and vulnerabilities, creating exponential threats that traditional thinking is simply not equipped to handle. Albert Einstein famously said, "We cannot solve our problems with the same level of thinking that created them".[34] The use of AI allows us to take our thinking on cybersecurity to that next level and identify more advanced problems. Cognitive technology gives us new insights into current and future threats, allowing for greater speed and precision in our response.

One of the biggest challenges is asking the right questions of the data. We collect so much, so fast, that soon we are overwhelmed. Inside that massive amount of data are the answers—the needles in the haystack. *Tomorrow's approach to asking the right questions will no longer be about making the haystack smaller to find that needle but, rather, about using AI and cognitive computing as bigger magnets to increase our chances of finding it.*[35]

Policymakers should collaborate closely with technical researchers to investigate, prevent, and mitigate potential malicious uses of AI. Researchers and engineers in AI should take the dual-use nature of their work seriously and proactively reach out to relevant actors when harmful applications are foreseeable. Best practices should be identified in research areas with more mature methods for addressing dual-use concerns. As AI capabilities become more powerful and widespread, the growing use of AI systems will result in an expansion of the existing threat landscape.

The costs of attacks can be reduced by the scalable use of AI systems to complete tasks that would ordinarily require human labor, intelligence, and expertise. New attacks may arise through the use of AI systems to complete tasks that would be otherwise impractical for humans, and malicious actors can exploit the vulnerabilities of AI systems being deployed by defenders. Attacks enabled by the growing use of AI have already become effective, finely targeted, difficult to attribute, and exploitative of the vulnerabilities in AI systems.

The use of AI to automate tasks involved in successfully completing cyberattacks will alleviate the existing tradeoff between the scale and efficacy of attacks. This may expand the threat associated with labor-intensive cyberattacks (such as spear phishing). Novel attacks that exploit human vulnerabilities (such as through the use of speech synthesis for impersonation), existing software vulnerabilities (like automated hacking), or the vulnerabilities of AI systems (through adversarial examples and data poisoning) will increasingly become the norm.

The use of AI to automate tasks involved in carrying out attacks with drones and other physical systems (for example, autonomous weapons systems) will expand the range of threats associated with these attacks. We should also expect that attacks that subvert cyber physical systems (causing autonomous vehicles to crash) or involve physical systems that would otherwise be infeasible to direct remotely (such as a swarm of micro-drones) will also become more prevalent with time.

The use of AI to automate tasks involved in surveillance, persuasion, and deception may also expand threats associated with privacy invasion and social manipulation. There will be a rise in attacks that take advantage of an improved capacity to analyze human behaviors, moods, and beliefs on the basis of available data. This would be of particular concern to the citizens of authoritarian states, and, as was seen in the US presidential election of 2016, it could also undermine public trust in government and erode the ability to sustain truthful public debates in democracies around the world.

Clearly, there is much that businesses and government can learn from professionals in the cybersecurity community about the intersection of cyberattacks and AI. As the dual-use nature of AI and ML becomes apparent, there is a growing need to reimagine legal standards and behavioral norms regarding the transparency of research and collective protection protocols. AI researchers and the organizations that employ them are in a unique position to shape the security landscape of our AI-enabled world. Positioning ourselves to be better able to combat malicious actions using AI and ML will be the result, not just for AI researchers and organizations active in the field but also for legislators, civil servants, regulators, security researchers, and educators.[36]

There is no alternative to changing the manner in which we are accustomed to thinking about the nexus between AI and cybersecurity, as well as the amount of resources devoted to staying a step ahead and creating new methods of combatting bad actors in cyber space. Along with the positive transformational changes that await us in the biometric, AI, and autonomous reality universe, lurks the dark vision that virtual terrorism plans to cast upon it. If we are successful in staying a step ahead of the virtual terrorists in the decades to come, that new universe may prove to be utopian. However, if we fail to stay a step ahead, it is sure to turn into a dystopian nightmare.

10. Spying and Fighting

<u>Spying</u>

While present day AI may not be anywhere near as advanced as it is often portrayed in fiction, the fear that it can produce, based on how and by whom it is deployed, is very real indeed. Some people worry about whether AI has the capacity to morph into a monster, as it was portrayed in the films *2001: A Space Odyssey* and *I, Robot*. That fear ranges from machines not obeying instructions to thinking for themselves and pursuing their own agenda. Fear is compounded with the knowledge that the designer of a neural network cannot know, once that network has been trained, exactly how it will perform. That is why AI's creators would like to better understand why the machines they create act the way they ultimately do. In no government sectors is this more important, or more advanced, than among intelligence agencies and armed forces.

In order to take advantage of the vast sensory and information collection apparatus the US has positioned throughout the world, the intelligence community must efficiently and effectively analyze *all* of the data flowing into each agency at any given moment, which is a tremendous challenge. Accurately relaying raw data and finished reports to other agencies, military commanders, and policymakers adds another layer of demand to that traditional dissemination requirement. One of the transformative AI capabilities being researched and deployed is Activity-Based Intelligence (ABI), an approach to geographical intelligence gathering that significantly reduces the data management role analysts at intelligence agencies face on specific assignments, improving the speed of the data exploitation process. For example, ABI emphasizes the development and understanding of patterns of life, which enable analysts to identify abnormal activities as well as new trends within the scope of what is considered normal. This is a direct application of SML algorithms focused on processing and then categorizing data inputs based on previously analyzed information.

To drill down on this concept a little deeper, if a persistent tactical Intelligence, Surveillance and Reconnaissance (ISR) platform over a small city in Iran routinely tracked 10 people entering a suspicious facility every day for a month, SML programs can note any instances of more or less than 10 people entering that facility. However, if over the next month eight people routinely entered the facility, UML programs would recognize a change in that pattern.

These programs all operate within the context of a specialized AI architecture that is able to extract the new pattern and implement it as the updated guiding template for the SML program classifying all new data from the ISR platform. This entire ABI process can occur with an analyst providing only an oversight and review role, demonstrating one capability of advanced AI that intelligence agencies can use to effectively exploit information and unburden the analyst from labor intensive data analysis.

Understanding how AI can significantly enhance the exploitation speed of incoming data and information, in addition to disseminating it more accurately, offers an opportunity to leverage and enhance intelligence collection capacity. From global ISR platforms extracting data from combat environments to developing analytical models that predict long-term global events, the more information that can be collected and properly analyzed in a timely manner, the more policymakers and military planners will benefit.

A 2017 Belfer Center study[1] acknowledged that the US intelligence community collects more data than its entire workforce can effectively analyze in their collective lifetimes. Recognizing that private sector funding for AI dwarfs that of the US Government, while it seems clear that future progress in AI has the potential to be truly transformative in national security technology, much will depend on how the public and private sectors work together to leverage their comparative advantages. A joint approach to enhancements of AI as it applies to national security should ideally focus on achieving information, economic, and military superiority concurrently.

AI SUPREMACY

Future AI advances will dramatically enhance capabilities for the collection and analysis of data, as well as its creation. Smaller countries with fewer resources that develop a significant edge in AI technology will punch far above their weight, as they already do in the cyber arena.

America's adversaries are providing their militaries and policymakers with information collected from highly advanced ISR platforms similar to American systems, so it is fair to conclude that the US has lost at least some of its comparative advantage in global technical information collection. In order to ensure that American policymakers remain equipped with superior intelligence, two general courses of action are possible: either seek to offset foreign technical collection capabilities with parallel US improvements or offset foreign capabilities with enhanced analytical capacity (or both).

The US intelligence community has heavily relied on collection superiority over foreign counterparts to support its importance within the national security apparatus. Having enhanced AI capabilities has provided the community with an opportunity to gain a new advantage that draws on enhanced exploitation and dissemination speeds to outcompete its adversaries. This enhanced capacity can strengthen the community's ability to influence American policymakers as competition inside and outside the US increases.[2]

Cracking Open the Black Box

The world's intelligence agencies and militaries are, not surprisingly, the furthest ahead in developing AI–spending vast sums of money attempting to better understand how and why intelligent machines end up operating the way they do. One of the first formal research programs to attempt to crack open this AI "black box" was the Explainable AI project, run by the DARPA. Monitoring places like North Korea from spy planes and satellites creates huge amounts of data; analysts looking at this data benefit greatly from tools that automatically alert them to suspicious activity, as well as explaining why they believe the activity warrants a closer look. Analysts from the US National Security Agency (NSA) have for some time been overwhelmed by the recommendations of old-fashioned pattern-recognition software. As AI adds to that deluge, it is more important than ever that computer programs be able to explain why specific pieces of data are worthy of a human operator's special attention.[3]

Some intelligence professionals admit that a lot of the work done in espionage is more boring than one might think, so the more such tasks can be assigned to AI, the better. Much of the essential work involved in spying *is* just plain boring and inefficient. Advances in AI that give algorithms the ability to analyze vast amounts of images and video to identify patterns, provide data about landscapes or identify unusual objects is critical for assessing national security concerns. The DARPA and the IARPA have been funding AI research for decades, and the US Central Intelligence Agency (CIA) routinely supports efforts to apply AI analysis to satellite imagery.

Since 1999, In-Q-Tel (IQT) has been the independent, strategic investment arm of the CIA (and the broader US intelligence community). IQT connects the government, startups, and the venture capital community to accelerate innovation. IQT Labs represent strategic initiatives established to explore new and disruptive technologies, uncover solutions, and demonstrate the art of the possible in a variety of technology areas. Within IQT, CosmiQ Works (CQW) supports the US intelligence community with analysis and insight to better leverage new products and services from commercial space startups. Operating as a technology challenge lab, CQW brings together participants from IQT, government, industry, and academia to address space-related challenges with the potential for high mission impact.[4]

As an example of how such collaboration works in practice, satellite operator DigitalGlobe has teamed up with CQW, Amazon, and chipmaker Nvidia to support its SpaceNet program, whose data will eventually include high-resolution images of half a million square kilometers of Earth (with a resolution of tens of meters). Amazon is making the SpaceNet data available via its cloud computing service and Nvidia is providing tools to help ML researchers train and test algorithms on the data. It will soon be possible to train software to do things like map the roads and buildings of shanty towns, track changes to urban infrastructure such as park benches and stop signs, and measure the materials used in roofs and other structures. That kind of information has intelligence value, but can also be commercially valuable, to provide information to health or aid programs.[5]

Another great example is Facebook. In order to be able to beam wireless Internet to the Earth's surface, Facebook was able to apply ML to satellite data to generate high-resolution population maps of 20 countries (within 15 feet of accuracy). The company's Connectivity Lab used maps containing 350 terabytes of information with 14.6 billion images. It used Facebook's AI algorithm to recognize structures that indicated where humans lived. To train the algorithm, the team pre-labeled 8,000 images of one country with information about whether a building was included in the image or not. They then showed this to the algorithm, which was able to use the information to learn which buildings were valid.[6] Google AI technology can similarly discern details in satellite imagery–such as the difference between snow and clouds–to be applied to accuracy-critical industries such as aviation and defense.[7]

In 2018, IQT invested in SigOpt, whose software enables consumers to automatically optimize the hyperparameters in ML systems–that is, control certain variables in the algorithm that determine how well the resulting model performs. Optimizing hyperparameters usually requires manual manipulation by experts through a mix of intuition and guesswork. SigOpt's software replaces that process with a system that observes how a model performs and uses that insight to suggest new hyperparameters that should improve its results. Offloading that task to the software allows data scientists to focus their efforts on other elements of enhancing an AI system.

The system works without observing a model's inputs or outputs, so users need not worry about sharing their data with a third party, or uploading it to a cloud service, if they would rather run the software only in their private datacenters. That ability to run in a private cloud setting helps set SigOpt's offering apart from some of its competitors' products, such as Amazon Web Services' SageMaker. Such systems were designed to simplify the entire process of creating an ML model–including the optimization of hyperparameters–but it requires users to perform some processing in the Amazon cloud, which is not viable for sensitive intelligence data. That is why SigOpt is selling its products directly to US intelligence agencies with IQT's blessing.[8]

Increasing Prominence in Government

In spite (or, perhaps, because of) the dramatic progress that is being made by integrating AI into the realm of government, and the degree to which AI is having an impact on such a broad range of industries and sectors, some practitioners and thought leaders worry about its future implications. Can governments maximize AI's potential benefits while successfully reigning in its potential dark side? Will AI ultimately put thousands of spies out of business because they may prove to be largely irrelevant in due course?

AI's prominence in the US government was already evident during the Obama administration; a white paper on its potential future impacts was released in its final month. The intelligence workforce is already concerned about the prospect of automation intruding in their tasks and potentially replacing them, but is it smarter to trust an algorithm over analyst?[9] AI will probably never possess the type of nuance, intuition, and gut instinct necessary to be a good intelligence analyst, so perhaps those working in the intelligence community can rest easier than those in manufacturing and other industries, but the truth is, the jury will essentially be out on that question until AI has been sufficiently developed to determine whether it is indeed capable of acting based on gut instinct and intuition.

An equally vexing concern is whether AI is being deployed with sufficient vigor to keep pace with how America's adversaries are doing the same. China and Russia have had a free hand to apply the technology unburdened by concerns about privacy, civil rights, or acceptability. China has excelled at developing AI via a vast number of well-trained computer engineers who program machines using an unrivaled amount of data, produced by the country's 1.4 billion people. Russia has little hesitation about what kinds of instruments of war it connects to AI, including the military's development of armed and unmanned vehicles (similar to a ground-based drones)– something the US has hesitated to pursue.

The US sees AI principally as a national security tool to be employed on the battlefield or to thwart terrorist attacks, and trails both countries in some aspects of the development of deployable machines. The manner in which it allocates and spends defense dollars tends to be slow and cumbersome– not conducive to speed or efficiency. That has hobbled the US from reacting swiftly, effectively, or proactively in response to cyberattacks, and has generally thwarted the development of cutting edge AI tools.

To counter that, the DIA started soliciting ideas from AI experts in a new way, through its Innovative Hub office. The office identifies problems it would like to solve, receives pitches from the private sector, and awards seed money to develop a technology for potential Agency investment. The Agency has spent millions of dollars this way to find tools that can quickly fight the wars of the future in cyberspace. Among the ideas pitched are: the ability to automatically identify a target across multiple closed-circuit surveillance cameras, explain cultural nuances such as the role religion plays in society, or accurately predict how an international incident will play out. Such uses could include anticipating how a "Black Hawk Down"-type incident might evolve or how the Iran nuclear deal might have looked in final form before the end of negotiations.

"Mind Mapping"—using computer programs created by conducting targeted interviews with people from a particular country—is helping determine how decisions are made and assisting in the prediction of reactions to specific provocations. This can be useful where winning over local support is as important as eliminating enemy combatants. An excellent example of this is when the US government originally presumed that the Iraqi people would greet American soldiers as liberators, showering them with rose petals following the end of what turned out to be the first phase of the Iraq War. The mind is quickly becoming the new battlespace. The ability to understand the psyche of a local population and accurately predict their likely reaction to a given set of circumstances is becoming ever more important. If your opponent has the ability to do that and you cannot, the potential consequences can be dire.

As is noted in Chapter 13, AI is so new that most countries do not have laws governing its use or addressing how involved humans should be in the decision-making process, particularly when a computer may decide on its own whether or not to carry out lethal action. Some critics question why governments would rely on AI technology before such questions have clear answers. However, if your adversary is doing something that does not take governance or ethical issues into consideration, you no longer have the luxury of doing so. That is certainly one impact of this AI "race" between nations—it is at times forcing fast tracking of tool development before an ideal amount of debate or testing has been completed.[10] That has raised questions about the credibility of the information being produced by AI in the intelligence arena, something that virtually ensures that human agents will remain bedrocks of the intelligence profession for some time to come.

James Bond on Steroids

In 2015, the CIA created its first new office since 1963—the Directorate for Digital Innovation—legitimizing the importance of technology, Big Data, and analytics, and has devoted substantial resources to it since then. As of 2017, the CIA had 137 pilot projects directly related to AI, which included everything from automatically tagging objects in video to better predicting future events based on Big Data and correlatable evidence. The objective is to identify and support correlations with cause and effect that enable better predictive powers. The Agency has significantly improved its "anticipatory intelligence", pitting sophisticated algorithms and analytics against complex systems to better predict the flow of everything from illicit cash to extremists around the globe. DL, and other forms of ML, can help analysts understand how seemingly disparate data sets could be linked, or lend themselves to, predicting future events with national security ramifications.

Intelligence analysts who incorporate open data sets may have a comparative advantage vis-à-vis those who do not. Doing so can enhance forecasting ability so as, for example, to be able to anticipate the development of social unrest or societal instability just days before they may occur. The objective is to leverage what is known from social science that is relevant to the development of instability, coups or financial instability, while adding historical perspective to predicting the future.

As an analogy: in the 1960s, James Bond carried around his classic Walther PP pistol and not much else. By 2015, Bond had progressed, in the film *Spectre*, to wearing an implanted transmitter, using a laser microphone, and accessing automated surveillance technology system that vacuums up all sources of intelligence. His pistol is now outfitted with an optical palm reader, which only he can fire.

Likewise, analysts are becoming more proficient in articulating the conclusions derived from cutting edge technology,[11] but given the margin for error in intelligence, and the potential costs associated with making mistakes, the stakes are extremely high.

A key function of artificial agents (both informational and cyber-physical) is the efficient manipulation of information, making such agents particularly suited to information warfare and cybersecurity applications. Adversaries can learn how to systematically feed disinformation to AI surveillance systems, creating unwitting automated double agents. One factor restricting intelligence in malware is the need for small malware payloads to prevent detection. For example, the payload for the intelligent malware Stuxnet was larger than most malware, but future developments in swarm or distributed AI may result in strategic botnets with small malware payloads packing devastating effects.

AI can crystallize any biases or falsehoods found in their training data. The alleged Russian interference in the 2016 US presidential election is perhaps the best-known example of the systemic vulnerabilities that exist. While this type of attack is detectable and recognizable (given the right information), more advanced artificial agents could make malevolent actors more effective and less detectable. The personalization of Internet-based information, combined with cognitive biases and the closed nature of online social media platforms, results in echo chambers that amplify misinformation. The artificial agents that feed our information consumption habits could be trained to make more strategic use of this vulnerability, or, alternatively, to nullify it via purposeful, systematic injection of noise or disinformation.[12]

Fake news

Misleading and deliberately false or malicious online content has become so prolific that humans may soon be forced to contemplate the possibility that only machines can discern fact from fiction. This has potentially serious implications in a whole range of areas–from political campaigns to law suits to the consumption and analysis of intelligence. One algorithm meant to shed some light on this is AdVerif.ai, which was specifically designed to detect phony stories, malware, and a host of other types of problematic content. AdVerif.ai scans content to spot telltale signs that something is not quite right–such as ensuring that headlines match article text–but it also cross-checks each story with its database of thousands of legitimate and fake stories, which is updated weekly. A variety of cybersecurity firms have added software to do something similar with their repertoire of services, but since ML is becoming more advanced each year, it is increasingly easy to generate realistic video and impersonate individuals with greater accuracy.

Identifying fake news has so many facets that the challenge of doing so accurately must be done in multiple steps. A first step might be "stance detection"–determining a story's relative worth based on what other news sites say about the topic, which would allow human fact checkers to spend less time checking individual pieces. The next challenge might be to address images with text overlay (i.e. memes, but with fake news), a format that is often promoted on social media, since it is harder for algorithms to break down and understand. AI could then be trained to integrate smarter ML techniques to dig deeper into recognizing what makes news fake.

Of course, just as AI can be used to detect fake news, it is also being used to create it, with incredibly convincing images and video. This has already raised serious concerns about attribution. In the absence of evidence to the contrary, who can say with any degree of certainty that one country disrupted the election of another by deploying fake news? Regarding the alleged Russian interference in the 2016 US presidential election, US intelligence agencies were publicly unanimous in their declaration of Russian involvement, but no proof was ever made public, and indictments were only produced in 2018. Doing so would, of course, reveal sources and methods, which any intelligence agency is loath to do. The anonymity by which actions can be taken on the Dark Web gives sophisticated individuals, and all leading intelligence agencies in particular, the ability to attribute their actions to another party. Fake news has already been deployed as a political and intelligence weapon.

In the coming decade, it is entirely possible–perhaps even likely–that the majority of people in advanced economies may consume and digest more false news than true information. Given the incentives and motivation to produce bogus content, the production of fake news is likely to outpace AI's ability to detect it, changing how and whether we will trust digital information in the future. Many people already do not, and with good reason. So, what AdVerif.ai and other similar platforms represent is just the opening round of an information arms race.[13] Smartphone apps such as FaceApp can already automatically add a smile to someone's face, swap genders, or make a person look younger. The app can also apply "beautifying" effects that include smoothing out wrinkles or lightening the skin. The Lyrebird platform can be used to impersonate another person's voice, as it demonstrated by posting clips of Barack Obama, Hillary Clinton, and Donald Trump—all supposedly endorsing the technology.

These are good examples of how powerful AI algorithms can be used to generate content rather than simply analyze data. Powerful graphics hardware and software, and recently developed video-capture technologies, are also propelling the sector forward. Researchers at Stanford University have unleashed a face-swapping program called Face2Face, which can manipulate video footage so that a person's facial expressions match those of someone being tracked using a depth-sensing camera. The result is often eerily realistic. These firms apply techniques that enable algorithms to go beyond mere learning so as to classify things and generate plausible data of their own.[14] Soon enough, AI may even be able to anticipate the objectives of fake news producers so they need not be bothered to even be involved in the process. That should really give intelligence agencies something to anticipate with trepidation.

Video and Imagery Analytics

Surveillance systems with video analytics consume video footage in real-time and detect abnormal activities that could pose a threat to an organization's security. The technology associated with video analytics help security software "learn" what is normal, so it can identify unusual and potentially harmful behavior that humans may miss. It does so by observing objects in a monitored environment (detecting when humans and vehicles are present) and by incorporating operator feedback about the accuracy of various events and integrating this intelligence into the system itself, improving its functionality. This interaction between operator and technology results in a "teachable" system wherein human oversight takes a backseat to the finely-tuned capabilities of AI.

Eliminating human error is the objective. Studies have shown that humans engaged in mundane tasks have a directed attention capacity for up to 20 minutes, after which their attention span begins to decrease. In addition, when humans are faced with multiple items to complete at the same time, their attention spans tend to decrease even more rapidly. Video analytics are therefore beginning to take the place of human judgment in an effort to increase operational efficiency. While a security officer may miss a person sneaking into a poorly lit facility, a camera backed with intelligent video analytics is designed to catch a flash on the screen and recognize it as a potential threat. It can similarly spot a person loitering at the perimeter of a schoolyard and alert security officials to investigate, while keeping close watch on many other cameras and locations. Rather than depend solely on human monitoring, AI-powered systems instead notify security teams of potential threats on a real-time basis, helping prevent break-ins or illegal activity, as well as increasing human accuracy.[15]

Taking this concept a step further, when Kim Jong Un prepared to launch a ballistic missile, analysts at the US National Geospatial-Intelligence Agency (NGA) would review satellite imagery looking for distinct signs on the ground in North Korea indicating that test preparations were underway. The NGA is increasingly using AI to conduct analysis typically done by humans. Some veteran imagery analysts fear the technology is not yet mature enough to replace human skill and analytic capability, believing that a machine can never possess the insight required to be a top-notch analyst. While the US National Reconnaissance Office is responsible for the satellites that collect earth imagery and data, the NGA plots the information on maps for use by the military and intelligence community in places where it is nearly impossible to send human sources. In 2011, analysts at the Agency helped locate the compound in Abbottabad, Pakistan where Osama bin Laden had been living prior to his assassination.

Some of those hard targets for the Agency's AI-focused imagery might include North Korea and Iran, where analysts must identify nuclear sites. The stakes are high, so the NGA's plan to shift toward the greater use of AI has sent shock waves among many in the intelligence community, who fear that machines will not be able to integrate intuition and gut feeling into analyses. Mistakes are certainly made by human imagery analysts—such as in 1999, when the US military mistakenly bombed the Chinese embassy in Belgrade—but significant errors tend to be relatively rare. In the intelligence business, there is no room for error.

It should be noted, however, that although mistakes are indeed made by AI—such as when Google was called out in 2015 for mistaking African Americans for gorillas—AI's analytical capabilities are improving over time in pattern recognition and statistical analysis by giving systems a set of examples or class of objects that make it easier to have intuitive sense—such as categorizing faces, vehicles, weapons, or vegetation. Machines are extremely well suited for performing rote tasks such as scanning endless numbers of pictures for specific objects, particularly given the attention span of humans. Machines are very good at identifying what is in an image; they are not yet so good at identifying what's not in an image, and that is where the real challenge resides.[16]

US researchers have trained DL algorithms to identify Chinese surface-to-air missile sites hundreds of times faster than their human counterparts. The algorithms proved capable of assisting individuals with no prior imagery analysis experience find the missile sites scattered across nearly 90,000 square kilometers of southeastern China. The neural network that was used matched the 90% accuracy of expert human imagery analysts in locating the missile sites, while helping humans reduce the time needed to analyze potential missile sites from 60 hours to just 42 minutes. This comes at a time when satellite imagery analysts are drowning in a deluge of Big Data. DigitalGlobe alone generates about 70 terabytes of raw satellite imagery each day.

DL models such as GoogleNet and Microsoft Research's ResNet were initially created to detect and classify objects in both color and black-and-white imagery. One major problem associated with applying AI to imagery analysis is the relative absence of large training datasets that include the hand-labeled examples needed to train DL algorithms to accurately identify features in satellite imagery. DL algorithms face their biggest challenge when trying to analyze smaller objects, such as mobile missile launchers, radar antennas, mobile radar systems, and military vehicles, because the available satellite imagery data has fewer pixels to work with when extracting identifying features. How well neural networks will work on a smaller scale will remain an open question for some time to come.

Imperfect AI tools can still prove incredibly helpful for intelligence gathering. For example, the International Atomic Energy Agency (IAEA) has the unenviable task of monitoring all declared nuclear facilities and searching for undeclared facilities among nearly 200 countries. DL tools can help the IAEA and other independent organizations use satellite imagery to monitor the development of nuclear power and related weapons of mass destruction.[17] Other applications include monitoring treaty compliance, troop and weapon movements, and human migration patterns. Clearly, AI and ML have a bright future in the intelligence space.

Fighting

A Commercial/Military Arms Race

A metaphorical arms race is in progress in the commercial sphere of autonomous systems development. The shift in R&D efforts and expenditures from the military sphere to commercial settings is proving to be problematic. Military autonomous systems development has been slow and incremental, and pales in comparison with the advances made in commercial autonomous systems. In a hotly competitive market for highly skilled roboticists and related engineers across the sectors most interested in AI (aerospace and defense), funding is far outmatched by the commercial automotive, information, and communication sectors, so is less appealing to the most able individuals in the marketplace. As a result, *the global defense industry is falling behind its commercial counterparts in terms of technology innovation*, with the gap widening as the best and brightest engineers move to the commercial sector.

The large disparity in commercial versus military R&D spending on autonomous systems development could have a cascading impact on the types and quality of autonomy that are eventually incorporated into military systems. One critical issue is whether defense companies will have the capacity to develop and test safe and controllable autonomous systems, particularly those that fire weapons. Fielding nascent technologies without comprehensive testing could put both military personnel and civilians at undue risk. However, the rapid development of commercial autonomous systems could normalize the acceptance of autonomous systems for the military and the public, and this could encourage state militaries to fund the development of such systems at a level that better matches investment in manned systems.

Although it is not in doubt that AI is going to be part of the future of militaries around the world, the landscape is changing quickly and in potentially disruptive ways. Given the challenge of feeding machines with knowledge and expert-based behaviors, as well as limitations in perception sensors, it will be many years before AI will be able to truly approximate human intelligence in high-uncertainty settings— as epitomized by the fog of war. Given the inability of AI to reason in such high-stakes settings, it is understandable that many people would want to ban autonomous weapons entirely, but the complexity of the field implies that its potential restriction must be addressed delicately.

For instance, does the term "autonomous weapon" describe the actual weapon (i.e. a missile on a drone) or the drone itself? Autonomous guidance systems for missiles on drones will likely be strikingly similar to those that deliver packages, so banning one could affect the other. Likewise, how will technologies emerging from the growing commercial marketplace be treated? Will the commercial sphere's emerging supremacy in the space change public perception about AI and its potential commercial applications? What are some of the global implications of this shift, and could it turn out to be a double-edged sword that will undoubtedly affect militaries in as yet unimagined ways?[18] These are but a few of the questions hanging on AI without clear, quick, or easy answers.

America has the Lead (for the moment)

Despite the US military's general lag over the US commercial sector in devoting resources and making bold advances in AI and ML, the US government is generally ahead of its adversaries (with some exceptions, as we will see) and has gotten the message that it must do a lot more, quickly, if it hopes to remain in the lead. In 2016, the DoD completed a study on military applications of "autonomy"[19] and concluded that the US needed to accelerate its exploitation of AI to realize its potential military value and remain ahead of its adversaries. The Study's recommendations focused on ensuring that AI systems were trustworthy and reliable, were properly understood to enable maximal advantage on the battlefield and expanded the range of technologies available for use on DoD missions. Among its recommendations were the establishment of early warning systems for understanding global social movements, the generation of autonomous "swarms"[20] that could exploit large quantities of low-cost assets, and the ability to detect cyber intrusions and enhance cyber resiliency for military vehicle systems.

America's military budget dwarfs those of its closest competitors (approaching $700 billion), with the US spending nearly 3 times more than China and almost 10 times more than Russia on defense. DoD spending on AI, Big Data and the Cloud reached $7.4 billion in 2017, a 32.4% increase over 2012. In Fiscal Year 2017, AI accounted for 33% of the spending total, while Big Data accounted for 48%, and the Cloud for 19%.

AI has contributed significantly to the overall growth in spending since 2012, however, with spending in the three AI segments—Learning and Intelligence, Advanced Computing, and AI Systems—growing the most during that period. The DARPA funded 60% of the Department's Natural Language Programs and 29% of DL programs.[21] In 2018, the Pentagon created the Joint Artificial Intelligence Center, intended to facilitate dozens of AI projects across the DoD, with a proposed budget of US$1.7 billion over its first five years of operation.[22] So, AI and its related components have garnered significant attention inside the Pentagon.[23]

AI and robotics–the forces that are ushering in the era of "hyperwar"–already allow for asymmetric responses that are inexpensive, resilient, and globally scalable.[24] AI technologies such as natural language-based dialog systems consume enormous amounts of information to augment human operators in non-combat situations, such as for maintenance and the remediation of equipment. Such capabilities will eventually be augmented by reality-based information-delivery technologies in combat scenarios. When deployed in an entirely autonomous fashion, the tactics and strategies of an AI system—its entire set of behaviors and corpus of acquired knowledge—can be easily reproduced between one system and another. This is the equivalent of having the most qualified veteran instantly transfer his or her experience and expertise to troops who have never been in battle.

Furthermore, an AI system's skills and specializations can be swapped instantly. The same autonomous aerial platform can be an expert "pilot" to suppress an enemy air defenses mission and, by quickly swapping the neural network controller, become the world's deadliest air superiority specialist. Other than the hardware, nothing would be lost; the "brains" of the pilot simply can be replicated on a different piece of hardware. The fusion of distributed machine intelligence with highly mobile platforms brings a speed and scale of concurrency never seen before. Such hyperwar applications have sparked a military revolution, with major implications for the security and defense arenas.

The speed of battle at the tactical end of the warfare spectrum will continue to accelerate, collapsing the decision-action cycle to fractions of a second, giving a decisive edge to the side with the more autonomous decision-action concurrency. At the operational level, commanders will be able to "sense", "see", and engage enemy formations far more quickly by applying ML algorithms to collection and analysis of huge quantities of information and directing swarms of complex, autonomous systems to simultaneously attack the enemy. At the strategic level, the commander supported by this capacity "sees" the strategic environment through sensors operating across the entire operational theater. The strategic commander's capacity to ingest petabytes of information and conduct near-instantaneous analysis—ranging from national technical means to tactical systems—provides a qualitatively unsurpassed level of situational awareness and understanding previously unavailable to strategic commanders.

AI-powered assistive technologies—such as intelligent assistants, advanced interactive visualizations, virtual reality technologies, and real-time displays projecting rapidly updated maps—will become conjoined and enable such situational awareness. This rekindles the perennial conversation about the nature and character of war. If, indeed, we are poised at the edge of hyperwar, we are compelled to explore the changes that will become necessary to embrace and adapt to this new conflict environment. It will require understanding the moral dimensions of these advances, educating a new generation of leaders and developing AI-powered analytical systems and autonomous weapons platforms.[25]

Project Maven

In 2017, the DoD created a task force dubbed the Algorithmic Warfare Cross-Functional Team (AWCFT), to develop more effective methods of sorting through the Department's vast amounts of archived and new signals and human intelligence, databases, and publicly-available material. Its objective was to modernize military processing, exploitation, and dissemination (PED) capabilities so that the vast reams of video, audio, and other data continuously gathered by the military can be made available to support real-time operational options.

The AWCFT accelerated the DoD's integration of Big Data and ML, creating actionable intelligence from the wealth of data it collects from the field. The Team's first task was to augment or automate PED for tactical unmanned aerial systems and mid-altitude full-motion video in support of the Department's anti-IS campaign. The AWCFT began focusing on developing, acquiring, and/or modifying algorithms to accomplish key tasks. It then identified the required computational resources, a path for fielding that infrastructure, and integrated algorithmic-based technology into the Department's strategy and planning efforts.[26]

The AWCFT's detection and classification effort involves literally putting boxes around objects on a viewing screen so that algorithms identify and classify them, determine a level of confidence about what they are, and enable a computer to do the analytical work that would previously have been completed by a human analyst. The end goal is to identify all the individuals in an entire city who are digging a hole and go straight to them without having to pore over thousands of hours of footage. The DoD is leveraging the AI technology that is already commercially available so as not to have to reinvent the wheel, which also supports to broader objective of enhancing public/private partnerships.[27]

The program's architects believed the initiative could overcome the array of challenges the military faces in the use of advanced intelligence equipment (particularly video sensors) and had the potential to dramatically improve PED outcomes for modern military purposes and warfare. The initial phase of the effort to transform the PED practice– code-named Project Maven– focused on developing computer vision algorithms for the fight against the IS. These AI-based programs were intended to detect and classify objects in full-motion video collected by surveillance drones and alert analysts when potentially noteworthy items or patterns were found. In essence, the AWCFT will transform PED by putting into practice a central tenet of data science: machines should do analytics and people should do analysis.

Project Maven adopted a developmental approach modeled after project management techniques in the commercial tech sector, wherein product prototypes and underlying infrastructure are developed iteratively and tested by the user community on an ongoing basis. Developers were encouraged to tailor their solutions to end user needs, and end users were similarly encouraged to prepare their organizations to make rapid and effective use of enhanced AI capabilities. Many of the key activities in AI system development–such as labeling data, developing AI-computational infrastructure, developing and integrating neural net algorithms, and receiving user feedback–were all run in parallel under Project Maven.

Though modern AI techniques for imagery analysis are extremely capable, building robust, DL AI systems requires huge data sets with which to train DL algorithms. Training data must not only be available, but categorized and labeled in advance by humans, and this phase of automation can be extremely labor-intensive. In Maven's case, humans had to individually label more than 150,000 images in order to establish the first training data sets, which quickly rose into seven figures in its first year. Such large training data sets were needed to ensure robust performance across a large diversity of possible operating conditions, including different altitudes, density of tracked objects, image resolution, and view angles. Having set the standard, every subsequent AI successor to Project Maven will need to adopt an effective strategy for acquiring and labeling large data training sets.

Once labeled data is ready, the algorithmic training process makes intensive computational demands. Traditional IT infrastructure has proven to be relatively useless for such computations. Many leading commercial tech companies have gone so far as to develop their own custom processors and cloud infrastructure networks to run AI computations. The DoD has spent years, and billions of dollars, trying to migrate its digital activity into the cloud, but none of that infrastructure was built with requirements for AI training and inference computation in mind. Project Maven therefore had to build its own AI-ready infrastructure, including computing clusters for graphics processing, from scratch. Fortunately, some of this capability can be leveraged for future algorithm training on other projects.

Even before the final versions of Maven's labeled data set and computational infrastructure were ready, the alpha and beta versions were used to develop algorithms that were shared with the user community to obtain feedback. Maven's team heard from users with full-motion video knowledge of counter-IS operations in the Middle East. From this, its developers were able to modify the course of their development in the direction of maximum operational effectiveness. Only such an approach could have provided high-quality, field-ready capability in the six months between the start of the project's funding and the operational use of its output. Just over six months from the start of the project, Maven's first algorithms were fielded to defense intelligence analysts to support drone missions in the fight against IS.

Project Maven delivered game-changing AI capability, demonstrating a level of technological innovation and programmatic agility that had been absent from most Defense Department digital initiatives. Maven's success was also clear proof that existing AI technology is ready to revolutionize future national security missions. Maven's secret sauce was not merely its focus on AI technology, but the fact that it was enabled and empowered by its organizational structure: a small, operationally focused, cross-functional team that was encouraged to: develop external partnerships, leverage existing infrastructure and platforms, and engage with user communities during development. Other divisions within the department will need to adopt project management structures similar to Maven's if they are to run AI acquisition programs in a similarly effective manner. Moreover, the DoD must develop operational concepts to use emerging AI capabilities and train its military officers and warfighters in their effective use.

The satellite imagery analysis community created its own version of Project Maven, migrating drone imagery analysis beyond the campaign to defeat the IS into other segments of the DoD that use drone imagery platforms. Maven copycats will likely be established for other types of sensor platforms and intelligence data, including analysis of radar, signals intelligence, and digital document analysis. Maven has successfully brought best-of-breed AI technology, along with project management practices from the commercial technology sector, to the US military.[28] That is a remarkable achievement, but undoubtedly one of many more to come.

A Broad Spectrum of Initiatives

The DoDs fiscal 2019 budget request illustrated where the Pentagon saw opportunities for future investment in AI:

- The Air Force proposed spending $87 million on its experimental program to further operationalize AI among several technologies, to test in war games, simulations, and field experimentation.
- The Navy was seeking $13.5 million to capitalize on rapid advances in terrain and environment collection to use AI to develop games to teach complex warfighting and decision-making skills, and to increase training tools for operation in Electronic Warfare and cyber-contested environments.
- The Army wanted to allocate $6.5 million to the Institute of Creative Technologies, its academic research laboratory at the University of Southern California that supports Army training and readiness through research into simulation, mixed, and virtual reality, AI, computer graphics, and learning sciences. The Army also sought an additional $6 million for medical training technologies that would use AI algorithms to assist in target recognition, next generation magnetometers, high resolution simulated three-dimensional terrain, and weapon orientation to enhance live training technology research.

- The Navy requested $49 million for its innovative Rapid Prototype Development program, which pursues a strategic focus on rapid prototyping of innovative combat system technologies and engineering innovations. AI is deployed in conjunction with directed energy weapons, hypersonics, ML, and computing techniques to improve the Navy's submarine combat systems.
- The Marine Corps requested $7.1 million for its own version of the Rapid Prototype Development program, which had already developed an unmanned swarm system that provides attack capabilities fused with AI to enhance situational awareness and decision-making.

In addition, the DoD requested funding for several programs to use AI to improve automated robotics. A $4.6 million request was intended to continue to improve robots' perception of their environments and their "intelligent control" abilities, enabling future systems to autonomously adapt and alter their behavior in dynamic tactical situations. Another program, for $4.2 million, would focus on better automating systems for robots, and a $9.5 million program would be dedicated to expanding the autonomous capabilities, utility, and portability of small robotic systems for military applications, with a focus on enhanced intelligence, biomimetic functionality, and robust mobility, to permit these systems to serve as productive tools.[29]

The US military already uses unmanned, remote-controlled vehicles such as the Predator Drone for aerial surveillance and attacks, and ground-based robots like the Mark 5A-1 for bomb disposal, but completely autonomous vehicles represent a new technological step. Critics argue that humans must be involved in life-or-death decision-making, and that such decisions should not be left up to machines, but it is unlikely that autonomous vehicles will remain in a benign, supporting role on the battlefield of tomorrow.

Several autonomous vehicles are either under consideration or are being used by the US military. Ironclad is a miniature tracked vehicle capable of operating in urban or off-road environments. Its modular platform implies that it can be fitted with different equipment, making it useful for reconnaissance, casualty evacuation, bomb disposal, and combat roles. It has a 50-kilometer range, runs almost silently, and is protected against explosives and small arms fire by its armored shell. While the Ironclad vehicle is not yet autonomous, the technology is being tested, and is described by BAE Systems as at a high state of readiness. Ironclad is also a step in the direction of the battlefield of the future, where fleets of unmanned air and ground vehicles work together, sharing situational awareness and pursuing combat objectives.

Lockheed Martin's Convoy Active Safety Technology system allows military drivers to switch their vehicles to fully autonomous operation. If a convoy were to come under attack, for example, drivers could press a button and their vehicles could stay in formation as the drivers took shelter or engaged targets. Lockheed has also developed a more advanced combat vehicle, called the Squad Mission Support System (SMSS), which can track and follow the movements of a single soldier, transport 1,000 pounds of supplies over rough terrain, or transport casualties. The remote-controlled type of SMSS has been in use in Afghanistan since 2011.

The Advanced Rotorcraft Sniper System is an autonomous helicopter drone that carries a human-operated sniper rifle. It can kill from hundreds of yards away based on software that corrects the operator's aim and steadies the .338 Magnum rifle attachment. With point-and-shoot simplicity, the remote operator can use an Xbox controller to squeeze off between seven and ten shots per minute, taking out targets at a fraction of the cost of a drone-carried Hellfire missile.

The US Navy and DARPA have been working on an anti-submarine drone, able to track and monitor modern, quiet diesel-electric submarines. Named the ASW Continuous Trail Unmanned Vessel (ACTUV), it was successfully tested in 2018 and will be able to patrol autonomously for between 60 and 90 days, while avoiding hazards like ship traffic, marine mammals, and fishing activities. While the ACTUV prototype is not equipped with weapons, once it detects an enemy submarine, it can mark the craft for other Navy units to take action if deemed necessary.[30]

The US Navy is expanding its ocean combat network—known as the Consolidated Afloat Networks and Enterprise Services (CANES)—to seamlessly connect its thousands of ships, submarines, and on-shore locations using AI. The expectation is that CANES will reduce the number of sailors required to maintain operations, potentially saving the Navy billions of dollars over years of use. The systems could also automate tactical information processing (such as threat and target information), making its fleets safer and more formidable. By 2020, the Navy expects to deploy CANES on 190 vessels and Maritime Operations Centers. Though CANES is not specifically intended to be weaponized, it certainly could influence future US military strategy[31] in a major way in the future. Clearly, the DoD understands the broad applicability of AI and is becoming aggressive about pursuing funding to make it a reality on a variety of battlefields.

The Race with China and Russia for AI Supremacy

Although the "big three" nations–China, Russia, and US–are clearly the leaders in the development of AI for intelligence and military applications, there are real contrasts between how each use and deploy AI. The US is believed to have a substantial lead in the application of AI in both arenas, but both China and Russia are investing heavily in the space, with the ambition of becoming the global leaders. China has already declared its intention to be the world leader in AI by 2030 (Chapter 11 is devoted to China's capabilities and ambitions in this area).

Russia's interest is already well established, with the Russian military having deployed AI capable of conducting independent military operations in Syria. Russia is preparing to fight on a roboticized battlefield in the near term, wielding anti-tank weapons, grenade launchers, and assault rifles. The US military's overall objective is to dominate AI and ML through its offset strategy, a research initiative meant to secure technological advantage over America's adversaries ("offsetting" similar gains by those nations.) But the US designs and deploys AI to be deliberately more human reliant than China or Russia, under the assumption that it has the most capable people, with comparative advantages to rely on.[32]

At the same time, America's adversaries have been betting that a new wave of weapons will negate technologies and tactics at the heart of US military might, among them aircraft carriers and high-altitude missile defense. In 2018, Vladimir Putin claimed that Russia's newest weapons were invincible against all existing and prospective missile defense and counter-air defense systems. In particular, Russia's Kanyon high-speed autonomous underwater nuclear torpedo can reach up to 115 miles per hour, making it difficult to defend against.[33]

Russia is clearly well advanced on the path toward developing the next generation of autonomous military weapons, and China is following a similar path, aggressively testing hypersonic weapons, unmanned aircraft, and advanced submarine detection, among other capabilities. The new Russian weapons appear to be intended less to pulverize than to provoke, meant to evoke a response that will further reinforce Putin's familiar narrative about being an encircled nation threatened by the North Atlantic Treaty Organization (NATO) and US missile defense systems.

It is no surprise that China and Russia are actively challenging the US for supremacy in space, cyberspace, AI, and other key technologies that have wide national security applications. The question is whether the US is taking this threat seriously enough. Some in the US Congress are alarmed that China continues to increase its R&D investments at an alarming pace, rapidly diminishing many of the technology gaps that exist between the two countries. There are, of course, strict laws and regulatory barriers in place to prevent China from accessing US technology or acquire American companies in sensitive sectors. Fear of Chinese theft of space technology has shaped some of the most stringent US export control policies, and the tariff war between China and the US, in 2018, included restrictions on AI and satellite components from the US side. A provision in the National Defense Authorization Act of 2017 bans the procurement of satellite communication by the US government if such systems use satellites or components designed or manufactured by China.

China has, nevertheless, built up a significant satellite manufacturing industry and has managed to develop quantum communications spacecraft with advanced encryption features. China is filling the need for affordable satellites so that countries like Nigeria, Bolivia, and Venezuela are able to buy satellites for a price that no other manufacturing country can compete with. China is also fast becoming a force to be reckoned with in the burgeoning small-satellite sector and should be expected to make significant inroads into the development of space-based weaponry in the not too distant future. Previously, China placed more emphasis on the legal and illegal acquisition of technology; it is now developing such technology on its own. China will have major advantages in translating private sector gains in the AI arena into national security applications,[34] given the heavy integration of government in all aspects of the Chinese economy.

Russia has also made great strides in designing, testing, evaluating, and fielding a variety of unmanned military systems. Until a few years ago, Russia had no official coordinated policy on how to marshal military robotics from concept and development through to testing, evaluation, and eventual acquisition by the country's armed forces. Multiple organizations within Russia's military-industrial complex were engaged in uncoordinated design of various platforms. Yet AI and ML constitute the most important aspect of Russia's military planning initiatives and the government created its own version of the DARPA, called the Foundation for Advanced Studies, tasked with developing unmanned and robotics projects for the military.

President Putin and top Russian military officials have in recent years called for the greater robotization of the Russian armed forces and increased investment in military robotics. These systems should account for a significant share of Russia's military strength in the coming decades. Numerous programs are at various stages of development, including mid to long-range designs, quadcopters, multi-rotor and tilt-rotor craft, micro-UAVs, and heavier models. Given Moscow's involvement in the Syrian civil war and the conflict in Eastern Ukraine, the Russian military has publicly acknowledged its use of unmanned platforms, resulting in the increased effectiveness of their fighting forces. Moscow used unmanned vehicles more often than piloted aircraft and is exporting the systems that have been battlefield tested.

The future trajectory of Russian unmanned military robotics is similar to that of the West: AI will continue to be a major focus of R&D. Russia's development of medium- and high-altitude unmanned aerial systems with long endurance will accelerate, along with further manufacturing and testing of unmanned ground and underwater combat vehicles. The Russian military understands that its robotics successes are limited, and that its defense industry lags behind China and the US. However, it is well aware that swarm technology powered by AI is a significant force multiplier, so the Russian government and its military establishment are devoting considerable resources to marshal the country's intellectual, industrial, and technological potential to create a modern military capable of projecting Moscow's power through unmanned systems in the future.

With ever faster concept development, testing, evaluation, and acquisition cycles, the world may discover that Moscow will deploy its military robotics at a more rapid pace than its Asian and Western counterparts. This is unlikely to automatically translate into a superior or more capable force than the US military, however, based on the budgetary resource constraints and bureaucratic inefficiency. Yet, it is clear that Russian development of military unmanned systems, in conjunction with the ongoing modernization of its armed forces, will result in a qualitatively different and capable force. Should Russia's success in Syria embolden it to act elsewhere in a similar manner, the US and other Western powers will need to consider appropriate courses of action. Russia has the advantage of being able to battle test its emerging AI-driven technology, which gives it an advantage that most other nations do not and will not have. Similarly, China's capabilities in this area may be less well-publicized, but should not be underestimated. The US (and other countries') reaction to their development will determine the nature of future conflicts for decades to come.[35]

Data Democratization

Since much of the technology that constitutes a nation's AI-oriented fighting power is typically acquired by the military in separate procurement programs, data tend to be stored in numerous individual repositories, each with its own set of cataloging procedures and proprietary technologies that effectively place the intelligence in silos. As a result, potentially advantageous communications among databases —for example, a fundamental but critical exercise such as immediately identifying a new building at a site by comparing the current video view with other flyovers in the past decade— are limited and, when these communications are possible, they are often not sufficiently timely to make a real difference. Instead of storing data in silos, the data is liberated, bringing together and integrating all the data available in one place at one time. "Democratizing" the data this way empowers intelligence analysts to address a much broader range of emergent problems. Analysts can view the larger context, see complex patterns and connections in the data, and understand how various forces impact each other.[36] That is clearly a comparative advantage on the battlefield.

A 2016 report[37] notes that, at their smartest, our most advanced artificially intelligent weapons are still operating at the level of insects—just armed with very real and dangerous stingers. Much remains to be done to more fully develop AI weaponry capabilities. The report looked at weapons in military arsenals around the world and found 284 systems at the time that included some degree of primarily standoff weapons that could find their own way to a target from miles away. Good examples are Aegis warships that can automatically fire defensive missiles at incoming threats or UAVs that shoot missiles from closer to a target. *Such autonomous weapons are not, in general, being developed to fight alongside humans on the battlefield but, rather, to displace them.*

At the other end of the technology spectrum are drones capable of loitering over an area, comparing objects on the ground against a database of images, and marking a target when a match comes up—all without human guidance. Such capabilities represent an emerging area of autonomy wherein a weapon does not have a specific target but a set of potential targets in an image library or target library (for certain signatures like radar) and it waits in the engagement zone until an appropriate target is detected. This technology exists in a relatively low number of deployed systems but is a major component of systems under development.

Cutting-edge experimental machines are being built— such as Dassault's nEUROn, BAE's Taranis, and Northrop Grumman's X-47B—which, unlike Predator and Reaper drones (which are designed to take out terrorist targets in places like Afghanistan), are more advanced and designed for war against countries that can actively engage them. These anti-access / area denial challenge environments require aircraft that use stealth to slip in under enemy radar and operate on their own over enemy territory, where UAV's must operate without communication.

Even if a military produces the best autonomic systems, other nations may advance AI further, to unexpected or destabilizing effect. As is the case with cyber terrorism, it may not be the world's strongest powers who tip the balance. Regional powers with greater abilities in autonomous weapons development, such as Israel, could, if it wanted to, destabilize a region through their use or choose to export their technology to other nations. It is, of course, also worth remembering that machines make decisions faster than humans. On the battlefield of the future, the fastest machines (those that make the best decisions with the least amount of human input) will offer the greatest advantage to their makers.[38]

Conclusion

As with prior transformative intelligence and military technologies, the coming national security implications of AI will be revolutionary, not merely different. Some governments will enact extraordinary policy measures in response, perhaps as radical as those considered in the early decades of nuclear weapons. The applications of AI to warfare and espionage are likely to be as impactful as the development of aircraft. Preventing expanded military use of AI is likely impossible and outright bans of AI applications in the national security sector are unrealistic. A more modest and reasonable goal is its safe pursuit, in which the technology is managed effectively.

Governments must both promote and restrain commercial activity, since a failure to recognize the inherent dual-use nature of technology can cost lives, money, reputation, and competitive advantage. Possessing the largest and most advanced digital technology industry in the world is an enormous advantage for the US, but the relationship between the government and some leading AI research institutions is fraught with tension. AI policymakers should support the interests of both constituencies, which is likely to be more difficult than many might imagine, if history is any guide.

More generally, governments should find a way to more meaningfully formalize goals for technology safety and provide adequate resources to support it. Safety results improve when governments create formal organizations tasked with improving the safety of their respective technology domains and appropriated adequate resources. This includes not only funding and materials but talented human capital. Governments should consider creating formal research and development organizations tasked with investigating and promoting AI safety across the entire public and private sector AI portfolio. Moreover, we call for the creation of a global entity whose objective is to oversee the creation and use of AI by establishing universally agreed guidelines, having the ability to monitor its broad development.

Countries have a strategic interest in shaping the cost, complexity, and balance between the offensive and defensive capabilities of national security technologies. As an example of how nations may "self-police" themselves in the area of weaponry, in the case of the US, the declining cost and complexity of bioweaponry led Washington to change its bioweapons strategy from aggressive development to voluntary restraint. As the development of America's stealth aircraft illustrates, targeted investments allow the US to strike a balance in the offense/defense domain and still build a long-lasting technological edge.

The same is true regarding how it can shape the technological profile of military and intelligence applications of AI. We stand at an inflection point; the pace of change in the AI sphere is advancing much faster than most experts had predicted. These advances will bring profound benefits to humanity, but such progress also entails risks, and AI has the potential to be every bit as fraught with risk as any prior transformative technologies. The stakes with AI are even higher than in the past, however, given the speed of technological progress in the 21st century and the ever more complicated relationship between government and business.[39]

While being sensitive to striking a balance between its own interest and that of the business world, the intelligence community's challenge is to improve source collection across platforms and domains without becoming overwhelmed. Likewise, the military's challenge is to exercise the same type of restraint it exercised in the US vis-à-vis the development of bioweaponry as it leaps further into the development of the next generation of AI-powered weapons. It is, perhaps, too fanciful to imagine that other nations would pursue their AI-driven ambitions in a similarly responsible manner, but the world would certainly be better off in the long-term if they did.

Daniel Wagner and Keith Furst

11. China's Quest for AI Supremacy

<u>Unrestricted Warfare</u>

In some countries, the pursuit of AI supremacy would be a simple, straight forward process–the government and national companies would compete on the global chessboard playing more or less by the same set of rules. But that is not how China plays the game. China does not merely aspire to be a leader in AI to become more economically competitive–it seeks to do so as part of a well-defined and executed plan to acquire what it needs from other countries in order to make up for any competitive deficiencies it may have so as to protect China, project its power, and prolong the power of the Chinese Communist Party (CCP). The acquisition of AI technology from other countries fits neatly into that framework.

Chinese military doctrine has long articulated the use of a wide spectrum of warfare against its adversaries. Much of what is known outside of China about its approach to asymmetric warfare is contained in a book first published in 1999 and translated to English with the title: *Unrestricted Warfare. The first rule of unrestricted warfare is that there are no rules and nothing is forbidden*. The book advocates tactics known as *shashou-jian* (Assassin's Mace), the concept of taking advantage of an adversary's seemingly superior conventional capabilities by "fighting the fight that fits one's own weapons" and "making the weapons to fit the fight". Having had nearly two decades to develop this philosophy, Chinese military strategists are, of course, prepared to use conventional weapons to fight their enemies, but, especially where it lacks a competitive advantage–such as not having a large or effective blue water navy–one of its tactics is to use cyberwarfare to make up the difference. *There is every reason to believe it is doing the same with AI.*

China is extremely adept at waging economic warfare. One way to estimate the damage done in this war is by the cost of intellectual property (IP) theft. The US Commission on the Theft of Intellectual Property has estimated that such theft costs the US $300 billion and 1.2 million jobs per year. Other organizations believe the cost is closer to $500 billion worth of raw innovation that is stolen from US companies annually, which would otherwise generate revenue, profits, and jobs. *The Chinese government's theft of IP for economic gain is just one piece of a larger strategy to fight a war while avoiding troop-to-troop combat.* The Chinese government's applications of Hybrid Warfare (that is, combining conventional means of warfare with unconventional means) are broad, enabling it to deploy a large array of tactics and methods to achieve its objectives.[12]

The Chinese government developed an elaborate system for processing and reverse engineering stolen designs decades ago. That has grown significantly larger than it was during the Cold War, morphing from a strictly military operation into a system permeating the entire Chinese government. *It is an elaborate, comprehensive system for identifying foreign technologies, acquiring them by every means imaginable, and converting them into weapons and competitive goods.* The departments in charge of reverse engineering acquired technology are officially called China's National Technology Transfer Centers (or National Demonstration Organizations) and became established by policy in 2007. Among their names are the State Administration of Foreign Experts Affairs (under the State Council), the Science and Technology Office (under the Overseas Chinese Affairs Office), and the National Technology Transfer Center (under the East China University of Science and Technology).

These organizations do not attempt to hide their purpose; their charters explicitly name 'domestic and foreign technology' as targets for 'commercialization'. The transfer centers play several roles, which include processing stolen technology, developing cooperative research projects between Chinese and foreign scientists, and running programs designed to 'encourage' Chinese nationals who have studied abroad to become part of the organizations. *China's meteoric economic rise can, in part, be attributed to this system of minimal investment in basic science through a technology transfer apparatus that worked to suck in foreign proprietary achievements while most of the countries which they were stealing from had no idea they were doing it.*

China could not have experienced the dramatic economic transformation it has experienced in the 21st century, nor have sustained its progress, without inexpensive and unrestricted access to other countries' technology. A 2010 report from the US Defense Threat Reduction Agency (DTRA) noted that modernization in the Chinese military depends heavily on investments in China's science and technology infrastructure, reforms of its defense industry, and procurement of advanced weapons from abroad. It added that the Chinese regime's theft of technology is unique in that, under the system, autonomy is given to research institutes, corporations, and other entities to devise collection schemes according to their needs.

State Theft as a Business

China's People's Liberation Army (PLA) is required to cover a portion of its own costs. Its decades-long focus on building external sources of cash has made its military leaders some of the most powerful people in China. With only 70% of its operating expenses covered by the state budget, the PLA must make up the difference and generate supplemental funds for its modernization. Just as is the case regarding the nexus between government and private business in China, the lines between the military and the state and the military and the private sector are thin. The PLA maintains thousands of front companies in the US, whose sole reason for existing is to steal and exploit US technology. According to the DTRA, the Chinese regime operates more than 3,200 military front companies in the US dedicated to theft.

Project 863 (also called the 863 Program) was started by former CCP leader Deng Xiaoping in 1986. According to a report from the US Office of the National Counterintelligence Executive, it provides funding and guidance for efforts to clandestinely acquire US technology and sensitive economic information. Project 863 originally targeted seven industries: biotechnology, space, IT, automation, laser technology, new materials, and energy. It was subsequently updated to include telecommunications, marine technology, and AI.

The Chinese government also runs the Torch Program to build high-tech commercial industries, the 973 Program for research, the 211 program for "reforming" universities, and countless programs designed to attract Western-trained scholars back to China. Each of these programs relies on foreign collaboration and technologies to cover key gaps, encouraging Western-trained experts to help China's technological development by returning to China or "serving in place" by providing needed information gained while working for their Western employers. Project 863 maintains a library of tens of millions of open source articles in scores of databases that contain more than four terabytes of information gleaned from American, Japanese, Russian, and British publications, military reports, and standard specifications.[3] There is no way to sugar coat this: *China's path toward AI supremacy is an extension of its well-crafted strategy to use unrestricted warfare and state theft to achieve its objectives.*

Harnessing the Power of Big Data

Given the size of China's population, it has an inherent advantage in harnessing the power of data into AI applications. Every time someone enters a search query into Baidu (China's Google), pays a restaurant bill with a WeChat wallet, shops on Taobao (China's Amazon), or catches a ride with Didi (China's Uber), that data can be fed back into algorithms, not only to improve their accuracy but, of course, to keep tabs on where people go and what they do. A similar phenomenon is occurring in the US and many other countries around the world, but China now has in excess of 750 million people online, and more than 95% of them access the Internet using mobile devices. In 2016, Chinese mobile payment transactions totaled $5.5 trillion– about *50 times* more than in the US.

China's State Council issued an ambitious policy blueprint in 2017 calling for the nation to become the world's primary AI innovation center by 2030, by which time, it believed, China's AI industry could be worth about *$150 billion*. China has been investing tens of billions of dollars in all aspects of information technology, from quantum computing to chip design, and Chinese government and industry have launched multiple simultaneous initiatives in an effort to commence its drive to achieve AI supremacy. By contrast, according to IQT, the US government's total spending on unclassified AI programs in 2016 was about *$1.2 billion*.

Facial recognition is the rage in China and is now used routinely for shopping and to gain access to some public services. In 2018, Alibaba led a $600 million investment in SenseTime, a Chinese firm that develops technology for tracking individuals, and was working with 40 local governments in China. It was estimated at the time that the company was the world's most valuable AI company, worth approximately $4.5 billion.

Throughout China, customers can authorize digital payment in some fast food restaurants via facial scan. Baidu's facial recognition systems confirm passenger identity at some airport security gates. Recent advances in AI have made it possible to identify individuals in video—a far more complex task than in photographs. The government has deployed facial recognition technology in Xinjiang, a Muslim-majority region in western China where tensions between ethnic groups erupted in riots in 2009. Facial recognition checkpoints operate throughout Xinjiang, from gas stations to shopping centers to mosque entrances. Even toilet paper in public restrooms is now being dispensed (in limited amounts, of course) after a facial scan.[4]

In its quest to achieve AI supremacy, the government is proceeding full throttle. As of 2016, Chinese researchers published more DL-related papers in journals around the world than researchers from any other country. China plans to expand the number of universities offering dedicated AI and ML departments. Chinese industry continues to bet heavily on AI. In 2017, Alibaba announced plans to invest $15 billion in research over 3 years to build 7 labs in 4 countries that will focus on quantum computing and AI. That same year, China's Ministry of Science and Technology (MST) issued a 3-year plan to guide AI development, naming several large companies as "national champions" in key fields; for example, Baidu, in autonomous driving, and Tencent, in computer vision for medical diagnosis.[5] China intends to be, and will indeed become, the world leader in AI and ML in the next decade.

Digital Leninism

China's transition to a 'great power" in AI has raised a host of fundamental issues the country's leadership must address: the transformation of labor markets, diminishing human control over critical decision-making processes, the increasing influence of those who develop and deploy AI, and potential changes in the power structure in a country where the CCP reigns supreme. Of course, any country with such ambitions would need to address a similar set of issues but, where China is concerned, *the potential of AI to dramatically transform Chinese society is a tiger being unleashed that the CCP may come to regret, for it has the potential to upend the command and control structure the Party has worked so hard to put in place since the 1940s.*

At the turn of the century, the Internet revolution posed a threat to authoritarian rulers, which they methodically turned to their own advantage by creating a surveillance state on steroids. The current Big Data revolution has been warmly embraced by the CCP as an opportunity, rather than a threat, and the country's leaders are reconfiguring many aspects of Chinese society to align with AI. In this reconfigured system, central co-ordination and control (fondly referred to as "top-level design" by the Xi Jinping) are intended to become an asset that propels technological innovation and enhances economic performance, while maintaining political stability.

The prospect of reaching a goal that has historically eluded Communist rulers—the perfection of centralized control—propels the CCP's intense and long-term commitment to digitization. All of China's major development strategies now rely on Big Data-driven solutions. "Healthy China 2030" seeks to expand online health services that can generate diagnoses and treatment advice with the help of Big Data. Local governments are investing heavily in the provision of public services by way of IT applications. China's "Social Credit System", announced in 2014, judges individuals' trustworthiness and aims to nudge citizens (and companies) into rule-abiding behavior by evaluating data ranging from payment morale and compliance with traffic rules to environmental regulations and opinions voiced in online chat rooms. Algorithms—rather than laws, policemen, or judges—will assume regulatory and enforcement powers in China's future.

AI SUPREMACY

China's leaders have more flexibility to experiment with the technologies of the future than elected leaders in the West, and they are doing so with vigor. *China's top-down authoritarian approach to governing meshes well with a compliant, bottom-up enthusiasm for new technologies in Chinese society.* From facial recognition to compulsory sharing of consumer data, China has become a laboratory for everything from mobile payment systems to online surveillance. *It is not incentivized to share new technology with the rest of the world because it seeks technological supremacy, which is one way of compensating for its lack of military supremacy vis-à-vis the US and West.* This applies in particular to AI and cyber-oriented technology and capabilities.

China's future success as a digital superpower is, of course, not guaranteed. The success of "smart" economic and social planning will ultimately depend on the quality, integration, and processing of the data that feed it, but also on the government responsiveness and agility that is essential to make use of feedback from data analysis. Xi Jinping has advanced his "digital Leninism" further than most had thought possible within a short time. Other authoritarian regimes around the world are watching the CCP's approach closely. If China manages to harness Big Data technologies to build a top-down yet responsive system of governance that turns out to be politically effective, economically productive, and socially stable, it has the potential to become a global model[6] that could further threaten democratic systems of government.

China's 2030 Vision

China's leadership sees technological innovation generally, and AI in particular, as a core element of its international competitiveness. Beijing has embarked on an agenda of "intelligentization" to take advantage of the transformative potential of AI throughout its society, economy, government, and military. As noted above, the Chinese government has not exactly been shy about declaring its intention to be the world's leading AI power by 2030. In 2017, it issued a bold AI plan[7] to proceed toward that objective in characteristic 5-year increments, which can be summarized as follows:

- By 2020, China will be on par with other advanced nations in the pursuit of AI supremacy. AI will become a focal point of economic growth, will improve Chinese citizens' livelihoods, and China will have made significant progress in developing a new generation of AI theories and technologies.
- By 2025, China will have achieved major breakthroughs in basic theories for AI and will become the primary driving force for China's ongoing economic transformation. A new generation of AI will be widely used in intelligent manufacturing, medicine, agriculture, defense, and other fields. By then, China will have established laws, regulations, and ethical norms governing the use of AI.
- By 2030, China's AI theories, technologies, and applications will lead the world and provide an important foundation for becoming a leading power in AI innovation.[8]

Despite the country's noteworthy rise in AI, the plan candidly acknowledges that, as of 2017, China had not produced major original results in AI and had lagged well behind leading AI nations in critical components such as high-performance chips for ML. At that time, there had not been systematic high-level design for research and development and Chinese research institutions and enterprises had yet to establish AI influence internationally.

The plan called for the leveraging of global innovation resources while acquiring what it needed to become competitive via overseas mergers and acquisitions, equity investments, the deployment of venture capital, and the establishment of research and development centers abroad. As a result, China will likely become less dependent upon foreign innovation resources as its becomes more capable in the space. Chinese investments in Silicon Valley AI startups prompted a debate in the US about whether to broaden the remit of the Committee on Foreign Investment in the United States (CFIUS) in order to expand reviews of Chinese high-tech investments, particularly in AI (discussed in more detail below).

The plan calls for progress in new "AI 2.0" technologies, including Big Data intelligence, cross-media intelligence, swarm intelligence, hybrid-augmented intelligence (human-machine symbiosis or brain-computer collaboration), and autonomous intelligent systems. China appears to be particularly focused on approaches that could enable paradigmatic changes in AI, such as high-level ML (self-adaptive or autonomous learning), brain-inspired AI, and quantum-accelerated ML. As the government enhances national competitiveness, of course, it also plans to bolster its capacity to ensure state security and national defense, in accordance with the CCP's imperatives. The country's AI will therefore have a range of applications intended to maintain public security, social stability, and national defense.[9]

While the plan formalizes and points to a national-level focus on AI, local governments and companies were, as noted above, already engaging in subnational planning on AI before the plan was announced. Additionally, crucial elements of the AI plan are rooted in previously adopted science and technology plans. So, while the central government plays an important guiding role in the plan's execution, bureaucratic agencies, private companies, academic labs, and subnational governments are all pursuing their own interests and staking their own claims to China's AI platform.

Although the rest of the world might be inclined to believe that ethics and safety considerations will be largely absent from Beijing's planning efforts, substantive discussions about AI safety and ethics are emerging in China. In fact, a wide range of Chinese AI researchers are participating in the Global Initiative for Ethical Considerations in AI. That said, it seems clear that China will not want to be part of any global arrangement which restricts its freedom of movement in the AI sphere. While there is no consensus about the endpoints of AI development in China—which will remain a work in progress for some time to come—it appears equally clear that the State Council's AI plan is not intended to be the totality of China's AI strategy.

The 2020 benchmark for the core AI industry's gross output (RMB 150 billion or approximately US$24 billion) represents a tenfold increase from 2017. Four factors are driving the overall development of China's AI: hardware (in the form of chips designed for training and to execute algorithms), data as an input for AI algorithms, research and algorithm development, and the commercial AI ecosystem. There are important similarities and differences between China's current approach to AI development and its past efforts to guide scientific and technological innovation in other areas. These include a strong degree of state support and intervention, transfer of both technology and talent, and investment in long-term, society-level measures. However, the plan differs from previous approaches because of AI's "omni-use" potential–the breadth of actors involved is much wider than for other technologies. As a result, international private tech giants and vigorous startups are leading players in driving innovation.

The Chinese government has adopted a "catch-up" approach to obtain the hardware necessary to train and execute AI algorithms. It has supported national corporate "champions" by providing substantial funding, encouraging domestic companies to acquire chip technology through overseas transactions, and making long-term bets on supercomputing facilities. Access to large quantities of data is an obviously important driver for AI systems. While China's data protectionism favors Chinese AI companies (which access domestic data), it simultaneously inhibits the cross-border pooling of data. This may prove to be in important impediment as China's AI industry matures.

The AI plan outlines a two-pronged "gathering" and "training" approach, wherein national and local-level talent programs accumulate AI researchers to work in China, while the country's tech giants set up their own overseas AI institutes to recruit foreign talent. The government has begun to take a more active role in funding AI ventures, helping promote the fourth driver of AI development, the commercial AI ecosystem. By disbursing funds through Government Guidance Funds established by local governments and state-owned enterprises, the government has already invested more than US$1 billion in domestic startups, with much of the investment directed toward AI. The government is also exploring ways to exert more influence over large technology companies, including through the establishment of party committees and "special management shares".[10]

<u>A Multiplicity of Purposes and Applications</u>

With such a diverse platform having been established to ramp up China's AI ambitions, AI serves a multiplicity of purposes for the Chinese government. While helping make China more competitive commercially and militarily, AI is also a vehicle through which the government can be perceived to provide better governance for the Chinese people, using AI to drive smart cities, manufacturing, government, and forming the infrastructure for a smart society. Consistent with the government's lofty aspirations, AI applications in agriculture, transportation, social security, pension management, public security, and a host of other government functions will enable the government to provide a range of new benefits to its people. Should the application of AI tools throughout Chinese society become as successful as the government envisions, *it could create a comprehensive new power source within the CCP's architecture, further deepening and strengthening the Party's power.* Doing so could also enhance the influence the technology industry within Chinese society and politics,[11] not unlike what is already the case in the US and other advanced nations.

The truth is, *China is particularly well placed to achieve its stated objectives and assume the global lead in AI. It has capital, people, and computing power in abundance.* In addition, China has two other resources that make China a promised land for AI. China already has approximately 40% of the world's trained AI scientists and most large universities have launched AI programs, meaning that number can only increase.

Then there is the data component: China's enormous population generates more data than any other nation, given its 750+ million daily Internet users. Almost all of them go online from smartphones, which generate far more valuable data than desktop computers, primarily because they contain sensors and are mobile. In its big coastal cities, cash has all but disappeared for small purchases: the Chinese make purchases with their devices using services such as Alipay and WeChat Pay.

Moreover, unlike in the West, where citizens are preoccupied with civil liberty protections, the Chinese are not necessarily concerned about privacy, which makes openly collecting data easier. The country's popular bike-sharing services not only provide cheap transport but what is known as a "data play"– when riders hire a bicycle, some firms keep track of renters' movements using a GPS device attached to the bike. This would likely be frowned upon in the West (even though, although most people do not think about it, nearly everyone riding a shared bike carries a smart phone that is broadcasting their location to their apps and service providers), but the average Chinese person may either care a great deal about location (or other forms of privacy) or could not really care less.

Young Chinese are particularly enthusiastic about their AI-powered services, while being relaxed about who uses their data. Xiaoice, a chatbot operated by Microsoft, has more than 100 million Chinese users. Most talk to the bot frequently about the problems they may have had during a given day. Each time they do so, the bot is learning from its interactions and becoming cleverer. Xiaoice no longer just provides encouragement and tells jokes—it has created the first collection of poems written with AI, which caused a heated debate in Chinese literary circles about whether there should be such a thing as artificial poetry.

While the country has more than 40 laws containing rules about the protection of personal data, they are rarely enforced. Technology firms are working closely with government agencies to develop AI and ML, so it is unlikely that the government will burden AI firms with overly strict regulation. Entrepreneurs are taking advantage of China's data strengths and lax regulatory environment. While many Chinese AI firms have only recently started up, many of them have progressed more rapidly than their Western counterparts. As a result, China already has a herd of AI unicorns (startups valued at more than US$1 billion). Baidu, Alibaba and Tencent, collectively referred to as BAT, are working on many of the same services as their smaller counterparts, including speech- and face-recognition, while also trying to become dominant in specific areas of AI, consistent with their comparative advantages.

Tencent is poised to develop a big presence in AI, given that it possesses more data than the other two. Its WeChat messenger service has almost 1 billion accounts and is the platform for thousands of services, from news to legal assistance to payments. It is also a world leader in games, with more than 100 million players for each of its most popular games. Alibaba is already a behemoth in e-commerce, investing billions of dollars to become number 1 in cloud computing. It already boasts traffic optimization software and AI-powered services to discover drugs and diagnose diseases from medical images.

However, Baidu's fate is mostly tied to AI in part because the technology may be its best opportunity to catch up with Alibaba and Tencent. It is putting most of its resources into autonomous driving, hoping to provide technology for fully autonomous vehicles by 2020. While its rivals—such as Waymo (Google's subsidiary) and Tesla—jealously guard their software and the data they collect, Baidu is sharing its data. The idea is to encourage carmakers that use Baidu's technology to do the same, creating an open platform for data from self-driving cars. That could shake up the autonomous vehicle marketplace globally.[12] Imagine that—a Chinese company leading the way in data transparency in one of AI's most significant sectors.

Don't Trust and Verify

The government began a nation-wide surveillance project— Skynet—in 2005. Today, as facial recognition has become omnipresent throughout the country, cameras track passengers at railway stations, identify homeless people on the streets, and monitor worshippers in state-approved churches. A 2017 documentary co-produced by the CCP claimed that the country had the largest network of closed circuit television cameras in the world (more than 20 million). That same year, some 55 cities became part of a plan called Xio Liange ("sharp eyes"), which entailed footage from surveillance cameras in public and private properties being centrally processed to monitor people and events. The intelligence collected from the video footage became part of what powers China's Social Credit System.

In 2017 alone, 530 patents related to video surveillance and surveillance cameras were published in China,[13] compared to just 96 in the US during the same period[14] China has also seen a rise in facial recognition patents, with more than 900 patents published in 2017. Applicants included government-supported academic institutions, big tech companies, and startups whose clientele includes government agencies. As the government adds a layer of AI to its surveillance operations, startups have come to play an important role in providing the government with foundational technology.

Although facial recognition has been in use for some time in many railway stations for ID verification, AI-enabled smart glasses developed by a startup are now being used to enable law enforcement authorities to identify criminals. Using Intel's Movidius Myriad vision processing chip, the technology matches faces with a database of known and wanted criminals stored on the device. By storing images on the device (as opposed to sending images to a central server on the cloud), it can function faster. One start-up led by the Chinese state government's venture capital fund had access to 1.3 billion face data records on Chinese citizens stored in the Ministry of Public Security's database.

As previously noted, China has a strong mobile payment market with WeChat (owned by Tencent) and Alipay (Alibaba's financial arm) having 1 billion and 500 million users, respectively. These payment giants have become digital ID repositories. At least three provinces in China are issuing electronic ID cards for their citizens using WeChat or Alipay's facial recognition technology. The mobile IDs can be used for authentication instead of carrying physical ID cards (mandatory for citizens at all times in China) for travel booking, name registration at Internet cafés, and for other security checks.

Alibaba and Tencent are also working with the government on smart city projects, offering their cloud processing and AI capabilities. The projects involve online and offline surveillance, in conjunction with smart city projects, to help monitor traffic, waste and water level management, and real-time monitoring of surveillance footage for crime prevention. Approximately 500 smart city projects were underway in China in 2017.[15]

AI with Chinese Characteristics

In some respects, the real race in AI between China and the US will be between the two countries' big cloud companies, which will compete to be the provider of choice for companies and cities that want to dive deeply into AI. China's tech giants are ready to compete with Google, Amazon, IBM, and Microsoft to serve up AI. While Alibaba's core business remains selling goods and providing a platform for business-to-business trade, it has spawned other lucrative operations, including a platform for logistics and shipments, an advertising network, and cloud computing and financial services. The company's ubiquitous mobile payments app, Alipay, is run by a sister company, Ant Financial, which also offers loans, insurance, and investing via a smartphone.

Alibaba announced in 2017 that it would spend $15 billion over the following three years on a research institute called the DAMO Academy (Discovery, Adventure, Momentum, and Outlook), dedicated to exploring fundamental technologies. This is evidence, as if it were needed, that China long ago dispensed with a reputation for simply copying Western innovations. According to the Organization for Economic Cooperation and Development, R&D spending in China grew tenfold between 2000 and 2016, rising to $412 billion[16]. The US still spends more ($464 billion in 2016), but its total has increased by only a third since 2000. Very soon, China will be the leader in that category, as well.

DAMO will effectively triple Alibaba's research budget, to more than $7 billion. This most likely means that Alibaba will overtake IBM, Facebook, and Ford, and will narrow the gap with the world's leaders—Alphabet and Amazon—which spent $16 billion and $14 billion on R&D in 2017, respectively. DAMO will include a portfolio of research groups working on fundamental and emerging technologies including blockchain, computer security, fintech, and quantum computing, with AI being the biggest focus with the greatest commercial potential.

Alibaba appears to be inspired by the way the DARPA funds different teams competing on the same project, and it is clearly learning from Alphabet and Amazon, as well. It has released a cloud ML platform, just as they have (the first from a Chinese company), which was launched in 2015 and has been upgraded significantly since then. The tools it offers are similar to those on Google Cloud and Amazon Web Services, including off-the-shelf solutions for such services as voice recognition and image classification. Developing these tools was a major technical undertaking for Alibaba and signals both how ambitious the company is to shape the future of AI and how big a role cloud computing will play.

Alibaba's cloud already supports several other companies' DL frameworks, including Google's TensorFlow and Amazon's MXNet. By supporting its competitors' frameworks, Alibaba gives developers a reason to use its platform instead, but Alibaba is creating its own DL framework. In 2017, the company released an AI program capable of reading a piece of text and answering simple questions about that text more accurately than anything ever built before. Alibaba has already used the program to improve its automated customer support on its online marketplace and it plans to deploy language understanding across all of its platforms and technologies.

Alibaba's AI researchers are working on other cutting-edge projects, such as generative adversarial networks, an ML approach developed by a Google researcher wherein two neural networks compete against one another, with one trying to generate data that seems as if it comes from a real data set, while the other tries to distinguish between real versus fake examples. Most Westerners may not realize it, but Alibaba is already exporting AI technology as the world's fifth-largest cloud computing provider (after Amazon, Google, IBM, and Microsoft). Alibaba is arguably already ahead of the competition in some areas. In 2017, it announced a collaboration with the Malaysian government to provide smart city services, including a video platform that can automatically detect accidents and help optimize traffic flow. Alibaba may have already done more to change the way business is done in China than any other organization.[17] It is ambitious on every front.

While the quantity of Chinese AI research has grown dramatically, researchers in the US remain responsible for a lot of the most fundamental groundbreaking work. *What Chinese researchers have been very good at doing is focusing on an idea and expanding on its different applications.* In fact, the Chinese have become prominent in adding value to existing research, with researchers in China wasting no time to produce papers on various applications, which can then be further developed. Chinese researchers usually speak English, so they have the benefit of access to all the work disseminated in English. By contrast, the English-speaking AI research community is much less likely to have access to work within the Chinese AI community, and the velocity of work is much faster in China than in most of Silicon Valley[18]–among the many distinct advantages China possesses.

AI SUPREMACY

A Plethora of Concerns

Baidu's call for data transparency aside, there are a number of additional concerns about how China will manage its rise to be a great AI power. Among them is the potential that the benefits of Chinese technological breakthroughs will be muted by data protectionism. In 2017, a cyber security law came into force requiring foreign firms to store the data they collect on Chinese customers within the country's borders; foreigners cannot use Chinese data to offer services to third parties. If data cannot be pooled, the algorithms that run autonomous cars and other products may not be the most efficient. There is also, of course, the risk of reprisals from foreign firms on Chinese firms and citizens outside of China.

On the subject of ethics and safety, in the US the technology giants of Silicon Valley have pledged to work together to ensure that any AI tools they develop will be safe. All the leading AI researchers in the West are signatories to an open letter from 2015 calling for a ban on the creation of autonomous weapons. Equivalent Chinese discussions about the limits of ethical AI research are more opaque, however. Chinese AI companies have incentive to think about some of these issues, since rogue AI would be a problem for the planet wherever it may emerge. There is a self-interest case to be made for the formulation of global safety standards, but it is hard to imagine that Beijing will lead the charge on that.

China's AI plan is clearly about maximizing AI's value to the state. AI techniques are ideally suited to identifying patterns in the massive amounts of data that Chinese censors sift through on a daily basis in order to maintain the government's grip on its citizenry. It is easy to imagine how the same data may be used to enhance the government's plans to create the previously mentioned Social Credit system that scores individuals based on their behavior and its perceived desirability. Once perfected, such algorithms would likely be of interest to autocratic regimes around the world. China's tech firms are in no position to prevent the government from taking advantage of such tools.

While Baidu preaches data transparency in public, it has been appointed by the government to lead a national laboratory for DL. That ought to cause a severe case of cognitive dissonance to AI ethicists in China and beyond.

Western firms and governments are no angels in the areas of data collection and espionage, but at least they are engaged in an open debate about the ethical implications of AI, and intelligence agencies in many other countries are constrained by democratic institutions.[19] Neither is true of China.

When, in 2014, the US Air Force wanted help making military robots more perceptive, it awarded Boston-based AI start-up Neurala the contract, but when Neurala needed money to finance the project, it got no help from the American military. The company ended up turning to China's Haiyin Capital[20], backed by state-run Everbright Group, for an undisclosed sum to support the Air Force contract. Everbright was initially owned by China's State Administration for Foreign Exchange[21], the wholly government-owned entity that manages the country's foreign exchange reserves. In 2008, it was sold to the China Investment Corporation,[22] the country's wholly government-owned sovereign wealth fund. So, *the Chinese government ended up partially funding the US Air Force's contract to make military robots more perceptive.*

In fact, Chinese firms have routinely become investors in American start-ups, particularly those working on cutting-edge technologies with potential military applications. These are companies that make rocket engines for spacecraft, sensors for autonomous navy ships, and printers that make flexible screens that can be used in fighter-plane cockpits. Many of the Chinese firms are owned by state-owned companies or have direct connections to Chinese leaders. According to the DoD, Beijing actively encourages Chinese companies with close government ties to invest in American start-ups specializing in critical technologies such as AI and robots to advance China's military capacity, as well as its economy.

US government controls intended to protect potentially critical technologies against countries like China have fallen short. It took a while, but such transactions eventually started ringing alarm bells in Washington. US lawmakers raised broad questions about the nature of China's economic relationship with the US well before the Trump administration started applying tariffs on Chinese products, in an effort to reduce the inherent inequity in bilateral trade between the two nations.

Neither the high-tech start-ups nor their Chinese investors had been formally accused of malfeasance, and some experts admit that much of the activity could indeed be perfectly innocent. Chinese businesses have money and are, after all, looking for returns, but the fund flows fit China's pattern of using state-guided investment to support its industrial policy, enhance its technology holdings, and acquire military-related technology. Yet, some start-ups—especially those making hardware rather than revenue producing mobile apps—have said that Chinese money has at times been the only available funding. If one is inclined to give them the benefit of the doubt, some Chinese investors appear to have a bigger appetite for risk and a willingness to get things done quickly, which is exactly what most start-ups need.

Although Neurala apparently made efforts to ensure that Haiyin Capital had no access to its source code or other important technological information, Haiyin's participation raised enough concern inside the Pentagon that some in the DOD argued that the US government should steer clear of the contract that it awarded to the company. To address concerns that it was not tapping a sufficient amount of innovation from start-ups, in 2015 the Pentagon set up Defense Innovation Unit Experimental, to enable investments into promising new companies. That same year, Haiyin Capital also invested in XCOR Aerospace, a US commercial space-travel company that makes spacecraft and engines and has worked with NASA. Haiyin Capital's founder later admitted that part of his firm's goal is to build Chinese industrial capabilities, noting the difficulty with which Chinese firms are able to obtain space technology from abroad because of American export controls.[23]

The Race for Faster Neural Network Processors

Part of China's race with the US for AI supremacy is in 5G wireless technology and includes Beijing's drive to achieve mass-production of neural network processors by 2020, with the intention of applying the chips to improve manufacturing, while putting into hyperdrive its move into smart cities. America's NVIDIA is the leader in neural processor chips, which are used by the world's largest tech companies to power everything from AI data centers to semi-autonomous cars. The company estimates that its total AI-related revenue will reach nearly $40 billion by 2025. China has taken notice of NVIDIA's

AI chip dominance and has specifically named the company as one that it would like its domestic companies to challenge. China's MST wants a chip that delivers performance and energy efficiency 20 times better than that of NVIDIA's M40 chip, branded as an "accelerator" for neural networks. Although it was first produced in 2015, the M40 is still used in a plethora of AI projects.

Chinese officials and tech companies each have good reason to target NVIDIA, which has provided chips for robots, drones, and autonomous vehicles, and has partnered with such auto makers as Toyota and Volvo. In response, in 2017, an investment fund owned by China's State Development and Investment Corporation led a $100 million funding round in Cambricon, a Beijing AI chip startup. Cambricon subsequently announced the creation of two server chips that could substitute for NVIDIA chips in some AI projects, while Huawei began collaborating with Cambricon to produce AI chips for phones and other devices. Plenty of other Chinese companies are jumping into the fray, with Horizon Robotics having raised $100 million and Deephi raising $40 million in 2017.

Although it wants to rely far less on foreign chip makers for commercial and military applications, China has struggled for years to make its chip industry more competitive. The US and other governments closely scrutinize proposed acquisitions of domestic semiconductor technology by Chinese companies. The US government has canceled multiple proposed purchases of US chip makers on national security grounds.[24] The concern is that cutting-edge technologies developed in the US could be used by China to bolster its military capabilities and gain a competitive advantage in strategic industries. The US government is strengthening the CFIUS, the inter-agency committee that reviews foreign acquisitions of US companies on national security grounds.

An unpublished 2017 Pentagon report warned that China was skirting US oversight and gaining access to sensitive technology through transactions that did not trigger CFIUS review, including joint ventures, minority ownership stakes, and early-stage investments in start-ups. The Trump administration recognizes that the CFIUS is outdated and inadequate for the new landscape the US faces with China and other countries. AI and similar technologies are so new that existing regulatory mechanisms related to export control and national security have not found a way to account for them. Legislation proposing revision of CFIUS guidelines would require the Committee to heighten scrutiny of buyers from nations identified as potential threats to national security. The legislation would provide a mechanism for the Pentagon to lead that ID effort, with input from the US technology sector, and the Commerce and Energy Departments.

As the legislation was making its way through Congress in 2018, a contentious issue was a provision that would give the CFIUS increased jurisdiction over a wide variety of transactions between US companies with "critical" technology (innovations that will sustain a US competitive edge in the future) and any foreign company. US technology company lobbyists tried to water down provisions that the industry deemed to be too restrictive on non-sensitive transactions, such as computer hardware sales and software licensing. Other industry lobbyists were concerned that stronger US regulations may not succeed in halting technology transfer and could trigger retaliation by China in the process.[25] The truth is that, in the absence of a lobbying effort by US technology firms, the CFIUS would have swiftly passed in Congress in 2018, lending credence to the argument that the Chinese have a distinct advantage when it comes to passing legislation and getting things done. They do not engage in a long drawn out debate about it–they just do it.

In 2018, President Trump signed into law the Foreign Investment Risk Review Modernization Act (FIRRMA), which establishes more vigilant reviews of foreign investments into American companies on national security grounds. Now that start-ups and minority investors are specifically included in the vetting process, countries such as China will no longer be given a free pass to slip under the radar undetected when making investments into sensitive areas of industry. According to the US Defense Innovation Unit Experimental, in 2015 alone, Chinese investors invested between $3 and $4 billion in early-stage venture transactions. Between 2015 and 2017, the Unit estimated that China contributed 13% of its total investment funds into US-backed companies, ranking second only to Europe as the largest foreign source of capital for start-ups.

Part of the reason this becomes so important is that China's sovereign, provincial, and local governments, along with state-owned enterprises, firms, and individual investors often form their own funds and pool their capital into each other's investment vehicles. Some have also adopted Western-sounding names, making it even more difficult to distinguish between Chinese and non-Chinese sources of investment. The US government believes that, in many cases, Chinese investments into US start-ups are not simply innocent investors seeking high returns on their capital but, increasingly, seeking information and insight into the inner workings of these start-ups.

Some analysts also believe that investing in start-ups may help prevent the ideas and technology they represent from becoming part of the US military. The DOD does not use technologies supplied by early stage companies with foreign investors, for fear that they could either share or steal information or clandestinely offer a back door into sensitive government computer systems. The implementation of FIRRMA will give the CFIUS enhanced discretion to review a host of new types of cross-border transactions in sensitive sectors and/or businesses. However, to fall within the purview of FIRRMA, investment must either include seats on a board of directors or access to sensitive material. And, just what constitutes "critical technology" will remain somewhat ambiguous. Some Chinese and other foreign investment that should fall within the purview of FIRRMA will remain outside its scope.[26]

There is certainly an argument to be made that *the US and other countries should focus at least as much on accelerating their own AI development as on restricting access to it*. Much of the work that is done on AI, ML, and DL takes place out in the open and is highly transparent. Companies such as Google and Facebook not only publish numerous papers detailing their latest ideas, but also the open source software and hardware that they use to do so. *America's best hope of staying ahead in AI is to keep alive the type of vibrant, open R&D culture that has made Silicon Valley the global hub for the blossoming of ideas and investment in the field.*[27] That may prove increasingly difficult with the tightening of protective legislation, in combination with a more competitive and combative tone having been set by Washington toward Beijing specifically, with respect to trade, and more generally.

Warp Speed

While the Obama administration sought to *increase* support for AI R&D in its final years and planned a range of initiatives to embrace AI in the future, in its first year in office, the Trump administration proposed *cutting* AI research to a variety of institutions supporting it. Funding for AI research at the National Science Foundation was cut by 10%, to a mere $175 million. By contrast, and at the same time, China is spending enormous sums on technology. In 2014, the Chinese government created a 1 trillion renminbi (US$150 billion) investment fund to turn the Chinese semiconductor industry into a global powerhouse. That was just an initial foray into the space.

While the US civilian sector races ahead of the US government in money spent and innovation achieved, and the government remains indecisive about how best to embrace the private sector to achieve AI supremacy, China has not been shy about pursuing AI in a manner deliberately designed to fuse together the most AI-capable aspects of its civilian and military sectors in AI. Beijing is proceeding full steam ahead, having created a Military-Civilian Fusion lab at its equivalent of America's MIT (Tsinghua University), to provide a formal platform for dual-use AI. In 2017, China also established its first national DL laboratory, through Baidu in partnership with three universities. Baidu has enthusiastically embraced AI technology, in essence reinventing itself around AI. As of 2017, the company had a 1,300-person strong AI team working on a range of cutting edge initiatives.

While the US government slowly ramps things up in the AI space, China's government is proceeding at warp speed, rapidly making up for any lag it has with Western firms and governments. It would not be surprising if it took the lead well before 2030. Chinese companies believe that, by rotating Chinese staff to Silicon Valley and American staff to Chinese campuses, they can accelerate the timeline for reaching parity with the US. For the time being, however, the size and experience of China's AI workforce is a fraction of that of the US. *Half of the top 10 employers of AI talent in China are US firms—including IBM, Intel, and Microsoft—who are integral to the development of China's human capital in AI.*

Governments around the world are beginning to grapple with the potential implications of AI on a range of societal sectors. As was the case under the Obama administration in the US, the Chinese plan also calls for government action to mitigate the economic pain and social instability of worker displacement. Given China's government-led transition of hundreds of millions of laborers from the agriculture to manufacturing sectors, it clearly has meaningful experience in this area, but its AI-related efforts have only just begun.

Unfortunately, under the Trump administration, the US is no longer attempting to plan for these challenges. The White House Office of Science and Technology Policy, which was instrumental in leading AI policy work during the Obama administration, has been depleted of 70% of its staff, depriving the administration of critical expertise and insights on AI, while China is using the Obama playbook to plan its own AI revolution.[28]

China's Game Plan

China lags behind the US and the UK in terms of fundamental research capability, which is a big reason why there is a shortage of talent. Despite its recent push to devote serious resources toward achieving AI supremacy, fewer than 30 university research labs in China are focused on AI, and they are unable to develop enough talent to meet the recruiting needs of China's AI industry. Chinese AI scientists have also disproportionately specialized in areas such as computer vision and voice recognition, creating gaps in some other areas. Beijing knows that, in order to turn the tide, one of the things it must do is increase the number of universities focused on AI, in order to produce more graduates prepared to contribute to the nation's AI effort.[29]

Another area where China has fallen behind, as previously noted, is its dependence on foreign suppliers for microchips. For some types of high-value semiconductors, China has had to rely on imports for virtually all of its needs. To address this, the Chinese government implemented a "Made in China 2025" policy, an initiative to comprehensively upgrade Chinese industry and become the global leader in manufacturing, while at the same time achieving self-sufficiency and reducing reliance on other countries. The policy outlines a wide-ranging strategy for harnessing and promoting the acquisition of foreign technology through outbound investment, including the use of industrial funds, state-owned capital dividends, and other channels to support the creation of advantageous manufacturing capacity to implement overseas investment acquisitions[30] and counteract China's comparative manufacturing disadvantages.

That said, many countries have become concerned that Made in China 2025 is not simply an effort by a country that lags behind to become more competitive. Governments around the world are increasingly concerned that such investments by Beijing, inside and outside of China, are not simply a product of market forces but are guided by the CCP (rather than the private sector), particularly where high-tech is concerned. Circumstantial evidence confirms this suspicion. For example, Chinese investment in the US and elsewhere has skyrocketed since 2015. *Between 1990 and 2015, Chinese investment in the US totaled $64 billion. In 2016 alone, Chinese investment in the US totaled $45 billion–triple the amount in 2015–with another $21 billion awaiting either regulatory of financing approval.*[31]

Such investments reveal a broader coordinated strategy by Beijing–to appear to have transparent and straightforward ambitions, and be playing by the rules, when not necessarily doing so. For example, Fujian Grand Chips is an ostensibly "private" Chinese company that attempted to acquire German machine maker Aixtron in 2016. Shortly before it attempted the public takeover, another Fujian-based company–San'an Optoelectronics–unexpectedly canceled a critical order from Aixtron on dubious grounds, sending its stock tumbling and presenting Fujian Grand Chips with an opportunity to purchase Aixtron for substantially less. It turned out that both Fujian Grand Chip and San'an Optoelectronics shared a common investor: a national semiconductor fund controlled by Beijing. Aixtron makes devices which produce crystalline layers based on gallium nitride that are used as semiconductors in weapons systems. The acquisition was stopped by the US and German governments at the last hour, but is illustrative of how Beijing can drive Chinese foreign direct investment in a highly coordinated manner. As the stakes associated with acquiring cutting edge AI technology become higher, such concerns can only become even more sensitive with time.[32]

Algorithmic Governance

Beijing is producing software to predict instability before it arises, based on the volumes of data mined from Chinese citizens about their friends, families, jobs, and habits. Taking a page from the film *Minority Report*, such predictive policing is used to deploy law enforcement or military units to places where crime (or anti-government political protests) is *likely* to occur. Predictive policing was actually first created and used in 1994 by the New York City Police Department, which embraced a pioneering and deeply controversial effort to pre-deploy police units to places where crime was expected to occur on the basis of crime statistics. It was highly controversial and ultimately deemed to be unconstitutional because minority youth in the wrong place at the wrong time were frequently targeted and harassed by the police. It did, however, reduce crime by 37% in 3 years.

China is putting predictive policing on steroids by rolling it out on a national level to create a unified information environment whose applications go well beyond simply sending police to a specific street corner. The government's control over the Internet places it in a unique position to extend the reach of surveillance and data collection deeply into the lives of citizens. Since Chinese authorities face far fewer privacy limits on the type of information they are allowed to gather on citizens, they can deploy police forces much more precisely. They might, for example, decide to target an individual who received and deposited a large payment in his or her bank account, or who reads pro-democracy news sites, or who is displaying a change in purchasing habits—labeling them as a "potential" threat to peace, stability, or the state.

Following the Arab Awakening in 2011, the Chinese government substantially increased spending on internal security, compelling 650 cities to improve their ability to monitor public spaces via surveillance cameras and other technologies. Collecting massive amounts of data inevitably leads to the question of how to analyze it at scale.

The Chinese government does not seek to keep its ongoing efforts to create a hyper-surveillance state hidden. On the contrary, it regularly updates (and sometimes overstates) its capabilities in an effort to maintain a sense of fear, uncertainty, and control over the Chinese people. It is based on the well-established notion that only strong authority can bring order to a potentially turbulent country.

Since President Xi assumed power in 2012, he has launched a significant upgrade of the Chinese surveillance state. China has become the world's largest market for security and surveillance technology, with analysts estimating that the country will have nearly 300 million cameras installed by 2020. Chinese buyers will purchase more than three-quarters of all servers designed to scan video footage for faces, and China's police will spend another $30 billion in the coming years on snooping-oriented hardware and software.

Government-sponsored contracts are promoting research and development into technologies that track faces, clothing, and even a person's gait. As the country spends heavily on surveillance, a new generation of start-ups has risen to meet the demand. Chinese companies are developing globally competitive applications in image and voice recognition. Start-up Yitu took first place in a 2017 contest to develop facial recognition algorithms held by the US government's Office of the Director of National Intelligence. The halls of Yitu's offices are filled with cameras searching for faces and tracking employees' paths from their desks to break rooms to exits.

China's public security market was valued at more than $80 billion in 2017 but will be worth considerably more as the country continues to build its impressive surveillance capabilities. China's national database of individuals it has flagged for surveillance includes up to 30 million people, which is too many people for today's facial recognition technology to parse. But when citizens do not know if they are being monitored, the uncertainty it creates tends to make them more generally obedient.[33] Those who are not can be quickly discerned from among the majority who are.

China has become a world leader in the use of AI and ML for national security and is leading the way in predictive policing research. It is developing systems to more easily recognize faces by compressing a Deep Neural Network to a smaller size, to enable cameras to recognize faces without calling up a distant database, and to use the datasets of thousands of disruptive occurrences to predict how and when others may occur. Beijing has the resources, will, data and inclination to turn predictive policing into something incredibly powerful.[34] No doubt this will prove to be another area of intense interest among governments from round the world.

Battlefield Singularity

As AI and robotics become more pervasive in warfare, some in the PLA anticipate battlefield "singularity"—when human cognition may no longer be able to keep pace with the decision speeds and tempo of combat in an AI-driven battlefield. That could result in humans shifting command and supervisory roles to machines. Limitations in the capabilities of current military AI systems preclude higher degrees of autonomy and automation for the time being, but there are, even now, missions and contexts in which they are either desirable or imperative—such as for air or missile defense operations. PLA planners have already highlighted the importance of human-machine collaboration, manned-unmanned teaming, and human-machine hybrid intelligence.

While the future trajectory of China-US strategic competition in AI remains uncertain, it is clear that the PLA has emerged as a true peer competitor to the US military in the AI sphere. As the PLA attempts to overtake (rather than simply catch up with or match) the US, *China's implementation of military-civil AI fusion could ultimately provide the PLA with a structural advantage by being able to rapidly adapt and integrate the latest advances in AI weaponry*. Accordingly, US competitive strategy and defense innovation initiatives should focus on a more nuanced understanding of the PLA's strategic thinking about, and development of, military applications of AI. The US (and other militaries) should also further explore the risks and advantages of developing "*counter-AI*" capabilities.[35] This is something that is likely to receive greater attention in the future.

In 2018, the PLA's Navy (PLAN) declared (presumably for propaganda or other purposes) that it was developing a nuclear submarine with AI-augmented brainpower, designed to give Beijing a comparative advantage in battle. The project involved updating the computer systems on nuclear submarines with an AI decision-support system that can think for itself, reducing commanding officers' workloads. The plan was for AI to take on "thinking" functions on nuclear subs, which could include, at a basic level, interpreting and responding to signals picked up by sonar, through the use of convolutional neural networks. The application of ML to acoustic signal processing has, at a basic level, been an active area of research in China for a number of years.

The PLAN is also pursuing the development and deployment of UUVs such as Sea Wing, a new generation of underwater glider using hybrid propulsion technology to support submarines engaged in military missions. In the future, the PLAN could seek to use such UUVs, in conjunction with submarines, in an attempt to advance its anti-submarine warfare capabilities and shift the undersea balance. As the deep-sea battlespace becomes even more complex and contested, the use of AI to support commanders for acoustic signal processing, underwater target recognition, and even more direct decision support seems plausible.

It is too soon to say whether putting "superintelligence" on nuclear submarines could ultimately unleash killer AI with nukes, but such a question must be asked and addressed. When (not if) such technology is fully developed, would there ever be a situation in which any country may consider ceding such decision-making authority to a machine? As any military seeks to use AI to enhance its response and fighting capabilities, there is always a risk that it may rely too heavily upon, or overestimate the superiority of, machine intelligence over the judgment of humans.

Past experience (such as with the Patriot missile system) has demonstrated that such complex systems can end up creating greater challenges for their operators, requiring nuanced understanding of their advantages and limitations through specialized training. In addition, the dynamic of "automation bias" can result in compromised decision-making when humans rely too heavily on automated systems, at the expense of their own judgment.

The use of AI on the battlefields of the future will necessarily create new and unexpected operational risks, which may include potential malfunctions, adversarial interference, or unexpected emergent behaviors. AI remains brittle and vulnerable to manipulation. As the world's major militaries start to rely more upon AI systems, the development of "counter-AI" capabilities to disrupt them will take on new importance. While the PLA's pursuit of decision support systems is not new, possessing the capabilities to develop intelligentized command decision-making capabilities may serve to fast forward the progress already made in developing cutting edge AI technologies. Beyond submarines, the PLAN also appears to be working on the development of systems to augment command decision-making and, at the tactical level, for the pilots of fighter jets. Having the capability to leverage AI-enabled support to command decision-making could become critical to achieving decision superiority and dominance[36] in the longer term, on land, in the sea, and in the air.

The Great Leap Forward in Quantum Cryptography

Quantum physics involves a discrete quantity of energy being produced in proportion to the magnitude of the frequency of the radiation it represents or, in plainer English, the minimum amount of any physical entity involved in an interaction. Quantum cryptography, which relies on individual photons to carry quantum information, allows communication that is guaranteed to be secure, thanks to the laws of physics. Physicists have long known that quantum computers can break almost all other types of cryptography and conventional encryption, but even the best optical fibers can carry these photons only so far (about 200 kilometers) before light absorption makes the process impossible.

That changed in 2016, when China launched the Micius satellite—the first satellite-to-ground quantum network. A year later, it smashed the record for the longest distance over which quantum entanglement had been measured, and China used its quantum network to teleport the first object from the ground to orbit. It then established the first intercontinental quantum cryptography service and tested the system by setting up a secure videoconference between Europe and China. For the first time, the security of this videoconference was guaranteed by the laws of physics, using a set of random numbers (a key) that could be used by two parties to encode and decode a message.[37] It all sounds like the realm of science fiction but China, the country which is lagging behind in semiconductor manufacturing and human talent, is breaking new ground in quantum physics. Can there be any doubt about its ability to do the same in AI in due course?

Conclusion

Many in the West may find China's way of doing business distasteful, but there is more than one way to achieve AI supremacy and, like it or not, China is well on its way to getting there. There are, after all, no real rules dictating how to do so. Similarly, those outside of China may detest the institutionalized state theft of intelligence, data, and industry secrets, but what is worse–to be stolen from or to give a thief the ability to steal? The Chinese government and Chinese companies undoubtedly see the world in these terms and would retort, if you do not want your secrets stolen, then do not allow me to steal them, for I will take what you cannot protect. If that is what they think, then they have a point.

It will do the West no good to simply complain about how China gets to the AI finish line. Perhaps it should play the game in a similar fashion. Maybe it already does and we are unaware of it. Regardless, the stakes are rather high and, in the end, it will not matter so much *how* a country achieves AI supremacy, but *that* it does so. China deserves credit for recognizing early on the importance AI fluency implies for national competitiveness in the coming decades. The West should either find a way to maintain its lead or get out of the way, because nothing will stop Beijing in its quest to achieve AI supremacy.

Daniel Wagner and Keith Furst

12. *International Relations*

<u>Playing Leapfrog</u>

International relations are a complex mosaic of bilateral and multilateral relationships among nations, often with high stakes and outcomes. How, and whether, countries choose to interact with one another has direct and indirect impacts on every one of us. International relations were, not so long ago, a rather straight forward proposition–policy makers took primarily political, economic, and security considerations into their calculations about what their national policies toward the world should be. That gradually changed in recent decades, as variables that either did not previously exist, or were deemed not significant enough to warrant inclusion in the crafting of such policies, became an essential component of national policy making. While rising global socioeconomic inequality, growing economic nationalism, terrorism, and unliteral political actions are all having a significant impact on the nature of how countries interact with one another, climate change and cyber risk have gradually worked their way into what diplomats and policy makers take into account to determine their moves and counter-moves on the global diplomatic chess board.

AI is similarly disrupting international relations today. China is developing an AI "policymaker" to support its crafting of foreign policy. China's system, also known as geopolitical environment simulation and prediction platform (GESPP), has already been used to help vet proposed foreign investment projects. GESPP has access to a variety of Chinese government databases and is equipped with DL and neural network tools to help predict and determine levels of risk for such events as political upheaval or terrorist attacks. GESPP remains unable to make a strategic decision by itself, but the next generation will have the support function to do so.

GESPP studies the strategy of international politics by drawing on a large amount of data that can contain information varying from cocktail-party gossip to images taken by spy satellites. The system provides policy makers with a range of options, along with recommendations for the best course of action. For Chinese leaders, the appeal of the GESPP is that it will be immune to fear, morality, passion, or other subjective factors that might "interfere" with strategic objectives.

The presumption, going forward, is that international actors who do not benefit from making decisions with the assistance of AI will be at a considerable disadvantage with respect to certain aspects of strategy selection, risk management, and execution efficiency. An early version of the AI system is already in use in the Chinese foreign ministry. The Chinese government does not expect that it will replace human decision-makers but will continue to support them. Researchers in China believe that US diplomats are using AI to advocate policy positions, promote awareness, and enhance transparency. Other advanced nations are believed to be doing the same.[1]

As was aptly illustrated in Chapter 11, some countries are pursuing AI more or less within the confines of international law and generally accepted principles of doing business, while others are choosing to do what is necessary to attempt to achieve AI supremacy outside those boundaries. In the process, AI is slowly altering the balance of power between global actors and among alliances in a number of ways.

Just as becoming adept in the cyber arena levels the playing field—giving countries such as Iran and North Korea the ability to go head to head with China, Russia and that US in cyber space—the pursuit of AI supremacy is providing an increased competitive edge in international business to some smaller, otherwise less competitive nations, enhancing their ability to secure preferential trade and investment arrangements with other countries, raising their global profile, and enabling them to progress into previously unimagined areas of international trade, investment, and diplomacy.

A 2017 study[2] took a deep dive into global competitiveness and noted that 46 countries account for approximately 80% of the global work force. It determined that the percentage of work activities that could be automated by adapting existing technology was highest overall in the developing world, with countries such as Kenya, Morocco, Thailand, Indonesia, India, Mexico, Costa Rica, Peru, and Colombia all theoretically the likeliest to be able to become more economically competitive by embracing cutting edge technology. Naturally, that comes with a trade off in terms of potentially higher unemployment rates and reduced tax revenue collection, but the point is that the rise in work place automation and greater productivity rates is not merely the domain of the developed world. Developing countries also want a seat at that table and will no doubt raise their collective voice in coming years to ensure they are not left behind.

AI also has the potential to shake up the established relations and practices of the world's leading security organizations. Some tech-savvy European NATO members' countries are already making efforts to compete with Silicon Valley to bridge the AI capabilities gap by combining smarter defense spending with the agile integration of available innovative technologies. These states are developing a better understanding of AI's potential, focusing on integration of AI technology, and supporting foreign-based start-ups that can produce customized technology that they may also use. Few countries that have the monetary resources and technological capability to bring AI to the forefront of their economies are failing to do so today—the question is the degree, speed, and impact with which they are doing so.

AI is also leveling the playing field for non-state actors, just as cyber and social media capability have done. The IS has turned virtual jihad into an art form through sophisticated use of social media to communicate with millions of adherents around the world and non-state actors have already proven their adeptness in conducting cyberattacks. Terrorists and extremist organizations can develop or acquire AI technologies for combat purposes or to attempt to influence the effectiveness of their opponents.[3] The use of AI-enabled drones to target individuals, or swarms of drones to target troops, is already being used on battlefields by non-state actors. In the 2017 battle for Mosul, the IS dispatched small drones armed with grenades to harass Iraqi forces attempting to retake the city.

On the streets and on battlefields, autonomous vehicles may well be used by non-state actors to conduct car bombings and hit and run assassinations, making terrorist groups more effective, and deadlier, in the future. The adept use of social networks built using AI capabilities will undoubtedly make it easier for such groups to conquer towns and cities by creating their own digital intelligence agencies. The ability to purchase AI technology on the Dark Web is already having an impact on the reach and potency of terrorist groups and rogue states alike, including extortion and kidnapping through the automation of social engineering attacks.[4]

While it is only natural for state and non-state actors to want to defend their turf and protect any technological or strategic advantages they may have developed in the AI realm, the temptation to try to maintain a "winner takes all" approach can be a trap. Many nations–large and small–will have comparative advantages, such as scale versus agility and deep pockets versus vast work forces. To the extent that allies can find a way to securely share AI technology for their collective benefit, they may end up being better off. Since China will surely not be sharing its AI competitive advantages but is likely to achieve AI supremacy in the next decade, it may behoove NATO nations to try to catapult themselves ahead of China by pooling their resources. Just as there can be no single winner in the cyber arena, there will not be only one in the AI arena. Those countries that understand this will find a way to marshal their resources to maximum advantage and leapfrog ahead of those that do not.

Data Warfare

How AI is deployed by governments can have severe consequences in international relations, particularly if the government in question has unusual capabilities in the AI arena. Access to, and the use of, autonomous weapons, for example, can potentially change the global balance of power. Challenges to a region's balance of power may occur as some states move to leverage AI technology to reverse historic military disadvantages vis-a-vis their neighbors. Those governments that choose to embrace AI responsibly and ensure that humans remain the ultimate arbiters of life and death decisions may be admired, but they may also be putting themselves at a strategic disadvantage by doing so. Pursuing AI on the military battlefield with ethics in the mix may prove to be a luxury few countries can afford, since not all countries will do the same.

There is, in addition, great danger that AI powered military systems and military led decision-making will eventually undermine existing approaches to conflict containment and de-escalation. The institutions and treaties designed to address 20th century foreign policy, arms control, and non-proliferation were never intended to apply to a world order incorporating AI. Perhaps the greatest threat from AI weapons will ultimately come from non-state actors (further explored below). *The cost of AI weapon deployment (via drone deployment) is already low enough to fall within the scope of even unsophisticated terrorists, which implies an increasing degree of symmetry between national militaries and non-state actors, going forward, in some aspects of warfare.*

Potential attack vectors are as diverse as there are uses for ML algorithms. As is the case in the cyber arena, attackers may have a strategic advantage over existing defense architectures. It is unlikely that there will ever be another conventional military conflict that does not contain components of information and cyber warfare; AI will play a central role in how these new forms of weaponry are deployed, enhancing tools of network penetration and exploitation. Future AI cyber operations will undoubtedly operate as if they are under a constant state of attack– seeking to penetrate as many networks as possible, then lying in wait for opportunistic exploitation. This could lead to the autonomous stockpiling of software vulnerabilities (e.g. zero-day attacks) and the proliferation of malicious code deployed in ways never previously envisioned (or subsequently controllable) by human designers.

AI-enhanced cyberattacks will have an asymmetrical advantage over exclusively human operators, automatically determining the most effective attack and defense vectors. In addition to expanded arenas of cyberattack, there will likely be a broader set of AI-driven information operations designed to deceive, disrupt, and distort public communications in enemy states. The effectiveness of AI-empowered disinformation campaigns, as exhibited[5] by Russia during the 2016 US presidential campaign, have already led to a significant re-evaluation of election security and the integrity of the public debate in democracies.

In the future, "data warfare" may include virtual battlefields between forms of AI seeking to disable one another, automatically infecting command and control systems with disinformation and/or malicious code. It may include sophisticated media forgeries developed by AI, designed to induce opposing populations into relying on falsehoods or acting contrary to their interests with maximum efficiency and impact. The "taste" of this that has been unleashed via fake news and disinformation campaigns in the US and Europe is nothing compared to what such actions may look like in the future—*damage to the integrity of democratic discourse and the reputation of state institutions and their representatives will be easier to inflict and more difficult to repair.*

This puts arms control and non-proliferation strategies in a whole new light and implies a need to align common foreign policies among allies in order to deter such actions by enemy states and nefarious non-state actors. There is too little understanding inside governments and among diplomats about how these technologies function, however, and what options are available to counter them. Governments have begun planning and investing for the AI future, but none has yet developed and articulated actionable red lines about how AI technologies may be used according to the norms of existing international law or in the context of human rights. Going forward, this should become a multinational effort, but there are more questions than answers about how, when, and in what manner a realistic and effective coordination of efforts may hope to achieve.

Even taking into account the breathtaking speed at which AI is being developed in a variety of realms, there remains a window of opportunity to begin coordinated efforts in earnest aimed at making progress in the arena of international relations (among others), to raise awareness about the pressing need for coordination on the ethical use of AI. For the foreign policy community, multilateral arms control for technologies utilizing AI will be at the top of the list. *A new set of guidelines governing its use in the context of dual-use, enforcement measures, and export controls are among the issues that must eventually be addressed.* This usually only entails government-to-government engagement but, given the nature of how AI is being developed internationally, government-to-business engagement to design safeguards will also have broad applicability in the future.

Given the growing challenge of disinformation and divisive propaganda, central to this notion will be the development of standards for what constitutes trustworthy communications among states and peoples. One objective would be to design open spaces for public communications that are free from the possibility of exploitation (assuming that is possible, of course), creating a widely accepted methodology for dispelling disinformation. This would require a willingness among states to share information about the use of AI-powered tools of digital deception and best practices to counter them. Perhaps this is nothing more than wishful thinking, for, as discussed earlier, comparative advantages using AI are, and will continue to be, as important as cutting-edge military technology is today.

The emergence of AI technologies poses serious challenges to the notion of strengthening democratic institutions and protecting social equality, since AI-enhanced surveillance practices can constrain civil rights and liberties, and sociocultural conflict can be exacerbated via the perpetuation of social bias and discrimination rooted in AI algorithms. It cannot be the task of foreign policy to design and implement checks and balances on the surveillance practices of security agencies, but it is within the domain of international diplomacy to communicate these policies with the rest of the world.

Research and analysis about the intersection of AI and foreign policy is embryonic and commentary on the role of diplomacy and statecraft is scarce. Given this, the best path forward in its development may be to build upon the successes (and learn from the failures) that "cyber" foreign policy has provided.[6] In most countries, foreign policy has already incorporated cyber issues to generate a foundation upon which to add an AI platform. Bringing statecraft up to speed in the Internet age is a significant achievement accomplished in a relatively short period of time. The same must be done for AI, but we do not have the luxury of waiting any longer to get started.

<u>Autonomous Weapons vs. International Law</u>

The development of lethal AI is posing some fundamental questions for the international legal regime and generating challenges for the stability of the international system as it begins to alter fundamental security dynamics. Leading scientists active in AI have long argued that the militarization and use of lethal AI would be a highly destabilizing development, as such weapons represent the third major transformation in the nature of war (the first and second transformations being gunpowder and nuclear weapons). They have warned that the deployment of such weapons will alter the future of peace and security for the worse, and they have advised that a lethal autonomous arms race could be prevented by imposing a ban on offensive autonomous weapons.

Their argument is a simple one: if any major military power pushes ahead in a significant way with AI weapon development, a global AI weapons arms race will be inevitable, and the technological trajectory of such a race is that such weapons could become the Kalashnikovs of tomorrow. Unlike nuclear weapons, they argue, *AI weaponry is neither costly nor derived from difficult-to-obtain raw materials, so they will become ubiquitous and inexpensive to mass-produce*. It will presumably only be a matter of time until they begin to appear on the black market and in the hands of terrorists, dictators, or warlords. They view autonomous weapons as being ideal for destabilizing nations, subduing populations, and carrying out assassinations.

Just as most chemists and biologists have no interest in building chemical or biological weapons, most AI researchers have no interest in building AI weapons—and do not want others to tarnish their field by doing so, potentially creating a public backlash against AI that curtails its future societal benefits in the process. Chemists and biologists across the world have broadly supported international agreements that have successfully prohibited chemical and biological weapons, just as most physicists supported the treaties banning space-based nuclear weapons and blinding laser weapons.[7] The same should be true of autonomous weapons, they argue, but has the genie already left the bottle? Can international law even address this in the context of international relations, and is it already too late to prevent the inevitable?

A 2016 review of 22 existing international treaties[8] noted that they were generated under a "precautionary framework", to establish new regimes to prohibit or control weapons systems that had been deemed to be destabilizing. These treaties—which ranged from establishing nuclear weapons free zones, to the prohibition of the placement of weapons on Antarctica (and on the moon), to the making of unacceptable weapons unlawful—achieved one or more of the following three goals:

1. Preventing further militarization or arms races,
2. Making weaponization unlawful, and
3. Stopping proliferation via the establishment of cooperative frameworks of transparency and common rules.

It is clear that a norm has been established with regard to lethal weapons systems: the utilization of disarmament and arms regulations as a tool to protect civilians. *The unchecked development of lethal autonomous weapons systems poses a potentially severe challenge to this global norm.*

This architecture of peace and security is ultimately built upon three pillars, all of which are the result of the intense codification of international law through treaties and cooperation among states. The first pillar is the regulation of war and the prohibition of the use of force and the peaceful settlement of disputes between states and non-state actors. The second pillar is composed of the complex network of global norms that constrain and guide the behavior of states. This includes an extensive array of treaties that form numerous governance regimes and comprise an elaborate framework of transparency, confidence building, and security mechanisms put in place to maintain peace and security. The third pillar is built upon a structure that fosters cooperation in cultural, economic, social, and environmental matters that affect humanity and address problems that must be solved collectively. Let us examine these in greater detail.

The first pillar is anchored on two broad norms that guide states: the peaceful settlement of disputes through courts and international organizations. The rise and use of lethal autonomous weapons raises two important questions: should a lethal autonomous arms race be stopped as a preventative measure before it is started, and should the world's people allow autonomous weapons to be empowered to kill within this framework? To simply contend that existing international legal frameworks are sufficient is a complacent response, since lethal autonomous weapons will make war easier to wage and more tempting to pursue.

The adoption of lethal autonomous weapons in warfare implies establishing a lower threshold for the use of force and has been carefully crafted upon international law since the founding of the United Nations (UN). The deterioration of such norms could lead to an increase in violence between states that do not have autonomous weapons systems and will make fragile conflict situations more precarious. The creation and use of such weapons may also give the leaders of countries that are technologically superior the mistaken belief that they can use such weapons in disregard of the long-established global rules that all actors must legally abide by. So, *the unlawful use of force, or the severe erosion of the existing thresholds for the exercise of force in international relations, will likely contribute to a weakened system of the rule of law, where the resolution of disputes through peaceful means should take precedence.*

The second pillar, the web of global norms that states must abide by in the pursuit of the maintenance of peace and security, is based upon the widely accepted understanding that some behaviors that were considered "acceptable" before the creation of the UN–such as piracy, torture, slavery, and the annexation of territory through extra-judicial means–are now considered illegal. Human rights law (HRL) and international humanitarian law (IHL) are primary areas of concern vis-à-vis the use of lethal autonomous weapons and how these weapons will disrupt the regulation of war and conflict under the rules of the UN Charter.

For both IHL and HRL, accountability for actions during violent conflict is a central premise under the law. The arbitrary deprivation of life and of basic human dignity is not acceptable under current international law. Being killed by an autonomous machine appears likely to violate that basic precept. The universal global norms protecting human rights through humanitarian law are foundational to post-war international relations and form a common legal code that is broadly adhered to. Article 36 of the UN Charter mandates that states conduct reviews of new weapons to assess where they are compliant with IHL; however, only a handful of states routinely conduct weapons reviews, which raises questions about just how seriously the world's nations would regard future restrictions on autonomous weapons in the context of HRL and IHL.

The third pillar, the pursuit of peace and security, is largely comprised of efforts to foster cooperation in cultural, economic, social, and environmental matters that affect humanity to address problems that can only be resolved collectively. Article 26 under the UN Charter calls on all states to create a system regulating armaments that focus on the promotion and maintenance of international peace and security and requires that they ensure that as little of the world's human and economic resources are used for the creation and maintenance of armaments as possible. Only the prohibition of the production and use of autonomous weapons would be consistent with this Article.[9]

Since international security depends in part on the international regulation of armaments, all states potentially benefit from their control or prohibition. This premise is embedded within the concepts of institutionalism, security regimes, and non-proliferation. Greater cooperation and transparency in the domain of new lethal weapons technologies implies greater peace and security, just as less coordination and the absence of clear rules applying to the world's nations implies a more insecure world.

Preventive Security Governance (PSG) is the codification of global norms, based on international law, which clarifies in advance what is considered universally agreed behavior regarding issues that have not been formally addressed by international bodies. Since there is no such agreement on the proper governance of autonomous weapons, PSG may be a way to establish a starting point for formally codifying such governance. The international community must ultimately address how the law protects civilians, who is to be held accountable and deemed responsible for actions that occur that are in contravention of the law, and how they will be prosecuted.

The best path for creating security governance frameworks addressing future lethal autonomous technologies is the formation of an international treaty that will preventively prohibit them, but, as noted in Chapter 13, the international legal regime is too far behind the curve to imagine that such a treaty will be created any time in the near future. PSG could prove to be a very useful stepping stone in that direction, since there is no clarity in the existing global legal/political architecture to ensure that autonomous weapons technologies coexist in a manner consistent with global peace and security.

As time goes on, the magnitude of the problem will become ever clearer. Since robots cannot discriminate between civilians and combatants, who will bear responsibility when non-combatants are killed by autonomous weapons– hardware manufacturers, software programmers, military commanders, the subordinates carrying out the commands, or nation-states? The issue of moral responsibility looms large in the background, even though none of the contributors to the existence and use of autonomous weaponry is addressing this in a meaningful way. Similarly, the absence of accountability is exacerbated by the absence of transparency in the development and potential use of such new technologies. Moreover, the vagueness and lack of precision in applying the existing international law to this subject are contributing factors in undermining the rule of law upon which international security depends.[10]

AI and Nuclear Strategy

AI has already disrupted a plethora of aspects of international relations–from national security policy to foreign affairs to nuclear strategy among the small handful of the world's nuclear nations. The effect of AI on nuclear strategy depends as much or more on adversaries' perceptions of its capabilities as on what is technically within its scope of action. Nuclear strategists have therefore added their opponents' uncertainty about just what is possible and what is not into their operational playbooks. A capability that is nearly effective could be perceived as even more dangerous than one that has been proven to work.

The pace at which AI is being developed, combined with complementary information technology, will clearly have a significant impact on nuclear-security issues in the coming decades. AI technology could continue to evolve at a rapid trajectory or it could plateau once current technology matures.

Some theorists envision the evolution of "superintelligence", wherein machines develop the ability to self-improve their own intelligence, resulting in abilities that humans can neither comprehend nor control. Others expect breakthroughs to be followed by setbacks, or that progress will remain incremental. Either way, maintaining strategic nuclear stability with the inclusion of AI and ML in the coming decades will require revisiting the foundations of deterrence theory.

In 2015, Russia revealed that it was developing the ultimate "killer robot": a nuclear-powered undersea drone designed to carry an enormous thermonuclear warhead. Shaped like a large torpedo and powered by a compact nuclear reactor, its Oceanic Multipurpose System Status-6 was claimed by Russian President Vladimir Putin to be able to overcome enemy defenses through a combination of speed and range that would enable it to outrun almost anything in the ocean. The drone was said at the time to be capable of being launched from submarines in the Russian arctic, traverse swiftly through the ocean, and autonomously circumvent antisubmarine defenses to deliver its deadly payload to the US coastline. Was the claim fact or fiction?

The nuclear drone claim was the manifestation of Russia's ongoing concern about the credibility of its retaliatory forces in the face of US targeting and missile defense capabilities. Unable to match these capabilities in kind, Russia hoped to be able to exploit AI to make up the difference. The claim, which many theorists believe was deliberately leaked by the Kremlin to raise question and/or elicit a response about America's own capabilities in this area, was an effort to maintain its post-war strategy of developing asymmetric responses to superior US capabilities. Status-6 was a stark warning that, if technological progress undermines nuclear capability, AI-operated technologies would overcome enemy defenses through a combination of speed and range.

While AI has the potential to exacerbate tensions among different aspects of nuclear strategy, it could prove similarly capable of alleviating the likelihood of conflict while actually enhancing strategic stability. For example, despite the prevalence of mutual distrust, nuclear states could become motivated by self-interest to share AI technology specifically to prevent the unthinkable from happening. While this may appear unlikely, given the desire to perpetuate competitive advantage in all spheres of military competition, AI systems may develop capabilities that would be less error-prone than human-driven alternatives, creating a stabilizing effect in the long term.[11]

As AI advances, ML systems might be able to control military exercises at superhuman levels. Once that capability has been demonstrated, the recommendations of robots may come to be treated on a par with humans, unleashing new forms of risk that have not even been considered at this juncture. It is already common for individuals to rely on AI to make routine decisions when driving, facilitate scheduling tasks, and respond to e-mails. Such successes may build sufficient confidence to prompt political and/or military leaders to rely on AI-based decision-making for the most complex and dangerous of tasks. While difficult to imagine today, based on what is publicly known about this technology, in the next decade it may seem like a perfectly natural next step.

NATO's Adaptation Initiative

In 2017, NATO acknowledged that, if it did not take bold steps, it risked falling behind the pace of political change and technological developments that could alter the nature of warfare, the structure of international relations, and the role of the Alliance within it. NATO acknowledged that it had failed to recognize the extent to which the human capacity to think, decide, act, and recover in a future war (a so called "hyper war") was rapidly changing the character of conflict and warfare. The Alliance concluded that it was in danger of rapid decline if it failed to embrace and exploit the full scope of technological advances in the art of waging war.

Hybrid warfare involves utilization of the full spectrum of warfare—both conventional and unconventional—so as to promote ambiguity and confusion about the scale and nature of the threat, as well as the necessary responses. This can include cyber warfare, fake news, and, of course, AI. The resulting inability to determine exactly where a threat emanates from, precisely what methods were used to inflict damage, and the scope of such damage, can create confusion that can easily paralyze decision-making and dislocate increasingly diverse societies, the cohesion and resilience of which may no longer be taken for granted.

AI SUPREMACY

In response, NATO's GLOBSEC Adaptation Initiative included creation of a bespoke Hyper War Center of Excellence (HWCE), to generate a coherent approach to future war, expand cyber defense capabilities, and enhance its AI capabilities. NATO recognized that it urgently needed a coherent approach to the development of AI, so the HWCE will become an important part of what the future NATO will become. NATO wants a future war strategy that fully-integrates hybrid warfare, cyber war, counter-terrorism, hyper war, and the continuum between them, so NATO decided to leverage the impact of new technologies into the security and battle spaces.

The idea is to use AI to help create Big Data solutions to assist NATO in streamlining its decision-making process. Humans would remain at the center of decision-making to ensure that virtual barriers are built into AI, but intelligent machines will play an ever more important role in the conduct of warfare. The command structure needed to be able to encompass operations across all the domains of contemporary warfare, including cyber warfare, information operations, and hybrid threats, while ensuring the heightened situational awareness needed for the Alliance to respond quickly enough to short-warning and ambiguous hybrid attacks.[12] This is admittedly a tall order and will take many years to attain the capability to do so effectively, but NATO is now firmly on that path.

Palo Alto or Paris?

In 2018, 24 European nations created the Joint European Disruption Initiative (with the cool Star Wars-esque acronym JEDI), because Europe knows it is also falling far behind in the race for AI supremacy. European countries acknowledge that, if the continent does not move forward swiftly with innovation, execution, and smart regulation, it risks becoming irrelevant with respect to AI. JEDI called for creating a €1 billion (US$1.2 billion) pan-European fund to promote projects leading to technological disruption in the AI space originating from Europe. The EU's unique form of bureaucracy and sclerotic approach to decision-making implies that it will take JEDI some time to become fully operational, but the important thing for European leaders is that a formal action has been taken to attempt to shift the balance of power away from the China and the US and toward Europe.

Europe's challenge is not to let petty politics get in the way of innovation and implementation, which will indeed test the continent's resolve. Can European countries prove that they can work together to transcend specific national initiatives for their collective benefit? While Europe has proven that it can marshal its resources politically, financially, socially, and military, what is at stake now is future competitiveness on a grand scale. Europe's inclination is to ensure that its efforts are consistent with its orientation toward ethics and governance—but who has time for that? The race is on, and countries such as China and Russia will not be factoring social justice into their AI equations. JEDI is calling for a DARPA-style effort focused on major technological breakthroughs—not ready-to-use or commercial innovations but, rather, the next generation of technology in areas such as biotech, quantum computing, and cyber-security.[13] To be successful will require maximum speed and agility in order for the necessary investment to be made, efficiently and effectively.

Some European governments are not waiting. In 2018, the French government announced that it will spend €1.5 billion (US$1.85 billion) over five years to support research in the field of AI, encourage startups, and collect data that can be used and shared by French engineers. The goal is to erase 30 years of underperformance in innovation and begin the process of catching up with China and the US in an effort to ensure that the smartest minds in AI choose to reside and work in France, over Palo Alto. President Macron wanted France to specifically embrace the disruption that comes with becoming a force to be reckoned with in AI development.

While it is clear that France itself has a lot of catching up to do, that it will be unable to match the fiscal resources China and the US are able to throw at AI development, and that venture capital exists on a much smaller scale in Paris than in Shanghai or Palo Alto, part of the plan is to lure French AI researchers back to the motherland. Macron's intention was to leverage the cash to develop sector-specific AI technology in areas where France has an edge, given its giant state-run agencies and troves of centrally collected public and private data. For example, medical records amassed by France's state-run hospitals are a data goldmine that has already fueled AI-powered medical technology.

Some of the money to be dedicated to this task will be drawn from a €300 million fund for radical innovation announced in 2017 by the French Finance Ministry. Financing would also come in part from the European Commission,[14] but France believes it has a number of major advantages which set it apart from the rest of Europe. Among them are dynamic ecosystems of innovation which have fostered numerous French high-tech start-ups and major industrial groups, a number of which constitute natural outlets for innovation transfers in the field of AI.[15]

The question is whether France's bold AI strategy can succeed as the Chinese and US goliaths are proceeding apace with their own ambitions for AI supremacy. Germany and UK are the only other European countries that could give it a run for its money, and it remains to be seen how their AI ambitions may unfold. That said, Macron obviously realizes that, going forward, AI fluency implies significant impact on France's ability to continue to be in a position to project its economic and political power and punch well above its weight, as it has for decades. Should Macron be successful in luring all that French AI talent back from Palo Alto (and elsewhere), France may well succeed at becoming, at a minimum, the preferred European center for AI research and development—to the extent that it chooses to share its technological advantages with its European brethren. That is far from certain, however.

Moreover, for Europe as a whole to become a leading AI power, it will have to find a way to transcend its own self-imposed constraints with respect to collective governance. While common legal and regulatory frameworks can enable companies to scale beyond their home countries in more conventional endeavors, such as manufacturing and communications, achieving open cross-border flows of proprietary technology is another matter entirely. Before Europe may hope to compete successfully on the global AI chess board, a way must first be found for it to harness its collective comparative advantages openly and transparently—otherwise, European countries will be no different than any other players striving to achieve AI supremacy.

Robotically Enhanced International Relations

Politics has always been inextricably linked with economics, finance, security, and other sectors of a society, but before the era of AI and ML, countries' fortunes, and therefore their politics, were never previously linked with robots. Today, AI is being used to predict economic trends and robots are being used to drive the cost of labor down, to enable nations to effectively compete in the global marketplace. That will ultimately have a profound impact on how countries craft their bilateral and multilateral strategies going forward, and global politics will be influenced by which countries think the most creatively, effectively deploying their comparative advantages to attain and retain AI supremacy.

We are accustomed to thinking about conventional issues in conventional ways, but what if, instead of thinking about robots merely as a means of streamlining production processes and cutting costs, we began to think of them as a way to potentially impact international relations? AI-enabled countries are gravitating toward more public-private partnerships in an effort to enhance their global competitiveness. In the process, some of them are (directly or implicitly) using AI to effectively marry their economic and political fortunes. Consider that robots could account for as much as 20% of global manufacturing by 2025.[16] The potential use of robots on a mass scale, to do things they have not previously done, deployed in a manner they have not been previously deployed, could, in time, have a profound impact on international economic and trade relations.

As an example, in 2015, the South Korean government made a US$15 million investment in Samsung to develop factory robots that can carry out complex tasks, normally reserved for nimble human fingers, specifically to compete with cheap labor in China. The robots will eventually make products such as mobile phones and other consumer electronics that require a level of precision normally associated only with human hands. Should the project prove to be a success (and there is no reason to believe it would not be), the company plans to provide the blueprint for mass-produced, six-axis vertical articulated robots, which would mean that South Korean companies will become less reliant on cheap labor to make competitive products.

AI SUPREMACY

AI-enabled China could decide to "retaliate" against South Korea (and other countries with similar initiatives) any number of ways. One of its most powerful tools is access to the global marketplace, so China may in turn choose to introduce a "robot tax" on products that were made outside of the country by robots. This could, in turn, trigger tit-for-tat tariffs against Chinese products. However, since approximately 25% of South Korean exports (worth approximately US$150 billion) in value are sent to China, which would become a costly path to go down for both countries. And yet, taxes or tariffs derived from robot-enhanced manufacturing could well become the new trade normal as soon as the coming decade.

Another alternative for South Korea might be to shift the core of its export base away from China and toward another equally populous country–India–which is a natural competitor and adversary of China. If Beijing chose to play hard ball against Seoul, it might offer to make India a new manufacturing hub for South Korean products, while opening its markets to Indian exports on a preferential basis. Doing so would also be consistent with Prime Minister Modi's "Make in India" policy[17], giving his government an incentive to capitalize on such an offer. AI, and the robotic manufacturing power it can support, may thus turn out to seriously impact multilateral trade relations, and by extension, regional and global politics.

Should AI-enhanced manufacturing, and the resulting robot taxes, become the norm, it could transform the ability of nations with aging populations and/or small labor forces to punch above their weight in the global marketplace. Nations whose work forces may have previously been limited to low added value textile production could find that producing electronic components is feasible and affordable. The flip side, of course, is the loss of low-end manufacturing jobs in some of the world's poorest countries, with a likely end result being labor strikes, opposition rallies, and unanticipated political change.

But it would be unreasonable to presume any given outcome without the benefit of knowing how specific governments would react to AI-enhanced manufacturing, what resources to influence positive change may be available to them, and whether other countries may wish to help them "transition" to an AI-dominated landscape. Would even freer trade and unanticipated prosperity turn out to be a result, or would the result simply turn out to be more protectionism, economic nationalism, and greater misery among the poorest of the poor?

The answer may seem obvious but may not be. What if wealthier nations chose to begin sending aid to countries that use AI to transform their economies, and what if that aid was predicated on ensuring that employees that were made redundant because of AI were retrained to give them skills that not only helped ensure they can get another job, but higher pay, which would mean more money to buy goods and services domestically while also implying that the government would have a better ability to raise its tax base in the longer term and become a stronger player in the global economy as a result? In this way, AI has as much potential to contribute to a country's well-being as it does to its downfall. Much will depend on how wealthier nations choose to respond to the coming AI manufacturing revolution.

Of course, there are many other potential applications for how AI can impact international relations. Algorithms already exist that give stock traders distinct advantages in timing trades or forecasting market movements in advance. In 2015, a Hong Kong-based hedge fund called Aidyia created an AI-based trading system that crunches data and makes trades based on suggestions from other AI platforms. If a government were to set parameters on such an AI fund (by, for example, restricting where it can trade or invest (i.e. only outside of Hong Kong, in this case)) and put a cap on the maximum amount it can invest, the countries in which the fund invests may choose to impose similar restrictions on its own AI-powered funds, artificially impacting who can trade, when, and for how much.

Apart from competing against one another's countries, such restrictions would have a (presumably) negative impact on AI-driven investment strategies, with the intention of limiting the impact of AI-driven investing at home. While human investors have biases, instincts, and emotions, AI does not. What would happen to bilateral relationships on a national level when an AI-driven trading program successfully crashes another country's stock market?[18] It seems unlikely that any national securities regulator has contemplated this and has a previously planned response ready to be swiftly implemented.

AI SUPREMACY

Predicting the Future

Diplomats, pundits, and political risk analysts have been trying to predict what will happen inside nations and between them for as long as the nation-state has existed. Political risk analysts have been trying to do so as part of the formal professional discipline of country risk analysis roughly since the 1970s, when public and private sector insurers began insuring against non-commercial risk by providing political risk insurance to traders, investors, and lenders. That has now become a multi-billion-dollar industry, complete with consultants, advisors, information providers, crisis management specialists, lawyers, and a host of other service providers all in the business of predicting when and how political change will impact cross-border trade and investment.

In recent years, the country risk analysis business has begun to gravitate toward using AI to enhance how the present is, or will, impact the future. Mostly new organizations, but also some established ones, are using predictive tools to enhance their craft. Although it is too soon to have established a track record based on long-term performance, what is clear is that AI is adding real value in the country risk analysis business— no easy task in a field where gut feeling and intuitive powers are as useful as historical precedent and past performance, if not more so.

While a number of AI-oriented start-ups are in the process of being launched, a few have already been in existence for a short time and have proven that they can receive funding from venture capitalists based on their ability to build on the strong foundation that the more conventional country risk businesses have established. Among them, GeoQuant quantifies political factors that were previously thought to be unmeasurable, identifying and comparing risks across markets and time. Its software scrapes the Internet to collect large volumes of data that are highly relevant to conducting political risk analysis. Doing so allows decision-makers to incorporate politics into investment strategies and asset allocations using real-time data rather than static indicators. Another recent entrant is Hozint, a threat intelligence platform providing political, security, and safety alerts in real-time, combining artificial and human intelligence. It has a global monitoring service that provides travel security and crisis management alerts through mobile applications. There are a host of other such service providers active in the space.

Some of the heavy weights of the industry are also moving in the direction of integrating AI into their platforms. Among them, Eurasia Group launched a subsidiary, in 2018, called egX, integrating its proprietary data-driven tools with strategic partners in the Big Data space to marry more traditional approaches to country risk analysis with emerging AI-technology. No doubt, other large players in the space will follow in its footsteps. This is clearly of great potential benefit to traders, investors, and lenders who commit large sums of money around the world. Their ability to peer into the future with less uncertainty and a greater degree of confidence is invaluable.

No matter what degree of future progress is made in this arena—and it should be considerable in the coming decade and beyond—country risk analysis appears to be one of the areas where the amount of potential value to be added by AI will be both enormous and limited in scope. That is because no machine can integrate human gut feeling, sixth sense, or instinct into its output. These are critical to getting predictions about the future direction, policy, or action of a country right. It is heartening, however, to know that the country risk business has taken a bold leap into its own AI future.

The Path Forward

With so many potential opportunities for AI to impact international relations, and with more questions being raised about its implications rather than answers, it seems surprising to some that AI has not featured more prominently in crafting foreign policy and influencing international relations. Others may think that, as an embryonic technology with emerging applications, too little remains known about how it may affect national relationships to warrant in depth study at this juncture. In most liberal democracies, the primary purpose of foreign policy is to promote and strengthen democratic institutions that protect social equality and representation around the world while pursuing a (human and civil) rights-based system of governance, commerce, and security within the international community.

AI SUPREMACY

The advent of the Internet era demonstrated the tension that exists between security considerations and the freedoms implied via connectivity (such as freedom of speech, movement, and congregation). AI has already aggravated this tension as enhanced government surveillance and censorship capabilities reach new levels of intrusiveness, especially in the name of national security. The challenge for foreign policy in the AI era will be to promote an enlightened agenda as such surveillance capabilities and hyper-competitiveness continue to rise. Foreign policies that leverage the existing tools of diplomacy while encouraging a responsible, thoughtful, and systemic adaptation of AI will, in the end, not only be more widely accepted and less resisted but stand a greater chance of achieving national objectives.

Although the current focus of AI centers on markets and weapons, there is growing interest in research and advocacy with respect to public policy and fundamental ethical questions about how societies should evaluate and manage the consequences of a world driven by automated decisions. At the same time, the nexus between AI and international relations is largely unexplored, even as there are clearly major implications for global economics, international security, and democratic ethics.

Since it is likely that transformational changes to diplomatic practices may eventually be required to meaningfully address that nexus, the most sensible path forward appears to be an incremental approach to crafting AI foreign policies that build upon the successes (and learn from the failures) of cyber foreign policy, which morphed into what it is today after numerous missteps and scandals. We do not have the luxury of merely thinking about whether or how to craft foreign policy in the AI age, for it will rapidly have a significant impact on international relations.[19] Fortunately, in many countries, cyber issues are now integrated into the practice of statecraft, which has become operationalized within existing diplomatic frameworks, giving policy planners and decision makers a firm foundation upon which to base the AI-enabled foreign policies of the future.

Some of the world's multilateral institutions have already created frameworks that will give them a better ability to integrate AI into their operations. As an example, the UN's Interregional Crime and Justice Research Institute launched a program in 2015 on AI and Robotics, in an effort to instill governance into its use. By utilizing experts in the field, the program seeks to educate and inform stakeholders (and in particular, policy makers) about how to build consensus and apply practical perspectives to use AI in a responsible manner.

A year later, the UN convened a gathering of states to attempt to create a legal framework, based upon the precautionary principle of international law, to prevent the further weaponization of AI. That same year, it formally opened its Center for AI and Robotics, to enhance understanding of the costs and benefits of AI through improved coordination, knowledge collection and dissemination, enhanced awareness, and outreach activities. Among its remit are risk assessments, stakeholder mapping and analysis, and training and mentoring programs.[20]

It is not too soon to consider creating a multidisciplinary International AI Agency (IAIA) that would include leaders from business, government, academia, law, and civil society to build on the early work of the UN so as to supercharge AI awareness across the globe.[21] Such an international body should include major players in all relevant disciplines and adopt a highly transparent operational philosophy whose starting point would be to share AI technology in an attempt to dampen the temptation to horde it. Existing international agencies adopted and gained wide acceptance for rules clearly defining the research for and use of lethal technology (such as the IAEA); given the potential lethal applications of AI, the IAIA must do the same.

In the race for achieving AI supremacy, those nations that quickly adopt good AI diplomatic hygiene stand the best chance of getting to the front of the line and staying there, but this must be done in a holistic manner that encourages a consistent approach throughout a country's ecosystem. For example, if the US hopes to be able to maintain its global competitive advantage in the AI space (especially with China nipping at its heels), it must do a better job of creating an open regulatory foundation upon which to build an even more robust entrepreneurial platform that encourages innovation in the private and public sectors. On one hand, America has an extremely open ecosystem from which to innovate–from its world class universities, to vibrant corporate laboratories, to the innovation corridors that have popped up around the country–

all designed to encourage creative thinking. On the other hand, that same openness is being used against the US to acquire secrets and technology to compete against it.

The same vibrancy that encourages American researchers to lead the world in innovation also pushes their discoveries quickly into public view, often before they are even granted a patent or other legal protection. When coupled with easily accessible online lectures from top universities, it is an invitation for competitors to copy American inventions. Yet, America's regulatory regime makes it difficult to produce sensitive items in the US and sell them to other countries, creating a market for foreign competitors. As demand for AI-related technology grows, the US, like all other countries, must strike a balance between protecting itself and hurting itself at the same time.[22]

One would have thought that a country like the US would not have waited until 2018 to address such basic issues, but this is emblematic of the reactive orientation so many countries adopt when deciding whether and when to take a deep dive in the AI pool. In 2018, a bill was proposed in the US House of Representatives (the National Security Commission on Artificial Intelligence Act[23]) to create a commission to conduct a review of the advances made in AI, ML, and associated technologies, to consider what the government should be doing to comprehensively address the national security needs of the nation in the context of AI and ML. Among the issues the commission will address are: how can the country stay competitive in AI, build a tech-savvy workforce, maintain data privacy standards, and develop technologies that comply with international law. Though the Act did not propose specifically allocating funds devoted to AI research, it did charge the Commission with identifying ways to push agencies, academia, and private companies to invest in research and development.[24]

Eventually, most countries will align their AI ambitions with AI-centered practice. One country is already doing so. The United Arab Emirates (UAE) and its commercial hub of Dubai have taken bold leaps to become the epicenter of futurism. In 2017, the UAE became the world's first country to release an official AI strategy and appoint a Minister for AI. Dubai has already deployed the world's first robotic policeman, started testing flying taxis, and is taking large steps toward becoming the "smart" city of the future. The objective was to make its economy hyper-competitive, while disrupting its own government to become more efficient and effective. By elevating AI to the front of its long-term objectives, the UAE intends to create future opportunities that will enable it to boost its GDP by 35%, reduce government costs by 50%, and make itself highly resistant to future financial crises after 2030.

To achieve this, the government created long-term objectives in line with its "Centennial 2071" plan, to make the UAE (unabashedly) the "best country in the world" by that time. AI will play a central role, using ML to boost government performance at all levels and create a marketplace with high economic value.[25] The UAE hopes its AI initiatives will encourage the rest of the world to consider how an AI-powered future can look, acknowledging that the future will be neither black nor white—much will depend on how AI is implemented.[26] The UAE is making AI the centerpiece of its growth and innovation strategy for the remainder of this century and is showing the rest of the world how it should be done.

13. Legal Safeguards

Research has always been central to practicing law, with lawyers and law firms in the US estimated to spend more than $8 billion annually to obtain the right information. The faster and more accurately an attorney can find information that is useful, the more time that can be spent developing an effective litigation or defense strategy. Part of the potential appeal of using AI in the legal arena is that ML algorithms are capable of building models that make sense of complex data by detecting patterns and inferring rules from it. The legal profession has already benefitted greatly by incorporating AI into various aspects of practicing law. There is no case–civil or criminal–that does not now involve the use of electronically stored information. Evidence has transitioned from what was once purely tangible and testimonial to being digitally stored and routinely used in the courtroom. From email, social media, and the cloud, to body cameras, GPS, and cellphones –it is all recorded and potentially applicable under the law.

While having started as a way to efficiently sift through mountains of data to find precisely the right reference or piece of information, AI has become an essential tool in interpreting the law. One of the basic cornerstones of the US common law system is that judges must explain their decisions in writing, establishing the reasons for their decision by referencing the law, facts, public policy, and other considerations upon which an outcome is based. ML can find correlations between opinion and other factors to determine whether there are any irregularities that impact a decision, while at the same time testing the system's strengths and vulnerabilities. It can also swiftly assist lawyers to find which judges could potentially be more sympathetic to their client.

There are a number of challenges and obstacles associated with using data analytics in the legal arena, however. To start with, the information that law firms analyze belongs to their clients and must be properly anonymized before conducting analysis. Many practitioners believe that practicing law requires cognitive abilities that are beyond the realm of ML and AI. Attorneys routinely use both abstract reasoning and problem-solving skills in the realm of legal and factual uncertainty. Modern AI algorithms have, by contrast, been unable to replicate most human intellectual abilities, falling far short in advanced cognitive processes such as analogical reasoning that are a foundation of legal practice.[1] So, is AI a net positive for the legal profession, and can it overcome the numerous obstacles that stand in the way of making ML even more widely adopted and accepted?

How AI is Transforming the Law

There are a number of potential applications for using AI in the legal domain, especially for those that relate to the automation of repetitive and routine tasks. Conducting legal research can be tedious, monotonous, and time-consuming, but performing timely and comprehensive legal research is critically important for lawyers. AI systems certainly aid lawyers by performing legal research on relevant case law and applicable statutes faster and more thoroughly than most lawyers may be able to do on their own. Such systems are proving powerful enough to use data to predict the outcome of litigation and enable lawyers to provide more impactful advice to their clients in connection with dispute resolution issues.

In 2016, law firm BakerHostetler became the first to publicly announce that it had "hired" a robot lawyer to assist with bankruptcy cases. The robot (ROSS) became the world's first artificially intelligent attorney. Powered by IBM's Watson technology, ROSS served as a legal researcher for the firm and was responsible for sifting through thousands of legal documents to bolster the firm's cases. ROSS highlighted relevant passages of law and shared them with attorneys, who could then choose either to enforce or question ROSS's hypothesis.

Prior to that time, lawyers had been using static pieces of software to navigate the law, which were limited in scope and required many hours of information retrieval tasks on the part of each lawyer. The use of complex software in the practice of law had become commonplace (for, say, purposes of discovery), but Watson assists legal professionals in making actual judgments. Using ML technology to fine tune its research methods, ROSS is simply accessed via computer and billed as a subscription service. Some legal research platforms—such as Bloomberg BNA, LexisNexis, and Thomson Reuters—come with a steep learning curve based on their current functionality and offerings, requiring training that is not built in to the billable hour model, making them less desirable from an operational perspective.

Since ROSS is built upon a cognitive computing system, it is able to sift through more than *one billion text documents per second* and return the exact passage a user needs. The days of manually poring through endless Internet and database search results are essentially gone for firms that use ROSS. It represents a transition from programming to learning and teaching, with ROSS becoming smarter with every transaction it completes. ROSS and Watson are not merely translating words and syntax into search results; they are learning to understand the law. Weighing data, drafting documents, and making arguments will remain the domain of humans but, by tackling the burdensome task of research, ROSS frees up lawyers to do what they do best.[2]

Although it remains early days for AI in the legal profession, ROSS and other cutting-edge AI tools that are under development should transform the legal landscape in a profound way. ROSS is not intended to be a way to replace attorneys, but it helps them learn faster and enhances their efficiency. Such tools stand to be so transformative in saving time and money that not finding meaningful ways to utilize AI throughout a law firm's operations will make some of them obsolete.[3] The legal sector's challenge remains to become forward leaning and imaginative enough to embrace the inevitable and adopt AI on a broad basis.

For many years, labor economists maintained that routine work (such as repetitive factory jobs) could be reduced to a set of rules that could be computerized, but they assumed that white collar jobs would not be in a similar domain because the nature of the work was steeped in thought and language. Advances in AI have clearly overturned that belief, so major law firms, sensing the long-term risk of not getting on board the AI train, are undertaking initiatives to understand the emerging technology, adapt to it, and exploit it.

As an example, in 2015, law firm Dentons established an innovation and venture arm–Nextlaw Labs–which not only monitors the latest technology but invests in legal technology start-ups. In 2017, Baker McKenzie set up an innovation committee composed of senior partners to track emerging legal technology and establish strategy. Most law firms are using AI primarily for electronic discovery, due diligence, and contract review. While more than 280 legal technology start-ups raised nearly $800 million between 2012 and 2017, progress in developing ground-breaking new technology is encouraging.[4]

Beyond helping to prepare cases, AI can also predict how they will hold up in court. Lex Machina, a company owned by LexisNexis, offers what it calls "moneyball" lawyering, applying natural-language processing to millions of court decisions to identify trends that can be used to a law firm's advantage. The software can, for example, determine which judges tend to favor plaintiffs, summarize the legal strategies of opposing lawyers based on their case histories, and determine the arguments most likely to convince specific judges. A company called Premonition takes this one step further, promising to predict the winner of a case before it goes to court, based on statistical analyses of verdicts in similar cases. A Silicon Valley start-up–Legalist–offers commercial litigation financing, meaning it will pay a lawsuit's fees and expenses if its algorithm determines there is a good chance of winning, in exchange for a portion of any judgment in your favor. Critics rightly fear that AI could be used to game the legal system by third-party investors hoping to make a profit.

Technologies such as ROSS and Lex Machina are intended to assist lawyers, but AI has also begun to replace them in some of the most straightforward areas of law. One of the most successful "robolawyers" (as of 2017), called DoNotPay, was developed by a British teenager. It is a free parking-ticket-fighting chatbot that asks a series of questions about a case–such as were the signs clearly marked and were you parked illegally because of a medical emergency–and generates a letter that can be filed with the appropriate government agency. The bot has helped hundreds of thousands of people beat traffic and parking tickets in London,

New York, and Seattle. Additional functions were subsequently added so that DoNotPay can help people demand compensation from airlines for delayed flights and file paperwork for government housing assistance. More such applications are being developed.

However, until there is a significant society-changing breakthrough in AI, robolawyers will not be disputing the finer points of copyright law or writing elegant legal briefs, but chatbots could be very useful for other aspects of the practice of law. For example, bankruptcy, deportation, and divorce disputes typically require navigating through lengthy and confusing statutes that have been interpreted in thousands of previous decisions. Chatbots could eventually analyze nearly every possible exception, loophole, or historical case to determine the path of least resistance.

While a court-appointed robolawyer may never actually be used for a criminal case, algorithms are changing how judges determine punishments. In many states, judges use software called COMPAS to set bail and decide whether to grant parole. The software uses information from a survey with more than 100 questions, covering topics including a defendant's gender, age, criminal history, and personal relationships, to predict whether he or she is a flight risk or likely to re-offend. However, with every enhancement of technology and applicability come pitfalls. The company that created COMPAS will not make its algorithm public, which means defense attorneys cannot bring informed challenges against a judge's decision, and a study by ProPublica also found that COMPAS appears to have a strong bias against black defendants.

There is, in addition, a possible downside to lowering barriers to legal services, as the range of automated services rise and costs fall. Imagine a future in which lawsuits are filed while standing in line for a latte, or are litigated with the swipe of a finger. Rather difficult to imagine today—but 5 or 10 years from now, perhaps not. Consider the possibility of a completely automated and ever-present legal system that functions based on sensors and pre-agreed-upon contracts. A company called Clause is creating "intelligent contracts" that can detect when a set of prearranged conditions are met or broken. If you were to agree with your landlord to keep the temperature in your house between 68 and 72 degrees, but you turn it off during the winter and a water pipe bursts, an intelligent contract could automatically deduct a penalty from your bank account.

At some point in the not too distant future, self-driving-car accident disputes may be resolved via simple reviews of the vehicle's logs and programming. A grievance against a local pizza shop's guarantee of a hot delivery in 10 minutes could be verified by a GPS sensor and smart thermometer, to ensure compliance with the terms of its service agreement. Imagine that divorce papers may be automatically drawn up when a smart phone detects, through location tracking and text-message scanning, that your husband or wife has been unfaithful. Your will could be executed as soon as your Fitbit detects that you have died.[5] What has been the world of science fiction is quickly becoming reality. The question is, can the legal profession—which is conventional and resistant to change—keep pace?

A Bold Unfolding Future

AI's impact on existing law is becoming increasingly significant, with a growing number of ML applications being used to interpret and/or enforce it. The range of tools that has emerged for general use in society has unleashed a broad range of potential applications for use in the courtroom. For example, Data mining involves discovering patterns or extrapolating trends from data. Data mining algorithms can make predictions about how a variable may be extrapolated to identify fraudulent entries or transactions.

ID or image tagging used in social media can provide facial recognition or obtain other data from a visual scan. Image processing uses algorithms to analyze images in order to get data or perform transformations. Optical character recognition uses algorithms to read handwritten text and convert documents into digital versions. Sentiment analysis identifies the mood of the person writing. Speech analysis applies similar skills used in text analysis and applies them to the spoken word (Apple's Siri and Amazon's Echo are examples of where this may be applied). Smart data streaming has become so advanced that it has drawn the attention of regulators. 3D environment processing enables an algorithm linked to a connected and autonomous vehicle or drone to understand its location and driving environment using image, radar or laser data, translating them in to three dimensional geometries.[6] Text analysis extracts information and applies classification to text-based data, such as used in social media and emails.[7]

AI is both enabling lawyers to establish new boundaries for how and where the law applies while establishing the boundaries of what AI is capable of. Typically, the legal system's interaction with software like robotics establishes liability where developers may have been negligent or could foresee harm. But certain aspects of AI, such as in RL–which involves establishing a balance between current knowledge and what is possible–makes it difficult to prove fault by humans and there is no way to accurately foresee injury. Traditional tort law would stipulate that a software developer would not be liable in such circumstances. The dangers posed by Terminator-like outcomes can therefore potentially proliferate without anyone having to assume responsibility.

While it is unlikely that outcomes associated with AI will result in a permanent state in which no one is held responsible for its actions, it is unclear how the law will evolve to address AI's every impact. Concepts such as liability or a jury of peers appear meaningless, unless AI's developers and users can be proven to have intended to create harm or damage. As is currently the case with cybercrime, wherein actors operate in an anonymous, boundary-less and largely lawless world, identifying AI's actors and proving their intent, can be a rather difficult task.

A useful starting point down that road would be to establish norms and standards to govern the AI world. Laws should be passed that force manufacturers and developers to abide by a general set of ethical guidelines–such as technical standards mandated by treaty or international regulation–to be applied when it is foreseeable that the algorithms and data can cause harm. This could be achieved by convening globally recognized AI experts[8] to create a framework that includes explicit creation standards for neural network architectures (containing instructions to train and interpret AI models). One of the biggest benefits of doing so would be the ability to substitute AI models for different applications without difficulty. At the present time, switching from one AI design that recognizes faces to one designed to understand human speech requires a complete overhaul of the neural network associated with it.[9]

Clearly, for such a standard to be enforced would require government intervention, and some governments have already initiated the process of forming teams to explore the concept. It will be a long time until such laws are routinely adopted, however, and even longer until a multilateral organization is created and tasked with passing and enforcing laws governing AI on a global basis. That is probably what will be needed to make the idea of universal enforcement and compliance a reality. As is the case with the cyber world, a large question mark hangs over the AI world, its various manifestations, and its future impact on the law.

Liability and Autonomous Vehicles

With more than 90% of all car crashes being attributed to driver error, the stakes are high. Fully and partially autonomous cars could improve that statistic substantially by reducing injury and death from speeding and drunk driving. Car insurance is big business, with insurers in the US alone generating more than $230 billion in premiums annually. The advent of autonomous vehicles could significantly impact that pile of cash, if driverless technology were to eliminate the human error that factors into so many automobile accidents.

The first known death resulting from the operation of autonomous vehicles occurred in 2018, when a pedestrian was hit by a self-driving Uber vehicle in Arizona. This event illuminated the question of who may be held legally responsible for accidents when a human is not at the wheel. At stake is the issue of liability between services such as Uber, versus the makers of such technology, versus the vehicle's manufacturer. Shortly after the Arizona fatality, a second fatality occurred in California, involving a Tesla. The driver's hands were not on the steering wheel for six seconds before the car collided with a highway median, although the driver had received several visual and one audible cues from the vehicle to grab the wheel earlier in the drive. The autopilot feature was not fully autonomous—it took control of some driving functions, but not all of them. Drivers were expected to remain engaged when the feature was activated.[10]

A subsequent investigation of the Arizona case by the US National Transportation Safety Board (NTSB) found that the vehicle's emergency braking system was disabled and that, while the vehicle's guidance system had seen the woman six seconds prior to hitting her, its emergency braking system was not enabled so as to reduce the potential for erratic vehicle behavior. Instead, the Uber system anticipated that the human back-up driver would intervene, even though the automated system was not designed to alert the driver of the impending danger. So, the vehicle's brakes were not applied until after the fatal impact. Uber ceased testing its vehicles in Arizona following the NTSB pronouncement, even though its report was not designed to assign responsibility for the crash.[11]

A question likely to arise in litigation is whether any companies that supply self-driving technology entered into indemnification agreements, in which an entity assumes another company's liability costs. Such agreements in the autonomous vehicle world are confidential and disputes over indemnification are generally arbitrated or mediated behind closed doors. If any litigation in the Arizona accident were to allege design defect claims, it could set legal precedent and be impactful on the autonomous vehicle industry.

Auto accident litigation is ordinarily determined by whether a driver acted negligently or failed to exercise a reasonable level of care. Most litigation prior to the Uber event involving non-fatal accidents with self-driving vehicles had been settled quietly and confidentially. But the stakes were raised with the Arizona case, raising question about whether the self-driving system had a design defect. Design defect claims do not require a finding of fault or negligence. To prevail, a plaintiff must show only that a product had an inherent design defect that would render it unsafe.

Automakers and software writers typically counter such arguments with detailed data gathered by on-board sensors recording how cars behaved during a collision, to show that it was impossible for the vehicle to avoid a collision and that all the systems functioned properly.[12]

In the Arizona case, the operator's judgment and response had been replaced by that of the machine. The fact that an autonomous vehicle caused the outcome might be enough to shift the liability and compensation to that of product liability on the part of Uber, but an individual *chooses* to ride in self-driving vehicles—so who is to blame? The Arizona case might neither be construed as vehicular negligence (because a person *was* and *was not* driving) nor product liability, because no product was leased or sold. The determinative question is whether a natural or legal person had a duty that was violated by acting unreasonably in a way that caused harm.

    A whole host of other unresolved issues arise, such as, if the safety driver were negligent, could Uber be vicariously negligent as an employer, and could Arizona be held accountable for allowing autonomous cars to roam public streets without sufficient oversight, including legal guidance for inevitable situations like pedestrian deaths? In the wake of the death, it became public that Uber had begun to test autonomous vehicles in 2016, without the public's knowledge. While states such as California had stringent regulations regarding the operation of such vehicles, Arizona, Florida, and Michigan did not. Eight days after the death, Arizona's governor, who had been so lenient on Uber, rescinded the company's right to operate autonomous vehicles on public roads in the state.[13]

    There is an argument to be made that a pedestrian death at the hands of an autonomous car—even one that would have been unavoidable—is no different from a human-driven car with a new, experimental combustion engine that malfunctions and blows up on a road. Human operators may be liable for any traffic infractions while he or she is in the vehicle, even if it is in fully autonomous mode. An operator could then be charged with vehicular manslaughter, although the courts would inevitably have to adjudicate such a matter. The entire subject is not possible to fully comprehend before legal precedent is set.

    Since autonomous Ubers can and do pick up ordinary passengers, such vehicles may also be subject to the common carrier doctrine—a law that requires common carriers (like buses and taxis, but also hoteliers and others) to be held to a higher standard of care than ordinary operators. The Arizona Supreme Court has held that common carriers in the state are only subject to "reasonable" care, the same as any other agent. It is also unclear whether ride-hail services even count as common carriers in Arizona or elsewhere.

    Federal legislation could push complaints arising from autonomous vehicle collisions or injuries into private arbitration. In addition to reducing citizens' rights, doing so could prevent courts from hearing some cases that may produce new statutory precedents. Given sufficient economic incentive to pursue public-private partnerships between municipalities and technology companies, cities, counties, and states might choose to adopt industry-friendly regulatory policy in exchange for changes to the urban environment. For example, should the use of autonomous cars become widespread, it could prove to be more expedient simply to close certain roads to pedestrians, bicyclists, and human drivers so that autonomous vehicles may operate at maximum efficiency.[14] That would, of course, defeat the purpose for the average user, however, who will want to use the vehicles to travel as they ordinarily do. It is, to say the least, complicated.

Many of the world's largest car and information companies— among them Apple, Audi, Ford, GM, Google, Honda, Mercedes, Tesla and Volvo—have incorporated advanced driver assistance systems into cars that can already be purchased. These are well developed initiatives and represent tremendous promise as well as potential peril. Among them is the question of how liability will be addressed by the legal and insurance professions. Two primary questions hover over automakers and tech companies in this regard: what, exactly, are the rules, and who is responsible when something goes wrong?

In 2015, Google, Mercedes, and Volvo stepped up to the plate to accept liability when one of their vehicles is in autonomous mode, but few others have stepped forward. For Volvo, the biggest threat to self-driving cars is not technology, but regulations. While the US is the most advanced country in terms of autonomous driving, its position could be eroded if a national framework for regulation and testing is not developed. Europe has suffered by having a patchwork of rules and regulations. Given that the US insurance industry is governed by 50 separate state jurisdictions and regulations, it is not hard to imagine that future autonomous vehicles will be governed by a disparate collection of rules and regulations. The fact that the rules governing cyber and AI risk are embryonic does not help. Without a clear set of rules, automakers cannot conduct credible tests to develop cars that meet the diverse guidelines of each of the 50 states.[15]

Then again, in two or three decades, the concept of conventional auto insurance may have disappeared entirely. It is not too much of a leap to imagine that, just as robots have dramatically decreased risks arising from human error in the medical profession, autonomous vehicles will undoubtedly become safer than human controlled driving, resulting in fewer losses and lower premiums. Insurance claims have not yet been made on behalf of autonomous vehicles, however, because manufacturers are being extremely cautious about releasing this technology into the marketplace, since there are enormous implications involved in getting it wrong. They only have one opportunity to make a good first impression, and they know it.

As an inevitable human-less driving reality awaits us, insurance companies are unlikely to change the way they do business for the foreseeable future. Their inclination is to wait until the technology has been proven beyond a reasonable doubt. The legal profession stands to benefit greatly as an entirely new area of the law emerges along with the new technology. Can it adapt quickly enough to get and stay ahead of the curve? Like the insurance industry, the legal profession is sometimes slow to change. In the AI world, time waits for no one.

AI and Data Stewardship

Data stewardship used to mean simply being a lawful collector and user of data, but in the AI era, an effective custodian must take into account stakeholder interests in a way that uses data to create maximum benefit for stakeholders with minimal implied risks to individuals and other stakeholders. In this context, accountability includes taking responsibility for the protection and appropriate use of information beyond legal requirements and being responsible for any misuse of information. In other words, users must take responsibility for the information they use. AI has forced legal practitioners to focus more intently on explicit data stewardship.[16]

In the US, social networks are considered to be public spaces and any information shared there is covered under the so-called 'third-party doctrine', which means that users have no reasonable expectation of privacy regarding the data their service providers collect about them. This exception to the Fourth Amendment of the US Constitution's prohibition on unreasonable search and seizure means that any data posted online in any format (regardless of privacy settings), or any data that is collected by the third parties with whom you may have an agreed-upon business relationship, is not considered private. As a result, any data that 'leaks' to places you did not explicitly authorize cannot be clawed back, and there is usually nothing that can be done about it.

The Apple Example

While most technology companies, telecommunication firms, and social media enterprises make no secret of their desire and intent to acquire and consume as much information as possible, and the law as it currently exists in the US more or less allows them to do just about anything they may want with that data, Apple portrays itself as the lone defender of user privacy. The company consistently rolls out hardware that makes it more difficult for Apple (or hackers, identity thieves, or governments) to access your data. Apple has traditionally limited data analysis so that it all occurs on the device instead of on Apple's servers.

According to Apple's website: "Personal data should always be protected on your device and never shared without your permission. We build encryption, on-device intelligence, and other tools into our products to let you share what you want on your terms. We also use techniques like Differential Privacy (DP) to improve user experiences while protecting the information you share with Apple. DP adds random information to your data before it is analyzed by Apple, so we cannot link that data to your device. Instead, patterns appear only when the data is combined with the data from many other users, because the random additions average out. These patterns help Apple gain insight into how people are using their devices without collecting information about an individual."[17]

That sounds nice on paper, but how can an average consumer really know if any of that is true? The truth is, we cannot know for certain and must take any company at their word (which can be an enormous leap of faith), but what we do know is that, like all other technology companies, Apple needs to know what its users are doing in order to improve its features, which presents a problem for a company that claims to put privacy first. DP uses hashing, sub-sampling, and noise injection, to permit crowdsourced learning while keeping the information of each individual user completely private. It is worth noting that Apple only began using DP with the rollout of iOS 10 update in 2017, so it may well have been guilty of the same approach to data aggregation and analysis as its brethren prior to that time.

DP works by algorithmically scrambling individual user data (so that it cannot be traced back to the individual) and then analyzing the data in bulk for large-scale trend patterns. The goal is to protect the user's identity and the specifics of their data while extracting general information to propel ML. iOS 10 randomized data on devices before sending it to Apple en masse, so that the data is never transported in an insecure form. Apple also limits the amount of data it can obtain from any single user. Although other companies have experimented with or actually used DP, Apple is believed to be the first organization of its kind to use it at scale.

The debate over privacy in Silicon Valley is often viewed through a law enforcement lens that pits user privacy against national security, but tech companies often see the debate as between user privacy versus features. Apple's introduction of DP could radically change that debate. Google and Facebook, among others, have been torn by the question of how to deliver feature-rich products that are also private. Neither Google's messaging app, Allo, nor Facebook's Messenger offer end-to-end encryption by default because both companies prefer to consume users' conversations in order to improve ML and allow chat bots to function. Apple also naturally wants to glean insights from user data but is unwilling to backpedal on iMessage's end-to-end encryption in order to do so.

Apple's move to DP should greatly enhance the debate about the relative tradeoffs between protecting privacy and improving ML. Apple's decision not to collect raw user data could encourage users to become more trustful of technology companies. It also happens to have the benefit of helping Apple become more resilient against government intrusion, which the company has not hesitated to defend against the US Federal Bureau of Investigation. Apple appears to have adopted a natural sweet spot between innovation and user safety. In doing so, it may have opened a pathway toward how other organizations collect and use private information,[18] for ML and other purposes.

Implications of Europe's General Data Protection Regulation

In 2018, the EU implemented the most dramatic law of its kind governing individual data protection. The GDPR strengthened and unified data protection for all individuals within the EU. It gave control of personal data back to citizens and residents, simplifying the regulatory environment for international business by unifying the regulation within the EU and replacing a data protection directive from 1995. The GDPR extended the scope of the EU data protection law to all foreign companies processing data of EU residents. It also provided for harmonization of the data protection regulations throughout the EU, making it easier for non-European companies to comply with these regulations. This came with a strict data protection compliance cost, with severe penalties of up to 4% of worldwide turnover (the UK Parliament's version contained increased fines of up to 5%) for non-compliance.[19]

Contrast that with the direction the US went on the subject of individual data protection. In the last months of Barack Obama's presidency, the then-Democratic-led US Federal Communications Commission (FCC) prepared guidelines that created unprecedented protections for the personal information that Americans put into cyberspace. The Internet data rules created under the FCC would have permitted ISPs to sell consumers' browsing data—but only with explicit permission by way of consumers "opting in" to their data being sold to third parties. In 2017, President Trump reversed all that and signed a bill that put *everything any American does on the Internet, and all the data generated in the process, on the table for Internet providers to sell to whomever they choose—with or without consumer consent*. If you call a car dealership, it is against the law for phone companies to sell information about your vehicle and your interest in selling it, but if you try to sell your car online, your ISP can now sell that information to whomever it wishes.

In the past, Internet users could selectively decide whether to use the services of Facebook or Google, or abstain from using them if they wished to safeguard their personal information.[20] That is no longer the case. The ToS we all unwittingly agree to when we in the US use Internet-related services gives these companies permission to basically do whatever they want with the information (and images) we voluntarily give to them—which was bad enough. Now, they have permission to sell it to whomever they please. The recipients of that information are, in turn, free to do with it whatever they choose. That process may now continue ad nauseam.

There is strange legal (and moral and ethical) dichotomy at play here, which is indicative of some of the larger questions that the law (to the extent that it exists on this subject) is at pains to address on the subject of cybersecurity as well as AI. This raises a number of issues, such as how to define the legal boundaries that apply to cyberspace and AI. In order for any law to be meaningful, the individual(s) perpetrating a crime must be identifiable and the law itself must be enforceable. In the AI and cyber worlds, boundaries are amorphous, so can any law be truly meaningful across the board? And how meaningful can a law against abuse of AI be if it addresses only a fraction of the lawbreakers?

Transforming the law to be more universally applicable to AI is a significant challenge, but the GDPR offers some guidance for how a global legal framework may be crafted in due course. The biggest change to the regulatory landscape of data privacy associated with the GDPR is its extended jurisdiction–it applies to all companies processing the personal data of individuals residing in the EU, regardless of the company's location. Previously, territorial applicability of the directive was ambiguous and referred to data processed "in the context of an establishment". This topic has already arisen in a number of high profile court cases.

The GPDR applies to the processing of personal data by controllers and processors in the EU, regardless of whether the processing takes place in the EU or not. The GDPR also applies to the processing of personal data in the EU by a controller or processor not established in the EU, where the activities relate to offering goods or services to EU citizens (irrespective of whether payment is required) and the monitoring of behavior that takes place within the EU. Non-EU businesses processing the data of EU citizens must also appoint a representative in the EU.

The conditions for consent have been strengthened under the regime. Companies will no longer be able to use long and illegible terms and conditions. Any request for consent must be given in an intelligible and easily accessible form, with the purpose for data processing directly attached to that consent. Consent must be clear and distinguishable from other matters, and provided in an intelligible and easily accessible form, using clear and plain language. It must be as easy to withdraw consent as it is to give it.

Breach notification became mandatory in all member states where a data breach is likely to "result in a risk for the rights and freedoms of individuals". This must be done within 72 hours of first having become aware of the breach. Data processors are also required to notify their customers "without undue delay" after first becoming aware of a data breach. Part of the expanded rights of data subjects outlined by the GDPR is their right to obtain from the data 'controller' confirmation as to whether or not personal data concerning them is being processed, where, and for what purpose. The data controller must further provide a copy of the personal data, free of charge, in an electronic format. This is a dramatic shift to data transparency and empowerment of data subjects.

Also known as data erasure, the "right to be forgotten" entitles the data subject to have the data controller erase his or her personal data, cease further dissemination of the data, and potentially have third parties halt processing of the data. The conditions for erasure include data that is no longer deemed relevant to the original purpose for processing, or a data subject withdrawing consent. It should also be noted that this right requires controllers to compare subjects' rights to "the public interest in the availability of the data" when considering such requests.

Privacy by Design as a concept has existed for years but it is only just becoming part of a legal requirement with the GDPR. At its core, Privacy by Design calls for the inclusion of data protection from the onset of the designing of systems, rather than as an addition to their design. Data controllers can hold and process only the data absolutely necessary for the completion of their duties, as well as limiting access to personal data to those needing to act in the course of processing. The GDPR also introduces data portability—the right for a data subject to receive the personal data concerning them and the corresponding right to have that data transmitted to another controller.

Controllers had only been required to notify their data processing activities with local data protection administrators (DPAs), which for large multinationals can be a bureaucratic nightmare, especially since most EU Member States have different notification requirements. Under the GDPR, it is not necessary to submit notifications or registrations to each local DPA of data processing activities, nor is it a requirement to notify or obtain approval for transfers based on model contract clauses. Instead, internal record keeping requirements and DPA appointments become mandatory only for those controllers and processors whose core activities consist of processing operations which require regular and systematic monitoring of data subjects on a large scale, or of special categories of data, or data relating to criminal convictions and offenses.

DPAs must be appointed on the basis of professional qualities and, in particular, expert knowledge on data protection laws and practices. They can be a staff member or an external service provider and must be provided with appropriate resources to carry out their tasks and maintain their expert knowledge. They are required to report directly to the highest level of management and cannot carry out any other tasks that could result in a conflict of interest.

As previously noted, under the GDPR, organizations in breach of its guidelines can be fined up to 4% of annual global turnover or €20 million (whichever is greater). This is the maximum fine that can be imposed for the most serious infringements, such as not having sufficient customer consent to process data or violating the core of Privacy by Design concepts. There is a tiered approach to fines (e.g. a company can be fined 2% for not having their records in order, not notifying the supervising authority and data subject about a breach, or not conducting impact assessments). These rules apply to both controllers and processors—meaning 'clouds' are not exempt from GDPR enforcement.[21]

There are some truly revolutionary aspects associated with the GDPR—both from the perspective of the data user and the data provider or processor. In time, either it will prove to be mostly successful and prompt global companies to modify their data handling methods in some fundamental ways, or it will turn out to be a Kafkaesque, bureaucratic enforcement nightmare. Global firms naturally took the implementation of the new rules seriously, and it cost companies a lot of money to be in full compliance with them. The real challenge was whether the EU had a monitoring and enforcement mechanism that was up to par with the GDPR, whether and how individuals will report infringements, and whether any of it will make a real difference at the end of the day. Only time will tell whether it has served its intended purpose, but it was a bold move on behalf of the EU to implement the GDPR. Its impacts will be felt for many years to come and should prove to have a permanently profound impact on data collection and use in the AI arena.

Direct Impacts of the GDPR on AI

Given how broad in scope the GDPR is—covering any information that can be used to directly or indirectly identify an individual (such as via names, photos, email addresses, financial details, posts on social networking sites, medical information, or a computer IP address), no matter when it was collected—it is only natural to wonder how it directly affects AI. While the bulk of the GDPR's language addresses how data is collected and stored, portions of the law focus on automated individual decision-making (including profiling) and potentially prohibits a wide swath of algorithms currently in use (such as in credit and insurance risk assessments, computational advertising, and social networks).

This is of direct concern to AI and ML because, in its current form, the GDPR's requirements could result in a complete overhaul of standard and widely used algorithmic techniques. The right of European citizens to receive an explanation for algorithmic decisions highlights the pressing importance of human interpretability in algorithm design. The subjects of data collection and analysis have the right *not* to be subject to a decision based solely on automated processing, including profiling. However, if the decision is (a) necessary for entering into (or in performance of) a contract between the data subject and a data controller, (b) authorized by an EU or Member State law to which the controller is subject and which also lays down suitable measures to safeguard the data subject's rights and freedoms and legitimate interests, or (c) is based on the data subject's explicit consent, then it may well apply.

While the 1995 Data Protection Directive that the GDPR replaces was subject to national interpretation and was only ever indirectly implemented through subsequent laws passed within individual member states, the GDPR requires no enabling legislation to take effect since it is the law. The scope of the GDPR is explicitly global; its requirements do not just apply to companies that are headquartered in the EU but to any companies processing EU residents' personal data. For purposes of determining jurisdiction, it is irrelevant whether that data is processed within the EU or abroad.

The GDPR pays specific attention to profiling that is intended to analyze or predict aspects of a person's performance at work, as well as their economic situation, health, personal preferences, interests, reliability, behavior, location, or movements. Given the breadth of inclusive categories, the GDPR's desideratum for profiling errs on the side of inclusion rather than exclusion. The law's definition of automated individual decision-making prohibits any decision based solely on automated processing that significantly affects a data subject.

Even in the case of the above-referenced exceptions, data controllers must provide appropriate safeguards, including the right to obtain human intervention to be able to express an individual's point of view or contest a decision. The law specifically prohibits automated processing based on special categories of personal data unless suitable measures to safeguard the data subject's rights, freedoms, and legitimate interests are in place. Since it is implied that the data is legitimately obtained, the provisions for algorithmic profiling are an additional constraint that apply even if the data processor has informed consent from data subjects.

All of this presents practical challenges related to the design and deployment of ML algorithms. The right to non-discrimination is deeply embedded in the normative framework that underlies the EU and the use of algorithmic profiling for the allocation of resources is inherently discriminatory. Concerns about discrimination are already central to a discussion about ethics associated with Big Data. Big Data claims to be neutral, but ML depends upon data that has been collected from society, and to the extent that society is unequal or discriminatory, so too is the data. ML can reinforce existing patterns of inequality and discrimination, making biased decisions out of what is thought of as an "objective" algorithm.

The GDPR explicitly requires data controllers to implement appropriate technical and organizational measures that prevent discrimination on the basis of processing sensitive data, which includes personal data revealing racial or ethnic origin, political opinions, religious or philosophical beliefs, trade-union membership, genetic data or biometric data for the purpose of uniquely identifying a natural person, data concerning health, or data concerning an individual's sex life or sexual orientation. The minimal interpretation is that this directive only pertains to cases where an algorithm is making direct use of data that is explicitly sensitive, which would include, for example, variables that code for race, finances, or any of the other categories of sensitive information referred to in the law.

A stricter interpretation takes a broader view of sensitive data to include not only those variables which are explicitly named, but any variables with which they are correlated. This puts the onus on data processors to ensure that algorithms are not provided with datasets containing variables that are correlated with categories of personal data. For relatively small datasets, it may be possible to identify and account for correlations between sensitive and non-sensitive variables, but removing all data correlated with sensitive variables could make the resulting predictor essentially useless. Postal codes can reveal racial information while at the same time providing non-discriminatory information on loan default rates or purchasing habits, and as data sets become larger, correlations will inevitably become increasingly complex and difficult to detect.

The link between geography and income may be obvious, but less obvious correlations—such as between race and the ISP —may exist within larger data sets and could lead to discriminatory effects. Given sufficiently large data sets, the task of exhaustively identifying and excluding data features correlated with sensitive categories is probably impossible. Since algorithms are by their nature risk averse, they will tend to make decisions based on predictions about which they are more confident, which could mean that predictive algorithms (i.e. for loan approvals) may favor groups that are better represented in the training data, since there should be less uncertainty associated with those predictions. Uncertainty bias illustrates the complex and multifaceted nature of algorithmic discrimination and suggests that appropriate solutions will require a clearer understanding of how it arises in practice, which highlights the need for human-intelligible explanations of algorithmic decision-making.

Standard SML for classification is based on discovering reliable associations and correlations to aid in accurate out-of-sample prediction, with no concern for causal reasoning or explanation beyond the statistical sense in which it is possible to measure the amount of variance explained by a predictor. The use of algorithmic decisions in an increasingly wider range of applications has led some to caution against the rise of a "black box" society and demand increased transparency in algorithmic decision-making.

Approaches to ML are unique in terms of the spectrum of their lack of interpretability. An algorithm can only be explained if the trained model that was used can be articulated and understood by a human. It is therefore reasonable to suppose that *any meaningful explanation would, at a minimum, provide an accounting for how input features relate to predictions*. This would enable its creators to answer questions such as: Is the model more or less likely to recommend a loan if the applicant is a minority, and which features play the largest role in prediction? It is unclear whether under the GDPR companies will be required to disclose their learning algorithms or training datasets, or whether such information will be made public.

While the GDPR presents a number of problems for current applications in AI and ML, the challenges described above imply that *algorithms should not only be efficient, but transparent and fair*. Research is under way to determine how to create algorithms that can do a better job of addressing the issues raised herein, but it remains to be seen whether any such enhanced techniques will be routinely adopted. For certain types of algorithmic profiling, it is possible to both identify and implement interventions to correct for discrimination, which is in contrast to cases where discrimination arises from purely human judgment.

If properly applied, algorithms should be able to not only make more accurate predictions but provide increased transparency and fairness over their human counterparts. *The GDPR is an important acknowledgement that, when algorithms are deployed in society, few if any decisions can be purely technical*. The ethical design of algorithms requires coordination between technical and philosophical resources,[22] but also the ability to comply with emerging laws governing the use of AI and ML.

AI and Copyrights

Creating works using AI could have very important implications for copyright law. In the past, the ownership of copyright in computer-generated works was not in question because the program was merely a tool that supported the creative process, very much like a pen and paper. Creative works qualified for copyright protection if they were original, with most definitions of originality requiring a human author. Using AI, a computer program is no longer simply a tool–it makes many of the decisions involved in the creative process without human intervention. AI is already being used to generate works in gaming, journalism, and music. These works could, in theory, be deemed free of copyright because they are not created by a human author. As such, they could be freely used or reused by anyone, which would not be welcome news for the companies selling the works.

Copyright law can either deny copyright protection for works that have been generated by a computer or attribute authorship of such works to the creator of the program. Conferring copyright in works generated by AI has never been specifically prohibited, but the laws of many countries are not amenable to non-human copyright. Granting authorship to the programmer has occurred in some countries–such as Hong Kong, India, Ireland, New Zealand, and the UK–by recognizing the work that goes into creating a program capable of generating works, even if the creative spark is undertaken by the machine.

The question of who the law would consider to be the person making the arrangements for the work to be generated remains open. Should the law recognize the contribution of the programmer or the user of that program? In the analogue world, this would be like asking whether copyright should be conferred on the maker of a pen or the writer. A similar ambiguity exists in the digital world. Microsoft developed Word software but the copyright resides with the user (the author who used the program to create his or her work). However, when AI algorithms are capable of generating a work, the user's contribution to the creative process may merely be to press a button so the machine can perform.

Some case law indicates that this question could be resolved on a case-by-case basis, but things are likely to become more complex as use of AI by artists becomes more widespread and as machines get better at producing creative works, further blurring the distinction between artwork that is made by a human versus a computer. Given sufficient computing power, we may soon be unable to distinguish between human-generated and machine-generated content. If (or when) we get there, we will have to decide what type of protection, if any, should be given to emergent works created by intelligent algorithms with little or no human intervention. Granting copyright to the person who made the operation of AI possible seems a sensible approach and may help ensure that companies will continue to invest in the technology, on the presumption that they will get a return on their investment. The next big debate may be about whether computers should be given the status and rights of people.[23]

One of the by-products of this ambiguity, and the legal minefield it represents, is that many AI researchers resort to using low-hanging, biased databases to train their algorithms. *The friction created by copyright law encourages AI bias by encouraging AI creators to use easily accessible, legally low-risk works as training data, even when those works are demonstrably biased.* As an example, more than one million emails sent and received by employees of the defunct energy company Enron were released by the Federal Energy Regulatory Commission in 2003, following a government probe into the corporation's fraudulent financial practices. The emails became a notorious treasure trove for computer scientists and academics, as they were easily accessible in machine-readable formats and were not believed to pose much legal risk. Former Enron employees were unlikely to sue for copyright infringement if their correspondence was used for AI research. The Enron emails might have seemed like an ML researcher's dream, but they were plagued with encoded biases. Implicit biases were embedded in emails sent among employees of the Texas oil-and-gas company that was investigated for fraud. Researchers used the Enron emails specifically to analyze gender and power bias.

AI SUPREMACY

Data in the public domain, like Shakespeare's plays, are not protected by copyright restrictions, can be readily found in machine-readable formats, and are entirely legal to use. The fair use doctrine is a legal provision that helps balance the interests of copyright owners with the interests of the public and potential competitors. The doctrine of fair use allows someone to quote a portion of a copyrighted work without authorization, meaning that AI creators and researchers can compete to create fairer AI systems by using copyrighted works. If the products associated with AI were to be classified as fair use, in most cases, computer scientists would be free to use any work to teach their algorithms. They could also disclose what they used without fear of legal repercussion, which allows academics and journalists to more easily check for bias in AI and give AI researchers access to a vastly larger set of data, which would help them to build even smarter algorithms.

Copyright law has historically exacerbated bias in artificially intelligent algorithms, but it also has the capability to vastly improve them. If companies knew they were protected legally, they would also be more likely to release the data their products were trained on. While incorporating fair use can promote the creation of fairer AI systems, an ongoing challenge is how to distinguish between what is legally permissible and ethically acceptable.[24]

AI and Ethics

Ethical questions abound with AI systems, raising questions about how machines recognize and process values and ethical paradigms. *Machine ethics* is explicitly concerned with the ethics of artificially intelligent beings and systems, such as Isaac Asimov's laws of robotics. *AI ethics*, by contrast, address broader social concerns about the effects of AI systems and the choices made by their designers and users. AI is certainly not unique among emerging technologies in creating ethical quandaries, but ethical questions in AI research and development present unique challenges in that they ask us to consider whether, when, and how machines should to make decisions about human lives—and whose values should guide those decisions.

Debates about whether AI systems should be used in sensitive or high-stakes contexts, who should be making important decisions, and what the proper degree of human involvement should be in various types of decision-making, are also commonplace. We have seen how Facebook mines user data to reveal teenagers' emotional state for advertisers, specifically targeting depressed teens. Cambridge Analytica, the controversial data analytics firm that claimed to be able to shift election results through micro-targeting before it became defunct following its scandal with Facebook, is reported to have had expansive individual profiles on 250 million adult Americans.

Fake news has been used to algorithmically filter news feeds and search rankings in order to influence elections. ML is being used to synthesize audio- and video-realistic representations of public figures and news events. Each of these examples shows how the interests of those deploying advanced data systems can overshadow the public interest, acting in ways contrary to individual autonomy and collective welfare, without this being visible at all to those affected.[25]

Given how relatively new AI is in the mainstream of society and how so many areas of society are struggling to become properly educated about it, there is a general absence of knowledge, oversight, and transparency in the AI space. Nowhere is this more apparent than in the criminal justice system, where the lack of proper safeguards risks eroding the rule of law and diminishing individual rights. Courts and corrections departments around the US use algorithms to determine a defendant's level of "risk", which ranges from the probability that an individual will commit another crime to the likelihood that a defendant will appear for a court date. Algorithms contribute to determinations made about bail, sentencing, and parole, with the objective of improving the accuracy of human decision-making, while resulting in a preferable allocation of finite resources.

Government agencies do not typically write their own algorithms; rather, they purchase them from private businesses, which often means that the algorithm is proprietary. Only the owners (and to a limited degree the purchaser) can see how the software makes decisions. *There is no federal law that sets standards or requires the inspection of these tools in a manner similar to the way the FDA approves new drugs.* This lack of transparency has real consequences. As an example, in 2016, Eric Loomis was found guilty for his role in a drive-by shooting in Wisconsin. Loomis had answered a series of questions that were entered into COMPAS and used by the Wisconsin Department of Corrections. The trial judge gave Loomis a long sentence in part because of the "high risk" score he received from this black box risk-assessment tool. Loomis challenged the sentence because he was not allowed to assess the algorithm. The state Supreme Court ruled against him, reasoning that knowledge of the algorithm's output was a sufficient level of transparency.

This was a worrisome precedent, since algorithms like COMPAS are opaque neural networks—DL algorithms meant to replicate the human brain—which cannot be transparent. Rather than being explicitly programmed, neural networks create connections on their own in processes that are constantly changing. This runs the risk of limiting a judge's ability to render a fully informed decision and defense counsel's ability to fully defend their clients.

Consider a scenario in which a defense attorney calls the developer of a neural-network-based risk assessment tool to the witness stand to challenge the "high risk" score that could affect her client's sentence. On the stand, the engineer could tell the court how the neural network was designed, what inputs were entered, and what outputs were created in a specific case. However, the engineer cannot explain the software's decision-making process. How could a judge weigh the validity of a risk-assessment tool without understanding its decision-making process? How could an appeals court know whether the tool used socioeconomic factors to determine a defendant's risk to society? Following the reasoning in the Loomis case, the court would have no choice but to abdicate a part of its responsibility to a hidden decision-making process.

The role of AI in the judicial system therefore creates two potential paths for the criminal justice and legal communities: *either blindly allow the march of technology to go forward unchecked or create a moratorium on the use of opaque AI in criminal justice risk assessment* until there are processes and procedures in place that allow for a meaningful examination of these tools. The legal community has never fully discussed the implications of using algorithmic risk assessments in court. Attorneys and judges are therefore forced to grapple with the lack of oversight and impact of these tools after they have already become a fact in the courtroom.

Implementing a preventative moratorium on their use would allow for the creation of rules governing how AI risk assessments should be examined during trial and give policy makers an opportunity to create standards and a mechanism for oversight. It would also permit educational and advocacy organizations time to teach attorneys how to manage AI in court.[26]

In 2017, the Asilomar AI Principles were created as a framework to govern how AI may be used ethically and beneficially. Thousands of AI researchers (and others) have signed on to these principles (which are listed at the end of this chapter). Some professionals in the field worry that regulations imposed in the future could prove to be unhelpful and misguided, or even stifle innovation and cede competitive advantage to individuals and organizations in countries where the Principles may not be adopted. Apart from clear examples, like the Loomis case, of the risks posed by AI in the real world, there is, naturally, disagreement among AI researchers about what the risks of AI are, just when that risk could arise, and whether AI could ultimately pose an existential risk.

Few researchers would suggest that AI poses no risk. The number of AI researchers who signed the Principles—as well as the open letters regarding developing beneficial AI and opposing lethal autonomous weapons—shows that there is strong consensus among researchers that much more needs to be done to understand and address known and potential risks of AI. The right policy and governance solutions could help align AI development with these principles, as well as encourage interdisciplinary dialogue about how that may be achieved. The Partnership on AI, which includes leading AI industry players, similarly endorses the idea of principled AI development.[27]

Accountability

Outside of Europe, debates on AI and the law tend to focus more on ethical design, education, and self-regulation than on individual rights. This divergence from the GDPR reflects the complexity of regulating AI and ML when compared with governing previous automated systems. It is clearly a more complicated legal landscape, with the issue of accountability being a critical ingredient, and largely absent in previous debates. In particular, the GDPR's "right to explanation" is potentially disruptive and technically challenging for AI, since some automated decisions must be explained to individuals under the law. Individuals are guaranteed "meaningful information" about the "logic" involved in certain automated decision-making under the law's "right of access". Although it fails to define the scope of information that must in turn be provided, only a general and easily understood overview of system functionality is likely to be required.

The European Parliament's 2017 civil law resolution on robotics struggled to define precise accountability mechanisms. Explaining the rationale or logic of robotic behavior and decision-making aided by AI are clearly called for, but left undefined. Even if future civil law rules for robotics are fully compliant with the GDPR's safeguards against automated decision-making, they will only apply to decisions based solely on automated processing, which may exclude many robotic systems. The outcome may be that robotic decision-making would not qualify as solely automated. Such hesitancy could, ironically, make systems less accountable by preventing the GDPR's safeguards from applying. Automated decisions must also have "legal" or "significant" effects in order for safeguards to apply.

Designing imprecise regulation that treats decision-making algorithms, AI, and robotics separately misinterprets their legal and ethical challenges as unrelated. Concerns about fairness, transparency, interpretability, and accountability are equivalent and ought to be addressed together, regardless of the hardware, software, or data involved. Security robots and predictive policing software identify threats with the same method (automated processing) and purpose (public safety). Such issues can only grow in importance with time. Regulatory and technical accountability mechanisms will be effective only if designed by taking into account the common functionality and diverse complexity of algorithms, AI, and robotics.[28]

In 2017, New York City unanimously passed a bill to bring transparency into the way that the city's government agencies use algorithms to make decisions. The Algorithmic Accountability Bill was the first-of-its-kind legislation to establish a task force to study how algorithms are being used by city agencies to make decisions that impact the citizens of New York. At the same time, the Bill mandated the exploration of how to provide greater transparency into algorithmic decision-making for the public. The task force was focused on investigating algorithmic bias and whether any of the AI being used by the City was discriminating against people based on age, race, religion, gender, sexual orientation, or citizenship status.

The driving force behind the Bill was to shed light on "black box" algorithms (i.e. without knowing how they actually work). By providing more visibility about how predictive models determine specific outcomes, it may be easier for decision-makers to either justify or refute results in the future. The Bill was an important step toward holding governments accountable from a legal perspective and likely set the foundation for future regulations that will be adopted by other governments and organizations.

There is, of course, an argument to be made that AI does not lend itself to being legislated at all. It is difficult to define, and what may constitute an applicable definition of AI or ML in one context may not necessarily fit a legal definition in another context. AI in the healthcare sector may mean something entirely different than in the financial services sector. Before government leaders and law makers may be in a position to pass legislation that is truly all-encompassing, a way should be found to segment AI by category, so that it may be properly defined. Doing so will enable leaders and decision-makers to drive desired outcomes while having a better ability to control possible dangers without putting all the technologies that constitute AI under a single umbrella.[29]

The Future

Following the onset of the Great Recession of 2008, a wave of budget constrained legal clients cringed at the idea of paying fees for research technology. At the time, law firms went from billing their clients for nearly all of their technology costs to recovering only a fraction of those costs, at the same time that clients began to demand alternative fee structures. Billable hours (as they had for decades been known) were quickly becoming obsolete. Law firms aspired to identify new ways to reduce costs while enhancing efficiency, opening the door for AI to enter the legal profession in a big way.

AI SUPREMACY

There is already a plethora of AI platforms in the legal arena. One of the most promising is Casetext, whose software uses natural language understanding to scan the text of legal briefs, providing access to 10 million court cases and statutes annotated by a community of litigators. Its most rigorous ML application is the Case Analysis Research Assistant (CARA), which enables attorneys to drag-and-drop a legal document and, within a few seconds, CARA reads and analyzes the document, returning relevant research to the attorney. This enables attorneys to ensure they are not missing any key precedents as part of their research or to catch opposing counsel that may have left something critical out. Attorneys who do not use CARA or a similar AI tool will be at a distinct disadvantage going forward.

Legal writing is based in large part on precedent. CARA can help predict the thrust of opposing counsels' briefs, since its ML model is focused on a network of citations between legal cases. It analyzes the relationships that shape the law, delving deep into literature ranging from the classics to science, and identifying anything that could be useful to legal research. Casetext develops "citation bundles"—examples related to case law because they are often cited together. CARA revised the way it conducted research to identify the topics specific briefs addressed, weighting the importance of each citation relationship and how recently certain cases had been cited, based on more than 100 factors. Yet, machines have a hard time deciphering ideology or a particular political philosophy that might underpin a decision.

One organization—FantasySCOTUS—tries to predict the outcome of US Supreme Court cases by using the predictive powers of tens of thousands of its members, compared with ML-driven predictions. Overall, humans have nearly always out-predicted AI systems, but the law is sometimes intentionally politicized or ambiguous, making accurate predictions even more difficult to achieve. Unless human minds can be read and the unique circumstances imbued in legal cases can be foreseen, there can be no way for humans or machines to accurately predict judicial outcomes.

Determining whether use of a copyright is permissible "fair use" requires a multiple-part test whose variables defy mathematical certainty. Any such decision will vary from judge to judge. While better defined areas and less obscure parts of the law (such as tax or accounting) may, in due course, prove to be computable, many other aspects of law are simply not computable,[30] which will limit just how far AI and ML will come to govern its outcomes.

Lawyers may never be replaced by machines, but they are already being augmented. The next frontier for law firms is to integrate NLP processing to carry out legal research. One such tool to achieve that is Legal Robot, which uses ML techniques such as DL to provide a suite of service offerings to understand legal language. Doing so permits comparison between the language with a specific contract containing boilerplate versus custom wordings, measuring the complexity and readability of the language and identifying the responsibilities, rights, and terms of an agreement. The information is then presented in a way that encourages exploration of a contract and is more easily understood, enabling lawyers to quickly interpret the components of the contract, quantify specific risks, and identify and resolve defects.

Legal Robot transforms complex legal language into numeric expressions so that ML techniques can derive meaning. Each layer of information is used to form higher level concepts, from words, to sentences, to themes, then continuously compares thousands of documents to build a legal language model that expands legal language analysis. Legal Robot either enhances or entirely replaces traditional legal processes (like contract reviews) with an automated intelligent assistant. By using its legal language model, the intelligent assistant flags issues and suggests improvements by considering best practices, risk factors, and jurisdictional differences.

The program's automated contract review also helps non-lawyers understand complex legal language and identify issues or problems before contracts are signed, without the time or cost required to hire an attorney. It evaluates contracts against standards across industries or jurisdictions, using algorithms that assess fairness and risk so that all parties may clearly understand contract terms and be confident about what they are signing. Legal Robot also provides instant error checking for contracts and uses advanced scoring to suggest changes that improve readability, internal consistency, and adherence to best standards and practices. It can even become integrated with Enterprise Contract Management systems to assist legal departments and auditors in analyzing portfolios of agreements in minutes, mitigating future legal expense by flagging agreements with questionable language for renegotiation.[31]

Some of the world's biggest law firms are getting on the AI bandwagon in a significant way. As part of its innovation program to reimagine the business of law, Baker McKenzie instituted an innovation committee that reports directly to the firm's executive committee. Part of its innovation framework involves employing "design thinking" to reimagine the services it provides and view them from the perspective of the client's needs, with technology being a key component of the firm's future suite of services. Other firms use key performance indicators to help them target resources on initiatives that make a strategic difference and will give them a comparative advantage. The challenge in doing so is to produce outcomes that are relevant, measurable, and forward thinking.

The integration of technology into the legal arena has already contributed to tens of thousands of jobs. Some predict that automation will cut jobs in the legal sector by another 40% over the next two decades. While there will be fewer jobs for traditional lawyers, there should be great opportunity for lawyers with a mix of new skills that incorporate science, forecasting, and social sciences into the law. Some universities recognize the shift and have already crafted programs designed to foster a new breed of lawyer. For example, Ulster University launched the UK's first Legal Innovation Center in 2017 to operate at the intersection between technology, innovation, and access to justice.[32]

Another product (on the Canadian market) is Tax Foresight, which can predict with greater than 90% accuracy how a court that ruled on a case would likely do so under new circumstances. It asks questions about the client's situation and analyzes thousands of cases produced by the Tax Court of Canada, Federal Court of Appeal and Supreme Court of Canada. Tax Foresight then provides a prediction, an explanation for why it believes in such an outcome, and a list of relevant cases for further research. Such technologies could well change the nature of litigation as it will increase the likelihood of settlements, while decreasing the likelihood that cases will go to court.

That said, the vast majority of law firms remain uncomfortable with the thought of being early adopters. According to a 2016 survey,[33] although more than half of all law firms (53%) reported larger tech budgets in 2016 than in 2015, the majority focused their efforts on bolstering cybersecurity, information governance, business continuity, disaster recovery or security compliance requirements; 87% of respondents said their firms were not evaluating or utilizing AI technologies. Law firms are notorious for being slow to adopt new technologies and many lawyers view AI as a threat rather than viewing it as an opportunity.[34]

Just as the insurance industry has had to be dragged kicking and screaming into the 21st century in the US by slowly adopting what has become standard practice for the industry in other countries—digitizing the purchase of insurance through the Internet, rather than through insurance brokers— the legal profession in the US and beyond faces a stark choice. Will it resist the inevitable and embrace the growing use of AI to streamline processes and save time and cost, or will it insist on doing things "the old-fashioned way"? Plenty of consumers will want to derive comfort from the belief that ML is not involved in the determination of their fate under the law but, as has already been noted, that belief is already incorrect. Law firms, attorneys, and other legal professionals can either be part of the solution by choosing to help craft the law's future in conjunction with AI, or they can be swept up in its outcome after the fact. The train is leaving the station.

The Asilomar AI Principles

1) The research goal should not be undirected intelligence but beneficial intelligence.

2) Investments should be accompanied by funding for research on ensuring its beneficial use, including difficult questions in computer science, economics, law, ethics, and social studies.

3) There should be constructive/healthy exchange between AI researchers and policy makers.

4) A culture of cooperation, trust, and transparency should be fostered among researchers and developers of AI.

5) Development teams should actively cooperate to avoid corner-cutting on safety standards.

6) AI systems should be verifiably safe and secure throughout their operational lifetime.

7) If an AI system causes harm, it should be possible to ascertain why.

8) Any involvement by an autonomous system in judicial decision-making should provide a satisfactory explanation that is auditable by a competent human authority.

9) Designers and builders of advanced AI systems are stakeholders in the moral implications of their use, misuse, and actions, with a responsibility to shape those implications.

10) Highly autonomous AI systems should be designed so that their goals and behaviors can be assured to align with human values throughout their operation.

11) AI systems should be designed and operated so as to be compatible with ideals of human dignity, rights, freedoms, and cultural diversity.

12) People should have the right to access, manage, and control the data AI systems generate.

13) The application of AI to personal data must not unreasonably curtail real or perceived liberty.

14) AI technologies should benefit and empower as many people as possible.

15) The economic prosperity created by AI should be shared broadly, to benefit all of humanity.

16) Humans should choose how and whether to delegate decisions to AI systems.

17) The power conferred by control of highly advanced AI systems should respect and improve, rather than subvert, the social and civic processes on which the health of society depends.

18) An arms race in lethal autonomous weapons should be avoided.

19) We should avoid strong assumptions regarding the upper limits on future AI capabilities.

20) Advanced AI could represent a profound change in the history of life on Earth, and should be planned for and managed with commensurate care and resources.

21) Risks posed by AI systems—especially catastrophic or existential risks—must be subject to planning and mitigation efforts commensurate with their expected impact.

22) AI systems designed to recursively self-improve or self-replicate in a manner that could lead to rapidly increasing quality or quantity must be subject to strict safety and control measures.

23) Superintelligence should only be developed in the service of widely shared ethical ideals, and for the benefit of all humanity, rather than one state or organization.[35]

14. An AI Future to be Feared or Embraced?

Throughout the course of this book we have explored a broad range of topics associated—directly or indirectly—with utilizing AI, benefitting from it, protecting against it, and racing toward a theoretical finish line. Achieving AI supremacy is as much about having fully embraced AI as it is about contemplating what your competitors are doing to achieve AI supremacy before you do. As AI and ML continue to evolve, new and ongoing challenges will arise, whether it is the loss of jobs, a continued erosion of privacy, or the integration of the human body with machines. In this chapter we explore some of the manifestations of the AI revolution currently sweeping the world, some implications, and some predictions about where we go from here.

The passage of time will raise more questions, many of which will either not have a single answer or perhaps any meaningful answer at all. Among the many questions AI raises are how bias can be eliminated from a work product, how we can be protected against unintended consequences, and how we can stay in control as we gradually hand the reins to machines. The pace of change in the AI and ML arena is already breathtaking and promises to continue to upend conventional wisdom and surpass some of our wildest expectations as it proceeds on what appears at times to be an unalterable and pre-ordained course. Along the way, much of what we now consider to be "normal" or "acceptable" will change. It is, of course, already changing.

Google's Vision of the Future

A 2016, video produced at Google provided a stunningly ambitious and unsettling look at how some people within the company envision using the information it collects in the future. Shared internally at the time within Google, the video imagines a future of total data collection, where Google subtly nudges users into alignment with the company's own objectives, custom-prints personalized devices to collect more data, and even guides the behavior of entire populations to help solve global challenges such as poverty and disease. Entitled "The Selfish Ledger", the 9-minute film maintained that the way we use our smartphones creates a constantly evolving representation of who we are, which it terms a "ledger", positing that these data profiles can be built up, used to modify behaviors, and transferred from one user to another.

This ledger of our device use—the data on our actions, decisions, preferences, movements, and relationships—is something that can be passed on to other users, much as genetic information is passed on through the generations. Building on the ledger notion, the video presents a conceptual Resolutions app by Google system in which Google prompts users to select a life goal and then guides them toward it in every interaction they have with their phone. The examples, which reflect Google's values as an organization, include, for example, urging users to try a more environmentally friendly option when hailing an Uber or directing them to buy locally grown produce. The concept is premised on Google having access to a huge amount of user data, insight into how a given user makes decisions, and in which policy-oriented direction a user may be inclined.

Neither privacy-related issues nor an exploration of the potentially negative aspect of this Google-governed reality are mentioned or explored in the video. The ledger's requirement for ever more data, and the presumption that billions of individuals would be just fine with a Google-governed world, are unnerving. The video envisions a future in which goal-driven automated ledgers become widely accepted; it is the ledger, rather than end user, which makes decisions about what might be good for the user, seeking to fill gaps in its knowledge in a *Black Mirror*-type utopian reality.

The vision of the future presented by the video sees the ledger as surpassing any concept of self-improvement; rather, it would plug gaps in its own knowledge while modifying human behavior, and not only that of an individual, but rather, that of the entire human species. The video's creator envisioned a future in which ledger users benefit from the preceding generation's behaviors and decisions, in essence "sequencing" it as the human genome was sequenced, while making increasingly accurate predictions about decisions and future behaviors. The ultimate objective is to develop a species-level understanding of complex issues such as depression, health, and poverty, modeled in Google's own image.

That it also happened to consume as much information as possible about as many people as possible in the process was taken as a given. In fairness to Google, it basically already does this, just as other Internet Service Providers, social media companies, smart device manufacturers, and a range of other firms in the Internet era already do–they just do not refer to what they do in that way. And, *every one of us who use the services these organizations provide has given them tacit approval to do so, because we all agreed to their Terms of Service (even though almost none of us have read them), which gives them the ability to consume our data and pretty much do whatever they want with it.* What the Google video implies is that, *going forward, the potential power the company will have, and the potential consequences of its actions, have not yet even been imagined.*

While the ledger concept would appear to be within the realm of science fiction, it aligns nearly perfectly with some of the approaches already taken by a variety of Google's existing products. For example, Google Photos already presumes to know what you will consider to be life highlights, and produces entire photo albums on the basis of its AI interpretations, before presenting it to users in a neatly arranged package. Google Maps and the Google Assistant already make suggestions based on information they have about your usual location and habits, and make suggestions about preferred routes or courses of action. The fact is, *Google is increasingly inquisitive about its users, assertive in how it wishes to interact them, and pressing existing limits about what is considered an acceptable level of intrusion into their lives–much of which may be welcomed, based on how we have already been "programmed" to accept the company's unsolicited overtures and now consider them to be perfectly normal and acceptable.*

As the ethical deployment of emerging technologies—and AI specifically—continue to be subjects of public discourse, Google appears to be unphased by the potential ethical implications of its current products, practices, and vision of the future,[1] or whether it is overstepping its bounds by proceeding apace to implement its that vision. *Google wants to understand and control the future before it occurs by in essence creating it, using AI and ML to help interpret and manage it.* That is both a welcome and a chilling proposition, but our collective technological future is unfolding at lightning speed, and no single government or company can control it.

So, is Google to be commended for attempting to contain and craft the future, or should it be feared and resisted at every turn? Is there a middle ground? Will the fact that most consumers do not know the difference, or necessarily care, enable organizations like Google to basically do whatever they want? Is our great leap into the AI unknown meant to be purely exhilarating, or should we be intuitively cautious and approach it with care? These are among the many questions you will undoubtedly have raised in your own mind throughout the course of this book. The truth is, there is no single answer to these questions, nor is there one that is necessarily a right or wrong answer.

Other Glimpses of the Future

What seems clear, however, is that human kind has never before had to contend with such blisteringly bold and mind-bogglingly complex potential implications for its own future. The idea that the future potential union of man and machine may not only be inevitable, but achievable sometime in our own lifetimes, is staggering. Based on some of the examples that follow, it should become clear that *we are already on the path toward what appears to be a convergence with our destiny in what should be considered baby steps. As "advanced" as we would like to believe our progress is and our capabilities have become in the AI and ML arena, we are merely at the beginning of the runway.*

That said, Google is also busy at work interpreting reams of data to train machines to predict when people will die. As an example, in 2018, a woman with late-stage breast cancer arrived at a hospital with fluid in her lungs. She saw two doctors, had her vital signs read, and received a radiological scan. The hospital's computers then estimated that there was a 9.3% chance she would die during her stay at the hospital. A Google algorithm subsequently reviewed more than 175,000 data points about the woman and predicted she actually had a 19.9% chance of death during her stay.[2] She passed away in a matter of days. Google has created a neural network tool that can forecast a host of patient outcomes, including, in addition, how long people may stay in hospitals and their odds of re-admission.

What made a real difference is Google's ability to sift through data that was previously out of reach to most medical professionals, such as notes buried in PDFs or scribbled on old charts. The neural network consumed vast amounts of information and processed them far faster and more accurately than existing techniques, even identifying for doctors which records led it to particular conclusions. Google's next step is to move this predictive system into clinics, to be able to link symptoms and disease with a level of accuracy that is being met with hope, as well as alarm.

Since Google declared itself an "AI-first" company in 2016, much of its work in this area has been devoted toward improving existing Internet services. The advances coming from its Medical Brain team give Google the chance to break into a brand-new market with real commercial potential. While software in healthcare is mostly coded by hand, Google's approach is to enable machines to learn to parse data on their own, which is enabling the company to leapfrog over most everyone else. Yet, for all the optimism over Google's potential in this arena, harnessing AI to improve healthcare outcomes remains a huge challenge.

Google has long sought access to digital medical records, which will only add to the vast treasure trove of data at its disposal. To support this foray into medical research, three US universities granted the company access to 46 billion pieces of anonymous patient data. Yet Google is treading carefully regarding patient information, particularly as public scrutiny over data-collection continues to rise.[3] Most people recognize that mass data collection is simply a fact of life now and are undoubtedly willing to live with it as long as some good comes from it.

Google's endeavors in this area, and advances in AI and ML more generally, are a leap into the unknown and a leap of faith. How *can* we actually know for certain that the data we provide to such organizations (wittingly or unwittingly) *will* be used for the purposes the data collectors claim? How will most of us ever know what the outcome of such data mining is? Can we say with any degree of certainty that the actions AI and ML developers are taking today will benefit humanity decades and centuries into the future? Some would say no one can know, and this is simply the price that must be paid to have the privilege of taking that great leap into the unknown. Others would say that the profit motive and human nature do not bode well for the ultimate outcome of such actions.

Still, one has to marvel at the sheer range of possibilities that AI represents. The world's leading militaries are rapidly developing cutting edge AI tools and weapons. In 2018, the DARPA selected teams to develop a "neural interface" that would allow troops to connect to military systems using their brainwaves while enabling those systems to transmit information back directly to users' brains. The Next-Generation Non-Surgical Neurotechnology (N3) program aims to combine the speed and processing power of computers with humans' ability to adapt to complex situations. The technology would let people control, feel, and interact with a remote machine as though it were a part of their own body.

The DARPA has already created a prosthetic limb that disabled veterans can control using an electrode implanted in their brains. The system gives users "near-natural" arm and hand motion while transmitting signals that mirror a sense of touch back to their brain. The N3 program is divided into two tracks: non-invasive interfaces that sit completely outside the body, and minutely invasive interfaces that could require users to ingest different chemical compounds to help external sensors read their brain activity. In both tracks, technologies must be "bidirectional", meaning they can read brain activity and also produce new information that is sent back to the user.

N3 technology is best thought of as means to use to a computer without a mouse, keyboard, or touch screen. The program is solely focused on designing an interface for humans to connect with technology. The interface might, for example, be used to help a pilot coordinate a fleet of drones with his or her thoughts, or permit a soldier to control a remotely deployed robot by using the brain's motor signals. In the future, neural interface technology can help militaries build a more intuitive interaction with autonomous weapons systems.[4]

China is developing smart, low-cost unmanned submarines that can roam the world's oceans to perform a wide range of missions, from reconnaissance to mine placement to suicide attacks against enemy vessels. The autonomous robotic submarines are expected to be deployed in the early 2020s. Apart from having built the world's largest testing facility for surface drone boats, China's military researchers are also developing an AI-assisted support system for submarine commanders to make faster, more accurate judgments in combat. While the new class of unmanned submarines will establish contact with the ground command for updates, they are designed to complete missions without human intervention.

The AI-powered subs are large compared to the standard UUVs. They dock as conventional submarines, their cargo bays can be reconfigured, and they can accommodate a wide range of freight, from surveillance equipment to missiles or torpedoes. As a result of their diesel-electric engines and other power sources, they may remain in continuous operation for months. The subs can gather intelligence, deploy mines, or be stationed at geographical choke points where armed forces are likely to pass, to ambush enemy targets. If necessary, they can also ram into a high-value target, like a conventional torpedo. The unmanned submarines would not be nuclear-armed. Their primary advantage is that they may be produced and operated on a large scale at a relatively low cost.[5]

Researchers at the University of Washington School of Medicine developed a robotic system, in 2018, to automate the production of human mini-organs, derived from stem cells. The ability to rapidly mass produce organoids promises to expand the use of mini-organs in basic research and drug discovery and, one day, perhaps human organ replacement. The system was tested in producing kidney organoids, including models of polycystic kidney disease, and the robots were also programmed to analyze the organoids they produced. The researchers also discovered a way to greatly expand the number of blood vessel cells in their organoids to make them function more like real kidneys.[6] This opens up exciting new pathways for treating a wide range of diseases in a cost and time-effective manner.

Researchers at Ohio State University have developed a nanotechnology-based chip that can successfully inject code into damaged skin cells to repair them. In the future, scientists will be able to use this process–Tissue Nanotransfection (TNT)–to allow *the genetic code of a robot to be passed onto its offspring, along with human genetic code. The robots of the future could ultimately be able to manipulate human skin cells to create human sperm eggs*. From the eggs, robots may be capable of *creating a human baby whose embryo can be nurtured and carried through pregnancy by a mother surrogate. In this way, humans and robots many be able to make babies together*. In due course, TNT could render in vitro fertilization and even sexual intercourse redundant.[7]

Swedish scientists have researched how to produce digital copies of humans who have already died, to create robots which resemble the deceased. In 2018, Sweden's Fenix funeral agency started looking for volunteers who would grant permission for scientists to replicate the images of their deceased relatives. The digital version of the deceased will be based on a chatbot that functions similarly to how humans currently chat with robots online. The idea is that the technology learns to communicate in the same way as the deceased person, by mapping and analyzing the person's social media and emails.[8]

The AI technology will reconstruct voices of the dead and teach the computer program to answer elementary questions. The program's creators envision that it may approach the level of sophistication of AI portrayed in the Black Mirror television series, in which the robots become sentient human clones, capable of not only conversing with, but coming into physical contact with, real people. The objective is to enable a robot to take the place of a photograph of a loved one to enable living relatives to relive memories of their past.[9]

Thousands of people from Sweden have taken a step toward the merger between humans and machine by inserting microchips into their bodies, which can function as contactless credit and key cards, as well as potential payment mechanisms. While the idea of carrying a microchip in one's body may feel dystopian to some, to many others it is simply a matter of practicality.

Swedes have a love affair with all things digital and they are more relaxed about sharing the personal details contained in microchips with businesses and government entities because of the way the Swedish social security system is structured.[10] It is relatively easy for Swedes to simply look up information about each other in publicly available databases, so, from many of their perspectives, they are giving up little in the way of personal privacy by doing so. Anyway, in a sense, we are already cyborgs since our smart phones and computers are already extensions of us and hackers can already access much of our most personal information. The only thing the rest of us lack is a chip inside our bodies and a willingness to voluntarily share such information.

      In 2018, Amazon Go was launched—a completely automated grocery store in Seattle with an automated checkout process. Sensors backed by ML monitor customers picking up and either returning a product to the shelf or placing it into their shopping carts. After swiping to gain entry into the store through the company's app, shoppers are able to freely remove items off the shelf and leave the store without needing to take out their wallets, because the app bills them and sends a virtual receipt. Human employees are still used to stock shelves and prepare fresh meals. Amazon Go is so confident of the reliability of its cameras and sensors that it has no real concern about shoplifting, and the company built in no safeguard against it. The company plans to open Go stores throughout the US.[11]

      Amazon also wants to expand its footprint from the home to the office, having launched its Alexa for Business product in 2017. The objective is to encourage businesses to feed Alexa (the voice-controlled virtual assistant in Amazon Echo devices) to book conference rooms, manage work calendars, order office supplies, control room temperature and lighting, and a range of other services. Amazon envisions a world wherein companies manage thousands of employees' shared Echo devices and employees routinely speak their requests out loud, enabling sales forecasts to be retrieved and presentations to be made without even clicking a mouse. A company's management will, of course, need to get comfortable with the concept that Amazon will hear everything that occurs in the company, and be satisfied with its security implications, before Amazon for Business is likely to be widely adopted.[12] In today's hacker-filled world, wherein cybersecurity protocols are successfully breached on a daily basis, that will be a true leap of faith.

In 2018, Scottish scientists genetically engineered pigs to be immune to one of the world's most costly animal diseases, an advance that could soon propel gene-editing technology into commercial farms. Pigs infected with porcine reproductive and respiratory syndrome (PRRS), a disease that is endemic across the globe, are safe to eat but the virus can cause pregnant sows to lose their litter and result in the deaths of piglets. There is no effective cure or vaccine. After deleting a small section of DNA that leaves pigs vulnerable to the disease, the animals showed no symptoms or trace of infection when intentionally exposed to the virus and when housed for an extended period among infected siblings.

Gene editing differs from older genetic modification techniques, which often involve transferring genes from one species to another. By contrast, gene editing uses precise molecular tools to remove small stretches of DNA or alter single letters in the genetic code–effectively speeding up processes that could otherwise occur naturally over many generations. In the trial, the animals showed no signs that the change in their DNA had any other impact on their health or fertility. PRRS is one of numerous other animal diseases being targeted through gene editing. Among them are E. coli, campylobacter, the oyster herpes virus, and amoebic gill disease in Atlantic salmon.[13]

Also in 2018, Alibaba's Cloud Unit joined forces with the Tequ Group–a Chinese food and agriculture conglomerate that raises about 10 million pigs annually–to deploy facial and voice recognition on Tequ's pig farms. Using image recognition, software will identify each pig based on a mark placed on its body, which will correspond with a file for each pig kept in a database, which in turn records and tracks characteristics such as the pig's breed type, age, and weight. The software can monitor changes in the level of a pig's physical activity to assess its level of fitness. It can also monitor the pigs' sounds on the farm, recording each animal's cough, for example, to assess whether or not the pig is sick and at risk of spreading a disease. The software will also use its data to assess which pigs are most capable of giving birth to healthy offspring.[14]

Xiao Long (or "Little Dragon") is a robot at China's first fully automated, human-free bank branch of the China

Construction Bank (the second largest bank in the world and the sixth largest company in the world as of 2015[15]). Xiao speaks with customers, accepts bank cards, checks accounts, and can answer basic questions. After first speaking to Xiao, customers pass through electronic gates where their faces and ID cards are scanned. On future visits, facial recognition alone will be sufficient to open the gates and retrieve customer information. Once inside, ATMs assist with services such as account opening, money transfer and foreign exchange. A staggering number of security cameras keep an eye on things. If someone loiters for too long or breaches security in any way, a human security guard appears.

Robots are taking over more and more aspects of everyday life in Chinese cities. They have been deployed in train stations for security purposes, programmed to scan travelers' faces and respond to common questions. Robots are being used to cook—both in restaurants and industrial kitchens—and robotic waiters have been a staple in restaurants across China for a number of years. Robotic waiters can be frustratingly slow and most must move along pre-programmed tracks, however, and some restauranteurs believe they may be more trouble than they are worth. Customers in the Hangzhou branch of Kentucky Fried Chicken can pay for their orders using only their faces, and retailers including Tencent have been experimenting with cashier-less stores[16] similar to Amazon Go.

In 2017, at London's Love and Sex Robot Conference, ethicists, computer scientists, and sex toy developers met to discuss how robots are already affecting our sex lives. It may come as a surprise to many, but the term "digisexual" (preferring to have sex with robots) is already a widely accepted sexual preference, with sex doll brothels having already proven popular in China and Japan. Europe's first robot brothel opened the same year in Barcelona and was staffed almost entirely by life-size sex dolls available for "appointments" at a rate of €80 an hour. Operating out of an apartment, the Lumidolls brothel offered sessions with one of its four life-like dolls. Upon arrival, clients would find their chosen companion in a private room, lit by candlelight.[17]

Some enthusiasts believe that, as robots become more intelligent and resemble human appearance more accurately, more people will fall in love and want to have sex with robots.

Sex robots have sensors hidden under their skin and can already respond to affection in a human-like manner. They have been designed to purr like a kitten or whisper endearments, if stimulated the correct way. A micro-computer inside their plastic skulls can also recall previous encounters, so as to enhance future experiences. They may also adopt "family", "romantic", "sex", and "extra naughty" settings. The sex robots may also be programmed to enter "dummy" mode when bored with a session or detect when touching becomes too aggressive or disrespectful, which will result in the robot in shutting off.[18]

Researchers have taught AI to recognize all sorts of patterns in peoples' actions and behaviors, allowing it to accurately predict their next move in advance. Built by a research team at the University of Bonn, its AI was taught to anticipate actions by watching hours of cooking videos. Using pre-recorded videos of people preparing a meal, the researchers were able to teach the machine to recognize each action being performed on-screen, including cutting tomatoes, adding salt, and flipping a pancake. Each recording was about six minutes in length and contained some 20 different actions. After four hours, the algorithm was able to recognize the sequence of events needed to prepare a dish.

The software was then presented with videos that it had not seen before, in which the machine was told what was happening for the first 20% or 30% of the clip. The algorithm was then asked to predict the next action before it took place on-screen. The machine drew upon on its knowledge of the recipe, and its understanding of how similar sequences have played out previously and was able to correctly anticipate actions in the near future with surprising accuracy.[19] It seems Google will not have a monopoly on AI's ability to predict the future. While predicting the course of cooking a meal is nothing like predicting a criminal act or the course of an election, it opens a world of possibilities, for good or ill.

An Expanding Malicious Attack Landscape

The cyber era heralded unparalleled opportunities for the advancement of science, technology, and communication, and also unleashed a range of new attack vectors for rogue elements, criminals, and virtual terrorists. The AI era is doing much the same, for the promise of advancement has gone hand in hand with a range of new perils and an expanded set of actors capable of carrying out attacks based on AI and ML. This flows naturally from the efficiency, scalability, and ease of diffusion of AI systems, which can increase the number of actors who can carry out attacks against civilian, business, and military targets utilizing AI.

      The typical character of threats derived from AI is likely to shift in some distinct ways in the future. Attacks supported and enabled by progress in AI will be especially effective, finely targeted, and difficult to attribute, as they have been in the cyber arena. Given that AI can, in a variety of respects, exceed human capabilities, attackers may be expected to conduct more effective attacks with greater frequency and on a larger scale. Attackers often face a trade-off between how efficient and scalable their attacks will be versus how finely targeted. For example: the use of drone swarms that deploy facial recognition technology to kill specific members in a crowd of people versus a mass casualty event.

      AI systems may be able to avoid detection by using a learning model that automatically generates command and control domains that are indistinguishable from legitimate domains by human and machine observers. Such domains can be used by malware to "call home" and allow malicious actors to communicate with host machines. Attackers are likely to leverage the growing capabilities of reinforcement learning to benefit from experience in order to craft attacks that current technical systems and IT professionals are not prepared for.

      For example, services like Google's VirusTotal file analyzer allow users to upload variants to a central site and be scrutinized by more than 60 security tools. This feedback loop presents an opportunity to use AI to aid in crafting multiple variants of the same malicious code to determine which is most effective at evading security tools. Additionally, large-scale AI attackers can accumulate and use large datasets to adjust their tactics, as well as modify the details of the attack for each target. This may outweigh any disadvantages they suffer from the lack of skilled human attention to each target, and the ability of defenders like antivirus companies and IT departments to learn to recognize attack signatures.

      Malicious AI actors and cyberattackers are likely to rapidly evolve in tandem in the coming years—in the virtual and physical arenas—so a proactive effort is needed to stay ahead of them. There is a growing gap between attack capabilities and defense capabilities more generally because defense mechanisms are capital-intensive and the hardware and software required to conduct attacks are increasingly not necessarily expensive and widely distributed. Unlike the digital world, where critical nodes in a network such as Google can play a key role in defense, physical attacks can happen anywhere in the world, and many people are located in regions with insufficient resources to deploy large-scale physical defenses. Some of the most worrying AI-enabled attacks may come from small groups and individuals who have preferences far removed from what is conventional, and which are difficult to anticipate or prevent, as with today's "lone-wolf" terrorist attacks.

Since the number of possible attack surfaces is vast, and the cutting edge of attack and defense capability is likely to be ever progressing, any equilibrium obtained between rival states, criminals and security forces, and competing organizations in a particular domain is likely to be short-lived as technology and policies evolve. Technology and media giants will in all likelihood continue to be the default technological safe havens of the masses, given that their access to relevant real-time data on a massive scale, and their ownership of products, communication channels, and underlying technical infrastructure, place them in a highly privileged position to offer tailored protection to their customers. Other corporate giants that offer digitally-enhanced products and services (in the automotive, medical, defense, and many other sectors) are coming under pressure to follow suit. This is due in large part to a growing trend, in which people routinely use the platforms provided by technology and media giants and interact less frequently with small businesses and governments.

Developed countries generally, and the leading countries in AI and cyber capabilities specifically, have a clear head start in establishing the control mechanisms to provide security for their citizens but maintaining that comparative advantage requires significant ongoing commitment from a plethora of resources. What is also required, of course, is the maintenance of forward thinking organizational strategic planning, which is not necessarily in abundance. Much more work must be done to establish the right balance between openness and security, improving technical measures for formally verifying the robustness of systems, and ensuring that policy frameworks developed in a world which was previously less AI-infused adapts to the new world we are creating.[20]

Integrated Strategy Machines

Although there are many more dislocations to be unleashed before the AI revolution has peaked, it appears unlikely that humans will be entirely replaced by machines, which lack gut instinct, emotion, and the five senses, among other things. In this new AI-inspired world, where comparative advantages have been transformed, strategic issues morph into organizational, technological, and knowledge issues—and vice versa. Structural flexibility and agility—for both man and machine—become imperative to address the rate and degree of change. In addition to reframing specific sources of competitive advantage, AI helps increase the rate and quality of decision-making. *Predictive analytics and objective data are in the process of replacing gut instinct and experience as a central driver of many decisions.* Rather than scrap traditional sources of competitive advantage—such as position and capability—AI reframes them.[21] Organizations therefore need to adopt a fluid and dynamic view of their strengths.

Big data, advanced analytics, and AI are at the forefront of the CEO agenda, an indication that the majority of companies view technology as a potential answer to many—perhaps even most—of their future challenges. Technology-enhanced strategy can be realized only in the context of an integrated strategy machine: a collection of resources (both technological and human) that act in concert to develop and execute business strategy. It comprises a range of conceptual and analytical operations—including problem definition, signal processing, pattern recognition, abstraction and conceptualization, analysis, and prediction—that connect into a seamless whole.

Since machines and algorithms can play increasingly large and important roles in strategy making and execution, the integrated strategy machine must, at least for now, be designed by humans, because we remain unique in our capacity to think outside the immediate scope of a task or problem. Machines cannot yet do that well. They are good at executing a well-defined task or solving a well-defined problem, but they cannot pose new questions or connect a problem to a different one they previously faced. Perhaps they will never be able to match humans in that regard.

Amazon provides an excellent example of an integrated strategy machine. The company has at least 21 data science systems, including several for supply chain optimization, an inventory forecasting system, a sales forecasting system, a profit optimization system, a recommendation engine, and many others. These systems are intertwined with one another and with human strategists to create an integrated, well-oiled machine. For example, if its sales forecasting system detects that the popularity of an item is increasing, it triggers a cascade of changes: the inventory forecast is updated, causing the supply chain system to optimize inventory across warehouses, the recommendation engine focuses on more profit optimization and adjusts pricing, and these changes, in turn, update the sales forecast. The integrated strategy machine can do for information technology what new factory designs did for electricity.

However, the increasing intelligence of machines will be wasted unless businesses reshape the way they develop and execute strategy. Business leaders must start thinking now about how they can integrate their two key assets—people and technology—or risk falling further behind.[22] They have come to realize that they must change how their organizations function and, as they do, they must be especially mindful of the failed technology transformations that have come before. Too many organizations are pigeon-holing AI talent, requiring them to think and work in silos. This is a mistake. *Just as risk managers should not be sent out for biscuits when decisions are being made, AI talent should be integrated throughout the organization to help ensure an appropriate amount of cross-fertilization.*

The organizational retooling process can be lengthy and even complex for many firms, but some of them will get it right, combining the correct amount of leaning forward with the right amount of visionary thinking. This "upskilling" of a workforce to become in sync with an organization's own AI future is not just about teaching and absorbing new skills—it is about a new mind set, new forms of collaboration, and a willingness to turn the pyramid upside down.

If it is harnessed correctly, AI will help decision makers to make smarter decisions while empowering employees to add more value to existing enterprises. This can be done by automating processes too complex for older technologies, identifying trends in historical data to create business value, and providing forward-looking intelligence to strengthen human decisions. Many organizations will find this a challenge. Some will face growing pressure from end users and regulators to deploy AI that is explainable, transparent, and provable. That may require vendors to share some of their secrets, and users of DL and other advanced AI to deploy new techniques that can explain things that were previously incomprehensible.[23]

Another great example of a company that has gotten it right is IBM, which has been in the AI arena for decades. The world is familiar with the firm's best-known incarnation of AI—Watson—which has been in the public eye since 2011. A wide variety of AI technologies sit under the Watson umbrella, including ML, DL, voice recognition, sentiment analysis, and NLP. The company's senior management decided to make Watson the centerpiece of its long-term strategy and created a business unit in 2014 around Watson, starting with a $1 billion investment and 2,000 full-time employees. To double down on its bet, IBM implemented Watson's Law, a visionary approach to thinking about the world, so that Watson can be applied in business, smart cities, consumer applications, and life more generally.[24] IBM has leapt to the front of the line in terms of AI investment, leadership, and product applicability—all dependent on IBM technology—which, the firm believes, gives it a significant competitive advantage.[25]

IBM is betting big on Watson and cloud computing, but it will be years before these services supplant the declining profit of the company's eroding legacy products. Unlike IBM's previous transformation into a consulting and technology company in the 1990s, its current transformation is proving more difficult, for two primary reasons. IBM is promoting radical transformation among its client companies using automated expertise on a large scale to efficiently solve problems too large and complex for humans to conquer on their own. IBM's cloud-based services will likely cannibalize three IBM mainstays: computer hardware, software, and data center services. By making such a bet, IBM is facing greater resistance internally (among business units protecting their turf) and externally (among investors decrying less than desirable short-term results and future implied risks).

To pull off this transformation IBM has to make a number of changes in thinking, practice, and culture—the types of pivots that come easily to start-ups but not so easily for decades old companies. Watson's progress in the healthcare market is a driver of this transformation, which also provides important lessons for any large organization seeking to deliberately disrupt its own businesses. Among the lessons to be learned is that rather than attempting to apply a cookie cutter approach to enterprise-wide change, it is important to develop a portfolio of capabilities with different risk/return profiles.

To achieve superior returns, the most successful venture capitalists (VCs) constantly analyze their portfolios to gauge this risk/return balance and focus on the companies with the best prospects. Why should established businesses be any different? Companies seeking to do what IBM is doing should think and act like a VC by quickly evaluating the effectiveness and return on investment of each of its ecosystem partners and acquisitions in accordance with the Watson's business unit's goals and the timelines, while having the courage to quickly modify or even eliminate those that do not work according to plan.

IBM developed Watson's base intellectual property (IP) largely based on its own research and development labs, and has been an active acquirer of companies with strategic assets and knowledge that complement Watson. IBM has favored buying and scaling firms with significant existing revenue, rather than buying only those that have IP. To ensure Watson's continued success, IBM has had to strike the right balance between buying large companies with market dominance and faster revenue impact versus purchasing additional startups with differentiated IP and strong teams that have a longer-term impact. It has also had to resist fueling the hype around Watson and AI too much, lest Wall Street pressure it even more to enhance its short-term results. Its message to Wall Street has been that Watson will still take years to really pay off.[26] In doing what it has already done, IBM has become well-positioned to stay at the cutting edge of the AI curve for decades to come.

New Forms of Risk and Reward

As discussed throughout the course of this book, AI and ML are transforming research, production, and logistics processes throughout the global economy. As previously noted, perhaps no company in the world better exemplifies the process redesign through the reinvention of the workflow and layout of its business than Amazon, following the introduction of robots and optimization algorithms in its fulfillment centers. But the truth is that AI is no longer the domain of the largest, most tech-savvy firms with the deepest pockets; ML systems are personalizing the way we purchase almost everything and how we are entertained. For example, instead of selling songs à la carte, online music subscription services now personalize stations by predicting the type of music a person is likely to prefer, even if he or she has never heard specific songs before. In the same vein, machine vision systems can now identify potential cancer cells, which allows radiologists to reallocate their time and focus on truly critical cases, while spending more time communicating with patients.

That said, humans often have difficulty interpreting how the systems arrived at their decisions. Deep neural networks may have hundreds of millions of connections, each of which contributes in a small way to the ultimate decision. As a result, these systems' predictions tend to resist simple, clear, or necessarily logical explanation. Machines do not always provide a rationale for why a particular applicant was accepted or rejected for a job, or why a particular medicine was recommended.

We face what is in essence a Polanyi's Paradox,[27] because we know more than we can tell. Human knowledge and capability have relied on skills and rule sets that are often transmitted to us via evolution, culture, and tradition, thereby depriving us of conscious appreciation of how we have gotten to where we are. In other words, many of the tasks we perform rely on tacit, intuitive knowledge that is difficult to codify and automate. As a result, *we may know more than we can tell. Similarly, machines undoubtedly know more than they can tell us.*

This creates three specific risks. First, as previously discussed, the machines often have hidden biases, not necessarily derived from any intent on the part of the designer but from the data provided to train the system. For instance, if a system learns which job applicants to accept for an interview by using a data set of decisions made by human recruiters in the past, it may inadvertently learn to perpetuate racial, gender, ethnic, or other biases. Moreover, these biases may not appear as an explicit rule but, rather, be embedded in subtle interactions among the thousands of factors considered.

A second risk is that, unlike traditional systems built on explicit rules of logic, neural networks deal with statistical truths rather than literal truths. That can make it difficult, if not impossible, to prove with complete certainty that a system will work in all cases, particularly in situations that were not represented in training data. Lack of verifiability can be a concern in mission-critical applications (such as controlling a nuclear power plant) or when life-or-death decisions are involved.

A third risk is that, when ML systems make errors, diagnosing and correcting the precise nature of the problem can be difficult. What led to the solution set may be unimaginably complex and the solution may be far from optimal if the conditions under which the system was trained happen to change. Given all this, the appropriate benchmark is not the pursuit of perfection, but rather, the best available alternative.

Of course, humans also have biases, make mistakes, and have difficulty explaining how we may have arrived at a particular decision. The advantage of machine-based systems is that their performance can be improved over time and they will give consistent answers when presented with the same data. Does that mean there is no limit to what AI and ML can do? No one can say that, but what can be said with some certainty is that *AI and ML will soon enough reach superhuman levels of performance* in most or all of these areas.

Perhaps the better question to ask is, what will AI and ML *not* be able to do? Some people continue to believe that AI will never be good at assessing human emotions, but some ML systems are already at or beyond human-level performance in discerning a person's emotional state on the basis of tone of voice or facial expression. Other systems can already infer when the world's best poker players are bluffing well enough to beat them, but computers are generally oriented to answer questions rather than ask them. There is also a big difference between passively assessing someone's morale or mental state and actively working to change it.

Going forward, those organizations that attempt to maintain barriers between humans and machines are likely to find themselves at an ever-greater competitive disadvantage when compared with rivals who prefer to tear such barriers down and put AI and ML to use in every way possible to effectively integrate their capabilities with those of humanity. A time of tectonic change has already begun. One of ML's greatest legacies is the creation of a new generation of business leaders whose vision, brashness, and willingness to smash through the status quo are propelling their firms, and the global economy, boldly into the future. Those organizations that can rapidly sense and respond to opportunities will seize the advantage in the AI-enabled landscape. In the near term, *AI may not replace managers, but managers who use AI will replace those who do not.*[28] The strategy that is most likely to be successful will therefore be to be willing to experiment and learn quickly.

Embracing or Fearing the AI Future?

Since data is the new oil, AI is the tool that enables it to be extracted and refined. Should we be exalted or mortified by the prospect that AI will become a dominant force in our lives as soon as the next decade? Can any government or company with the capability to become AI literate afford not to dive into the deep end in pursuit of AI supremacy, or is it bordering on lunacy to imagine that any country other than China or the US, or any company other than Alibaba, Baidu, Tencent or Amazon, Facebook, and Google stand any kind of reasonable chance of ultimately winning the race? And if another country or company cannot win the race, or remain the victor for long, is there really any point in even running the race?

One of the dangers in deciding to dive in is that, as more organizations jump on the AI bandwagon, a "shoot first and ask questions later" type of mentality is developing. The perceived need to get in the race (and do so quickly) is prompting many participants to skip steps they may otherwise have taken in the product development process. A 2017 survey of 260 global organizations[29] found that 80% of enterprises were actively investing in AI, but many of these firms anticipated significant barriers to adoption of AI internally and were in the process of creating new C-suite-level positions to streamline and coordinate AI adoption. You might think this is a good thing, but, ordinarily, a Chief AI Officer would have been appointed, strategic plans would have been adopted, and necessary allocations of resources would have been made prior to actually proceeding down the AI path.

Furthermore, the massive increase in AI-oriented startups and established vendors is confusing consumers, who are unable to distinguish between one service provider, service, or product, and another. Just as when some companies raced to join the "environmentally friendly" band wagon and engaged in "greenwashing" (exaggerating the environmental-friendliness of their products, services, or practices), many technology vendors are now "AI washing" by applying the AI label somewhat indiscriminately.[30] No wonder consumers are confused. Eventually, an equilibrium will be established between firms offering AI products and services and consumers who purchase them, wherein product and service providers will be clearly differentiated and consumers will become knowledgeable and experienced enough to recognize, and reward, the difference—but we are many years away from that at this juncture.

Ironically, although data is the food that feeds AI[31], it remains one of the most neglected intangible assets *not* on companies' balance sheets. As previously discussed, very few companies treat data as a balance sheet asset, either because they do not think of it as an asset or because there is no standard methodology for attributing tangible value to data. As the race toward AI supremacy marches on, this is becoming an increasingly important omission. The failure to accurately quantify the Enterprise Value of Data[32] may woefully undervalue not only a firm's stock value and brand equity but the potential value of its AI-related assets and investments.

Definitions for what constitutes EvD, and methodologies intended to calculate it, remain in their infancy. Many existing means of determining the impact of data-related risk on the bottom line—such as evaluating the degree of cyber security in an organization—miss the real potential value of data. Companies such as Apple, Coca Cola, and Disney have recognized this and have begun to quantify what brand equity, which includes a valuation for data, is actually worth, by including it as an asset on their balance sheets.

To unlock the hidden value of data, firms should begin to treat data itself an integral part of their supply chain, because data impacts the entire ecosystem in which a firm operates. The challenge is to attribute value and quantify it in the context of what 'data equity' actually means to a firm, so that economic value can be ascribed to this asset class over time.

AI is already a fact of life whose potential will grow exponentially, along with its applicability and impact. Just as manned-flight could only have occurred once combustion engines technically enabled it, the use of graphics cards, creation of custom hardware, the rise of cloud computing, and the growth in computing capabilities—all occurring at the same time—have made AI a force to be reckoned with. Being able to rent cloud space or outsource computational resources means relative costs have come down to earth, and will continue to do so. The widespread use of open source Internet-based tools and the explosive growth in data generation have also made a big difference. *So much data is now generated on a daily basis globally that only gigantic infusions of data are likely to make a difference in the growth of AI going forward. That implies that only the largest, most technically sophisticated firms with the capability to consume and process such volumes of data will benefit from it in a meaningful way in the future.*

Some of the greater thinkers of our time are pondering what our AI future may imply. Henry Kissinger sees AI as dealing with ends rather than means, and as being inherently unstable, to the extent that its achievements are, at least in part, shaped by itself. In his view, AI makes strategic judgments about the future, but *the algorithms upon which AI is based are mathematical interpretations of observed data that do not explain the underlying reality that produces them*. He worries that, by mastering some fields more rapidly and definitively than humans, *AI may diminish human competence and the human condition over time*, as it turns them into mere data.

AI makes judgments regarding an evolving, as-yet-undetermined future, and Kissinger argues that *its results are imbued with uncertainty and ambiguity, which leads to unintended results and a danger that AI will misinterpret human instructions*. By achieving intended goals, AI may change human thought processes and human values or be unable to explain the rationale for its conclusions. *By treating a mathematical process as if it were a thought process, and either trying to mimic that process ourselves or merely accepting the results, we are in danger of losing the capacity that has been the essence of human cognition.* While the Enlightenment began with philosophical insights being spread by new technology, the period in which we are living is moving in the opposite direction, for *it has generated a potentially dominating technology in search of a guiding philosophy.*[33]

In a world filled with unintended consequences, what else might be lost along the way? Will our collectively shared values fall by the wayside in an effort to reach AI supremacy? Will we lose our ability to distinguish between victory and a victory worth having—in business as well as on the military battlefield? Will the notion of human accountability eventually disappear in an AI-dominated world? Some military strategists view an AI-laden battlefield as "casualty-free" warfare, since machines will be the ones killing and at risk. Could the commercial AI landscape evolve into a winner takes all arena in which only one firm or machine is left standing?

New forms of threats are evolving as AI becomes more widely utilized, so it is important that we regain agency over it. Just as Microsoft proposed a Digital Geneva Convention that would govern how governments use cyber capabilities against the private sector, in 2017,[34] *an international protocol should be created to govern not only how governments project AI onto one another, but how they do so with the private sector, and how the private sector will do so with itself.*

Yet, there are no such "rules of the road" in existence for AI right now. While AI remains in an embryonic state, it would be a perfect time to establish rules, norms, and standards by which AI is created, deployed, and utilized. We should ensure that it enhances globally shared collective values to elevate the human condition in the process. While there will probably never be a single set of universal principles governing AI, by trying to understand how to shape the ethics of a machine, we are at the same time forced to think more about our own values and what is really important.

Attempting to govern AI will not be an easy or pretty process, for there are overlapping frames of reference and many of the sectors in which AI will have the most impact are already heavily regulated. New norms are emerging, but how will the two be merged? It will take a long time to work through the various questions that are being raised. Many are straight forward questions about technology, but many others are about what kind of societies we want to live in and what type of values we wish to adopt in the future. *If AI forces us to look ourselves in the mirror and tackle such questions with vigor, transparency, and honesty, then its rise will be doing us a great favor. History would suggest, however, that the things that should really matter will either get lost in translation or be left by the side of the road in the process.*

We may see a profound shift in agency away from man and toward machine, wherein decision-making could become increasingly delegated to machines. If so, our ability to implement and enforce the rule of law could prove to be the last guarantor of human dignity and values in an AI-dominated world. Yet, as we continue to grapple with such fundamental issues as equality and gender bias with great difficulty, what should be on the top of the AI "values" pyramid? How can we even know what human compatible AI is or will become?

Some would argue that we are getting ahead of ourselves by imagining a world dominated by AI and worrying about its potential implications, particularly since, in many respects, we are at the beginning of the runway. Much of what has been accomplished in the AI arena remains rudimentary. How excited should we actually be getting by robots that can master repetitive tasks or a robot that can be programmed to jump up and down or answer basic questions? *The truth is, we have only just begun to understand how to actually "build" AI. Mostly, what we know how to do is collect and utilize statistics from Big Data.* AI cannot be "produced" simply by collecting data, any more than weather may be predicted with complete accuracy.

We might just be approaching a wall in terms of how much further we will advance meaningfully using AI in the near term. Longer term is a completely different proposition. To achieve even close to our potential in scientific discovery will require a much larger effort than teams of people sequestered in a room to think about it all. Only those organizations and countries that commit *massive* resources toward solving many of AI's inexorable challenges and creating a competitive edge *today* have a chance of achieving AI supremacy in the next decade. That implies adopting a mindset that makes AI an integral part of the long-term planning process, with clear objectives and benchmarks in view. That will be much easier said than done, of course, no matter how large the organizations or how committed the government.

Sprinting to a Mythical Finish Line

Those organizations that will ultimately prevail in the race for AI supremacy will master how to use AI technology *and* Big Data. That means they will not necessarily be the industry leaders in *providing* data and technology; rather, they will be the leaders in successfully *integrating* data and technology into their organizations, using data and technology to maximize operational and financial value. Consider the examples of Alphabet, Apple, Amazon, and Facebook. Sure, these firms all have a rather large finger in the technology pie, but they are all exceedingly busy *dominating* the markets they operate in, and are pushing aggressively into new markets that they hope to one day dominate.

But achieving AI supremacy is about much more than industry domination or sprinting to some mythical, unseen finish line somewhere out there in the mist, for *there is no real end to this race*. AI will continue to evolve, with amazing new technological *advances* that will occur with regularity, establishing a perpetual state of "new normal" in the process and redefining what it means to succeed in the AI arena. *There can be no single victor, and "victory" will be a relative term, for it will always be a matter of time until another entity surpasses whatever the victor of the moment may have been able to achieve. Any victor's time in the winner's circle will, by definition, be short lived.*

Numerous organizations and governments will achieve evolutionary states of "supremacy" in this realm, and they will all have certain basic things in common. It is a given that they will have invested heavily in AI and ML–in terms of financial, human, and technological resources. But *they will also have adopted a defensive and offensive posture, a strategic planning process that is aligned with AI-oriented objectives, and a willingness to fail along the way, dust themselves off, and get back to it.*

You might say that sounds like what any organization must do to stay in the ring, and you would be right. The difference is, *only those organizations that have made the decision to do so at the beginning of the AI race stand a chance of competing effectively in it or, indeed, ever achieving AI supremacy. This is not a race that one chooses to enter at any time in the future, for the others who have been at it for years will be too far ahead for the laggards to ever catch up. That is likely already true.*

A number of eminent scientists, engineers, thought leaders, and visionaries fear that, once we have succeeded in building an AI smarter than we are, our own demise may be the natural result. Elon Musk has warned against summoning the demon, envisaging an immortal dictator from which we can never escape. Just prior to his death, Stephen Hawking declared that AGI could spell the end of the human race. While such dire predictions are not new (they have been made since AI came into being in the 1950s), it has taken on new meaning as we grow ever closer to making AGI a reality, even though it remains a somewhat distant goal.

As a result of advances in chip design, processing power, and Big Data hosting, advances in AI's capabilities have grown so ubiquitous that we rarely notice it. It is simply a matter of fact that Siri schedules our appointments, Facebook tags our photos, and GPS plots our routes for us before we even know where we are going. We may not even notice when something that would have seemed a revolutionary technological advance a decade ago is released as a new product to the general public. We have become so conditioned to expect rapid AI advance that we are disappointed if a new smart phone is released without some major new AI-oriented gadgetry embedded inside it.

So, what is left for humans? As Larry Tesler, the computer scientist who invented copy-and-paste, has noted, *human intelligence has become whatever machines have not done yet.* Some argue that the relationship between human and machine intelligence should be understood as synergistic rather than competitive. AI in large part helps us do what we can do better; working alongside cobots augments human productivity. That is, for now, at least. What about when most applications of AI surpass human capability in more areas than it complements it?

The real risk of AGI may stem not from malice, or emergent self-consciousness, but simply from autonomy—for intelligence entails control. A recursive, self-improving AGI may not be smart like Einstein, but rather in the sense that an average human being is smart compared to a beetle or a worm.[35] No matter how intelligent machines may become, they will lack human emotion and instinct, but instead of seeing this as a disadvantage, perhaps we should be thinking of it as a net plus. AI will presumably not adopt a course of action that it can calculate will fail, while humans may proceed anyway, by letting their emotions guide them or by following their gut instinct.

Perhaps, rather than sprinting toward a mythical finish line, we should slow things down, take a deep breath, and ponder where we, and AI, are going. Most businesses and governments are, after all, really just getting their feet in the starting blocks to begin their version of "the race". Given all the unknown unknowns, should we not figure out where we are going, and why, first?

That may fly in the face of conventional wisdom and runs counter to business culture in our hyper-competitive world, but, in the end, if you recall in Aesop's Fables, the tortoise surpasses the hare. AI supremacy will not be achieved simply by virtue of devoting necessary resources to the task. Strategy, direction, resources, vision, and application will ultimately determine who gets to stay in the race and who will have their chance to join the winner's circle.

AI SUPREMACY

Notes

1. Creeping Irrelevance

[1] Oscar Li, "Artificial Intelligence is the New Electricity — Andrew Ng," Medium. April 28, 201, accessed March 25, 2018, https://medium.com/@Synced/artificial-intelligence-is-the-new-electricity-andrew-ng-cc132ea6264

[2] Klaus Schwab, "The Fourth Industrial Revolution: what it means, how to respond," World Economic Forum, January 14, 2016, accessed March 25, 2018. https://www.weforum.org/agenda/2016/01/the-fourth-industrial-revolution-what-it-means-and-how-to-respond/

[3] "Cognification," Wiktionary, accessed March 25, 2018. https://en.wiktionary.org/wiki/cognification

[4] "Gartner Says By 2020, Artificial Intelligence Will Create More Jobs Than It Eliminates," Gartner, December 13, 2017, accessed March 25, 2018, https://www.gartner.com/newsroom/id/3837763

[5] "Local Pay Reports," Glassdoor, February 2018, accessed March 25, 2018, https://www.glassdoor.com/research/local-pay-reports/

[6] Jeremy Hsu, "Africa's Drones are Zipping Past the US," Wired, September 13, 2017, accessed March 25, 2018, https://www.wired.com/story/africas-delivery-drones-are-zipping-past-the-us/

[7] "Comparing the Performance of Artificial Intelligence to Human Lawyers in the Review of Standard Business Contracts," LawGeex, February 2018, accessed March 25, 2018, https://www.lawgeex.com/AIvsLawyer/

[8] "Sizing the prize: What's the real value of AI for your business and how can you capitalize?" PWC, 2017, accessed March 25, 2018, https://www.pwc.com/gx/en/issues/analytics/assets/pwc-ai-analysis-sizing-the-prize-report.pdf

[9] "DeepMind AI Reduces Google Data Centre Cooling Bill by 40%," DeepMind, July 20, 2016, accessed March 25, 2018, https://deepmind.com/blog/deepmind-ai-reduces-google-data-centre-cooling-bill-40/

[10] "What DeepMind brings to Alphabet", The Economist, December 15, 2016, accessed March 25, 2018, https://www.economist.com/news/business/21711946-ai-firms-main-value-alphabet-new-kind-algorithm-factory-what-deepmind-brings

[11] "Jobs lost, jobs gained: workforce transitions in a time of automation", McKinsey Global Institute, December 2017, accessed March 25, 2018, https://www.mckinsey.com/~/media/McKinsey/Global%20Themes/Future%20of%20Organizations/What%20the%20future%20of%20work%20will%20mean%20for%20jobs%20skills%20and%20wages/MGI-Jobs-Lost-Jobs-Gained-Report-December-6-2017.ashx

[12] Jathan Sadowski, "Why Silicon Valley is embracing universal basic income", The Guardian, June 22, 2016, accessed March 25, 2018, https://www.theguardian.com/technology/2016/jun/22/silicon-valley-universal-basic-income-y-combinator

[13] Gil Press, "Artificial Intelligence (AI) Defined", Forbes, August 27, 2017, accessed March 25, 2018, https://www.forbes.com/sites/gilpress/2017/08/27/artificial-intelligence-ai-defined/#5bb75e777661

[14] "Electroencephalogram (EEG)," Medical Discoveries, accessed March 25, 2018, http://www.discoveriesinmedicine.com/Com-En/Electroencephalogram-EEG.html

[15] Pamela McCorduck, *Machines Who Think: A Personal Inquiry into the History and Prospects of Artificial Intelligence*, (2nd ed.) (USA: A. K. Peters, Ltd, 2004), p. 204.

[16] Steve Paulson, "The systems theorist explains what's wrong with standard models of intelligence", Nautilus, April 23, 2015, accessed March 25, 2018, http://nautil.us/issue/23/dominoes/ingenious-david-krakauer

[17] Alan Turing, "Computing Machinery and Intelligence", Mind, 1950, accessed March 25, 2018, https://www.csee.umbc.edu/courses/471/papers/turing.pdf

[18] Saundra Young, "For amputees, an unlikely painkiller: Mirrors", CNN, March 19, 2008, accessed March 25, 2018, http://www.cnn.com/2008/HEALTH/03/19/mirror.therapy/index.html

[19] Michael Graziano, "How Phantom Limbs Explain Consciousness", The Atlantic, February 4, 2016, accessed March 25, 2018, https://www.theatlantic.com/science/archive/2016/02/phantom-limbs-explain-consciousness/459780/

[20] Rick Strassman, *DMT: The Spirit Molecule: A Doctor's Revolutionary Research into the Biology of Near-Death and Mystical Experiences*, (1st ed.) (USA: Park Street Press, 2001), p. xv.

[21] Darold Treffert, "Is There a Little 'Rain Man' in Each of Us?", Wisconsin Medical Society, accessed March 25, 2018, https://www.wisconsinmedicalsociety.org/professional/savant-syndrome/resources/articles/is-there-a-little-rain-man-in-each-of-us/
[22] Aatif Sulleyman, "AI is highly likely to destroy humans, Elon Musk warns", The Independent, November 24, 2017, accessed March 28, 2018, http://www.independent.co.uk/life-style/gadgets-and-tech/news/elon-musk-artificial-intelligence-openai-neuralink-ai-warning-a8074821.html
[23] "Chinese Room Argument", Internet Encyclopedia of Philosophy, accessed March 25, 2018, https://www.iep.utm.edu/chineser/
[24] Amir Husain, *The Sentient Machine: The Coming Age of Artificial Intelligence*, (1st ed.) (USA: Scribner, 2017), p. 21.
[25] Adam Gibson, and Josh Patterson, *Deep Learning: A Practitioner's Approach*, (2nd ed.) (USA: O'Reilly Media Inc., 2017) p. 7-10.
[26] Julian Wong, "Unsupervised Analytics: Moving Beyond Rules Engines and Learning Models", DataVisor, August 1, 2016, accessed March 25, 2018, https://www.datavisor.com/technical-posts/rules-engines-learning-models-and-beyond/
[27] Adam Conner-Simons, "System predicts 85% of cyberattacks using input from human experts", MIT, April 18, 2016, accessed March 25, 2018, http://news.mit.edu/2016/ai-system-predicts-85-%-cyberattacks-using-input-human-experts-0418
[28] "The story of AlphaGo so far", DeepMind, accessed March 25, 2018, https://deepmind.com/research/alphago/
[29] Ethem Alpaydin, *Machine learning: The new AI*, (1st ed.) (USA: MIT Press, 2016), p. 126-128.
[30] "DeepFace: Closing the Gap to Human-Level Performance in Face Verification", Facebook, 2016, accessed March 25, 2018, https://research.fb.com/wp-content/uploads/2016/11/deepface-closing-the-gap-to-human-level-performance-in-face-verification.pdf?
[31] "Deep Learning", Investopedia, accessed March 25, 2018, https://www.investopedia.com/terms/d/deep-learning.asp
[32] Nahua Kang, "Multi-Layer Neural Networks with Sigmoid Function— Deep Learning for Rookies (2)", Towards Data Science, June 27, 2017, accessed March 25, 2018, https://towardsdatascience.com/multi-layer-neural-networks-with-sigmoid-function-deep-learning-for-rookies-2-bf464f09eb7f
[33] Adam Gibson, and Josh Patterson, *Deep Learning: A Practitioner's Approach*, (2nd ed.) (USA: O'Reilly Media Inc., 2017) p. 4-6.
[34] "List of public corporations by market capitalization", Wikipedia, accessed March 25, 2018,

https://en.wikipedia.org/wiki/List_of_public_corporations_by_market_capitalization

[35] "The world's most valuable resource is no longer oil, but data", The Economist, May 6, 2017, accessed March 25, 2018, https://www.economist.com/news/leaders/21721656-data-economy-demands-new-approach-antitrust-rules-worlds-most-valuable-resource

[36] "Data Age 2025: The Evolution of Data to Life Critical", IDC, April 2017, accessed March 25, 2018, https://www.seagate.com/files/www-content/our-story/trends/files/Seagate-WP-DataAge2025-March-2017.pdf

[37] Anasse Bari, "Deep learning and Big Data: Wall Street and the new data paradigm", International Business Times, November 22, 2017, accessed March 25, 2018, https://www.ibtimes.co.uk/deep-learning-big-data-wall-street-new-data-paradigm-1648448

[38] John Kay, "The parable of the ox", Financial Times, July 24, 2012, accessed March 25, 2018, https://www.ft.com/content/bfb7e6b8-d57b-11e1-af40-00144feabdc0

[39] Alexis Madrigal, "Does Anne Hathaway News Drive Berkshire Hathaway's Stock?", The Atlantic, March 18, 2011, accessed March 25, 2018, https://www.theatlantic.com/technology/archive/2011/03/does-anne-hathaway-news-drive-berkshire-hathaways-stock/72661/

[40] Chris Moyer, "How Google's AlphaGo Beat a Go World Champion", The Atlantic, March, 28, 2016, accessed March 25, 2018, https://www.theatlantic.com/technology/archive/2016/03/the-invisible-opponent/475611/

[41] Cade Metz, "As China Marches Forward on A.I., the White House Is Silent", The New York Times, February 12, 2018, accessed March 25, 2018, https://www.nytimes.com/2018/02/12/technology/china-trump-artificial-intelligence.html

[42] Neil Connor, "Chinese police using facial recognition glasses to identify suspects", Telegraph, February 7, 2018, accessed March 25, 2018, https://www.telegraph.co.uk/news/2018/02/07/chinese-police-using-facial-recognition-glasses-identify-suspects/

[43] Cheryl Pellerin, "Deputy Secretary: Third Offset Strategy Bolsters America's Military Deterrence", U.S. Department of Defense, October 31, 2016, accessed March 25, 2018, https://www.defense.gov/News/Article/Article/991434/deputy-secretary-third-offset-strategy-bolsters-americas-military-deterrence/

[44] James Vincent, "Putin says the nation that leads in AI 'will be the ruler of the world'", The Verge, September 7, 2017, accessed March 25, 2018,

https://www.theverge.com/2017/9/4/16251226/russia-ai-putin-rule-the-world

[45] "Exposed: Undercover secrets of Trump's data firm", Channel 4, March 20, 2018, accessed March 25, 2018, https://www.channel4.com/news/exposed-undercover-secrets-of-donald-trump-data-firm-cambridge-analytica

[46] Hannes Grassegger and Mikael Krogerus, "The Data That Turned the World Upside Down", Motherboard, January 28, 2017, accessed March 25, 2018, https://motherboard.vice.com/en_us/article/mg9vvn/how-our-likes-helped-trump-win

[47] Carole Cadwalladr and Emma Graham-Harrison, "Revealed: 50 million Facebook profiles harvested for Cambridge Analytica in major data breach", The Guardian, March 17, 2018, accessed March 25, 2018, https://www.theguardian.com/news/2018/mar/17/cambridge-analytica-facebook-influence-us-election

[48] "CA Political", Cambridge Analytica, accessed March 25, 2018, https://ca-political.com/ca-advantage

[49] "Forecasting Presidential Elections since 1912", Primary Model, accessed March 25, 2018, http://primarymodel.com/

[50] "Facebook to launch two smart speakers in July 2018", Digitimes, February 14, 2018, accessed March 25, 2018, https://www.digitimes.com/news/a20180214PD207.html

[51] Christi Olson, "Just say it: The future of search is voice and personal digital assistants", Campaign, April 25, 2016, accessed March 25, 2018, https://www.campaignlive.co.uk/article/just-say-it-future-search-voice-personal-digital-assistants/1392459

[52] CB Insights, "Top AI trends to watch in 2018," 2018, accessed March 4 2018, https://www.cbinsights.com/research/report/artificial-intelligence-trends-2018/

[53] Kevin McLaughlin, "Facing New Rivals, Amazon May Open Up Alexa Data for Developers", The Information, July 12, 2017, accessed March 25, 2018, https://www.theinformation.com/articles/facing-new-rivals-amazon-may-open-up-alexa-data-for-developers

[54] Rebecca Robbins, "The Sound of Your Voice May Diagnose Disease", Scientific American, June 30, 2016, accessed March 25, 2018, https://www.scientificamerican.com/article/the-sound-of-your-voice-may-diagnose-disease/

[55] "How to survive in the 'Digital Amnesia' world", Kaspersky Lab, July 1, 2015, accessed March 25, 2018, https://usa.kaspersky.com/blog/digital-amnesia-survival/5548/

[56] Stephen Dinan, "Drones become latest tool drug cartels use to smuggle drugs into U.S.", The Washington Times, August 20, 2017, accessed March 25, 2018, https://www.washingtontimes.com/news/2017/aug/20/mexican-drug-cartels-using-drones-to-smuggle-heroi/

[57] Christopher Woody, "The Coast Guard is detecting a new trend among high-seas narco smugglers", Business Insider, September 20, 2017, accessed March 25, 2018, http://www.businessinsider.com/coast-guard-is-detecting-a-new-trend-among-high-seas-narco-smugglers-2017-9

[58] J.P. Buntinx, "Autonomous Submarines are the Next Frontier for the US Navy", The Merkle, September 17, 2017, accessed March 29, 2018, https://themerkle.com/autonomous-submarines-are-the-next-frontier-for-the-us-navy/

[59] Robert Beckhusen, "Mexican Cartels enslave engineers to build radio network", Wired, November 1, 2012, accessed March 25, 2018, https://www.wired.com/2012/11/zeta-radio/

[60] "Fourth Industrial Revolution for the Earth: Harnessing Artificial Intelligence for the Earth", PWC, January 2018, accessed March 25, 2018, https://www.pwc.com/gx/en/sustainability/assets/ai-for-the-earth-jan-2018.pdf

2. AI Globalization

[1] Mark Minevich, "Globalizations second act is the digital boom", Forbes, accessed August 2, 2018 https://www.forbes.com/sites/forbestechcouncil/2017/07/06/globalizations-second-act-is-the-digital-boom/#dbf0e5264bf6

[2] Aridam Bhattacharya, Hans- Paul Burkner, and Aparna Bijapurkar, "What we need to know about globalization's radical new phase", BCG, accessed August 2, 2018, https://www.bcg.com/publications/2016/globalization-growth-what-need-know-globalization-radical-new-phase.aspx

[3] Martin Reeves, and Johann Harnoss, " An agenda for the future of global business", Harvard Business Review, accessed August 2, 2018, https://hbr.org/2017/02/an-agenda-for-the-future-of-global-business

[4] Jeffrey Immelt, "Preparing for What's Next", GE, accessed August 2, 2018, https://www.ge.com/reports/jeff-immelt-preparing-for-whats-next/

[5] Susan Aaronson, "Artificial intelligence is trade policy's new frontier", Centre for International Governance Innovation, accessed August 2, 2018,

https://www.cigionline.org/articles/artificial-intelligence-trade-policys-new-frontier

[6] Jennifer Nesbitt, "4 ways artificial intelligence is transforming trade", Trade Ready, accessed August 2, 2018, http://www.tradeready.ca/2017/topics/import-export-trade-management/4-ways-artificial-intelligence-transforming-trade/

[7] "AI and cargo shipping full speed ahead for global maritime trade", Microsoft News, accessed August 2, 2018, https://news.microsoft.com/apac/features/ai-and-cargo-shipping-full-speed-ahead-for-global-maritime-trade/

[8] William Cassidy, "Artificial intelligence in global trade – HAL 9000 it is not (so far)", Journal of Commerce, accessed August 2, 2018, https://www.joc.com/artificial-intelligence-global-trade-%E2%80%93-hal-9000-it-isn%E2%80%99t-so-far_20180206.html

[9] "IBM begins introducing AI-enabled scanner for global trade", Port Technology, accessed August 2, 2018, https://www.porttechnology.org/news/ibm_begins_introducing_ai_enabled_scanner_for_global_trade

[10] "Blockchain insurance platform piloted by Maersk launches" Port Technology, accessed August 2, 2018, https://www.porttechnology.org/news/blockchain_insurance_platform_piloted_by_maersk_launches

[11] Marcos Troyjos, "We should make ai and blockchain boost global trade", Huffington Post, accessed August 2, 2018, https://www.huffingtonpost.com/entry/we-should-make-ai-and-blockchain-boost-global trade_us_59cf7646e4b0f58902e5cc65

[12] Bruce Love, "The lawyers using AI to keep tabs on new global sanctions", Financial Times, accessed August 2, 2018, https://www.ft.com/content/0c2bda9e-4caf-11e8-97e4-13afc22d86d4

[13] "Revenues from the artificial intelligence (AI) market worldwide from 2016 to 2025 (in million U.S. dollars)", Statistica, accessed August 2, 2018, https://www.statista.com/statistics/607716/worldwide-artificial-intelligence-market-revenues/

[14] Alison Denisco Rayome, "AI investments will hit $232B by 2025, but businesses don't plan to cut jobs", Tech Republic, accessed August 2, 2018, https://www.techrepublic.com/article/ai-investments-will-hit-232b-by-2025-but-businesses-dont-plan-to-cut-jobs/

[15] "How artificial intelligence is disrupting the ETF industry", The Street, accessed August 2, 2018, https://www.thestreet.com/technology/how-artificial-intelligence-is-disrupting-etf-industry-14638427

[16] David Thomas, "Artificial Intelligence investing gets ready for prime time", Forbes, accessed August 2, 2018, https://www.forbes.com/sites/greatspeculations/2017/10/25/getting-ready-for-prime-time-of-artificial-intelligence-investing/#3d2ba4572010

[17] World Economic Forum, "Extreme automation and connectivity: The global, regional, and investment implications of the Fourth Industrial Revolution: UBS White Paper for the World Economic Forum Annual Meeting 2016", accessed August 2, 2018, http://www.tadviser.ru/images/b/b7/Extreme_automation_and_connectivity_The_global%2C_regional%2C_and_investment_implications_of_the_Fourth_Industrial_Revolution.pdf

[18] Nicholas Wright, "How artificial intelligence will reshape the global order", Foreign Affairs, https://www.foreignaffairs.com/articles/world/2018-07-10/how-artificial-intelligence-will-reshape-global-order

[19] Tim Dutton, "An Overview of National AI Strategies — Politics AI — Medium," Medium, June 28, 2018, accessed September 05, 2018. https://medium.com/politics-ai/an-overview-of-national-ai-strategies-2a70ec6edfd

[20] Carlos E. Perez, "How Artificial Intelligence Enables the Economics of Abundance," Medium, Augmenting Humanity. February 18, 2018, accessed March 4, 2018, https://medium.com/intuitionmachine/artificial-intelligence-and-the-economics-of-abundance-92bd1626ee94

[21] "The Impact of the Technological Revolution on Labour Markets and Income Distribution," 2017, Department of Economic & Social Affairs, https://www.un.org/development/desa/dpad/wp-content/uploads/sites/45/publication/2017_Aug_Frontier-Issues-1.pdf

3.   Jobs and Manufacturing

[1] Klaus Schwab, *The Fourth Industrial Revolution*, Crown Business, 2016.

[2] François Barbier and Global Operations, "5 Trends for the Future of Manufacturing," World Economic Forum, accessed March 25, 2018, https://www.weforum.org/agenda/2017/06/what-s-going-on-with-manufacturing-b013f435-1746-4bce-ac75-05c642652d42/

[3] Sam Ransbotham, David Kiron, Philipp Gerbert, and Martin Reeves, *Reshaping Business With Artificial Intelligence: Closing the Gap Between Ambition and Action*, PDF, MIT Sloan Management Review, October 2017,

https://www.bcg.com/Images/Reshaping%20Business%20with%20Artificial%20Intelligence_tcm30-177882.pdf

[4] David Autor and Anna Salomons, *Does Productivity Growth Threaten Employment?* PDF, MIT Department of Economics, June 2017, https://www.ecbforum.eu/uploads/originals/2017/speakers/Speech/D_Autor_A_Salomons_Does_productivity_growth_threaten_employment_Final_Draft_20170619.pdf, pp. 46-47.

[5] David Autor, "Will Automation Take Away Our Jobs?" TedX Cambridge, September 2016, accessed September 7, 2018, https://www.ted.com/talks/david_autor_why_are_there_still_so_many_jobs/transcript?language=en

[6] David Rotman, "How Technology Is Destroying Jobs," MIT Technology Review, September 01, 2016, accessed March 25, 2018, https://www.technologyreview.com/s/515926/how-technology-is-destroying-jobs/.

[7] "How Robots Will Redefine Competitiveness," https://www.bcg.com, September 23, 2015, accessed March 25, 2018, https://www.bcg.com/publications/2015/lean-manufacturing-innovation-robots-redefine-competitiveness.aspx.

[8] *A Future That Works: Automation, Employment, and Productivity*, PDF, McKinsey Global Institute, January 2017, accessed March 25, 2018, https://www.mckinsey.com/~/media/McKinsey/Global%20Themes/Digital%20Disruption/Harnessing%20automation%20for%20a%20future%20that%20works/MGI-A-future-that-works_Full-report.ashx

[9] "What's Now and next in Analytics, AI, and Automation," McKinsey & Company, accessed March 25, 2018, https://www.mckinsey.com/featured-insights/digital-disruption/whats-now-and-next-in-analytics-ai-and-automation

[10] Jacques Bughin, Eric Hazan, Sree Ramaswamy, Michael Chui, Tera Allas, Peter Dahlstrom, Nicolaus Henke, and Monica Trench, *Artificial Intelligence: The Next Digital Frontier?* PDF, McKinsey Global Institute, June 2017. accessed March 25, 2018, https://www.mckinsey.com/~/media/McKinsey/Industries/Advanced%20Electronics/Our%20Insights/How%20artificial%20intelligence%20can%20deliver%20real%20value%20to%20companies/MGI-Artificial-Intelligence-Discussion-paper.ashx.

[11] Joao-Pierre Ruth, "6 Examples of AI in Business Intelligence Applications," TechEmergence, August 12, 2018, accessed March 25, 2018, https://www.techemergence.com/ai-in-business-intelligence-applications/, accessed March 25, 2018,

https://www.techemergence.com/ai-in-business-intelligence-applications/

[12] Jon Walker, "Machine Learning in Manufacturing - Present and Future Use-Cases," TechEmergence, May 30, 2018, accessed March 25, 2018, https://www.techemergence.com/machine-learning-in-manufacturing.

[13] *Human Amplification in the Enterprise: Automation. Innovation. Learning,* PDF, Infosys, 2017, accessed March 25, 2018, https://www.infosys.com/human-amplification/Documents/human-amplification-enterprise.pdf

[14] "Manufacturing Sector Transforming with AI, Automation," Industry Today, August 02, 2017, accessed March 25, 2018, https://industrytoday.com/article/manufacturing-sector-transforming-ai-automation/

[15] *Technology and Innovation for the Future of Production: Accelerating Value Creation*, PDF, World Economic Forum, March 2017, accessed March 25, 2018, http://www3.weforum.org/docs/WEF_White_Paper_Technology_Innovation_Future_of_Production_2017.pdf, pp. 4-5, 32.

[16] "The Onrushing Wave," The Economist, January 18, 2014, accessed March 25, 2018, https://www.economist.com/news/briefing/21594264-previous-technological-innovation-has-always-delivered-more-long-run-employment-not-less?spc=scode&spv=xm&ah=9d7f7ab945510a56fa6d37c30b6f1709

4. Financial Services

[1] Niall Ferguson, "The Ascent of Money: Dreams of Avarice", *YouTube*, 2009, accessed April 3, 2018, https://youtu.be/X1sjFHEI5GI

[2] Jessica Resnick-Ault, "Goldman Sachs buys personal finance start-up Clarity Money", *Reuters*, April 15, 2018, accessed May 2, 2018, https://www.reuters.com/article/us-clarity-m-a-goldman-sachs/goldman-sachs-buys-personal-finance-start-up-clarity-money-idUSKBN1HM0Z7

[3] Daniel Wagner and Dante Disparte, "Do You Know What Your Company's Data Is Worth?", *Harvard Business Review*, September 16, 2016, accessed May 2, 2018, https://hbr.org/2016/09/do-you-know-what-your-companys-data-is-worth

[4] "Some hedge funds deliver double-digit gains for 2017", *Reuters*, January 4, 2018, accessed April 29, 2018, https://www.reuters.com/article/us-hedgefunds-returns/some-hedge-funds-deliver-double-digit-gains-for-2017-idUSKBN1ET29Y

[5] Ian Allison, "Fintech: Can Machine Learning Be Applied To Trading?", *Newsweek*, October 16, 2017, accessed May 4, 2018, http://www.newsweek.com/business-technology-trading-684303

[6] Nishant Kumar, "How AI Will Invade Every Corner of Wall Street", *Bloomberg*, December 5, 2017, accessed April 22, 2018, https://www.bloomberg.com/news/features/2017-12-05/how-ai-will-invade-every-corner-of-wall-street

[7] Robin Wigglesworth, "Quant hedge funds set to surpass $1tn management mark", *Financial Times*, January 8, 2018, accessed April 23, 2018, https://www.ft.com/content/ff7528bc-ec16-11e7-8713-513b1d7ca85a

[8] Cade Metz,, "The Rise of the Artificially Intelligent Hedge Fund", Wired, January 25, 2016, accessed April 29, 2018, https://www.wired.com/2016/01/the-rise-of-the-artificially-intelligent-hedge-fund/

[9] "Will the "black boxes" cause the next financial crisis?", *13D Research*, December 23, 2017, accessed April 29, 2018, https://latest.13d.com/will-the-black-boxes-cause-the-next-financial-crisis-algorithmic-revolution-asset-management-d8827c29b537

[10] Lindsay Whipp, "'Flash-crash' trader Navinder Sarao pleads guilty to spoofing", Financial Times, November 9, 2016, accessed April 9, 2018, https://www.ft.com/content/a321031a-a6cb-11e6-8898-79a99e2a4de6

[11] Arjun Kharpal, "Traders used poker games and food menus to make a killing off illegal trades and cover it up but then AI found out", CNBC, March 27, 2017, accessed April 15, 2018, https://www.cnbc.com/2017/03/27/behavox-rogue-traders-compliance-artificial-intelligence.html

[12] Andrew Griffin, "Facebook's artificial intelligence robots shut down after they start talking to each other in their own language", *The Independent*, July 31, 2017, accessed April 29, 2018, https://www.independent.co.uk/life-style/gadgets-and-tech/news/facebook-artificial-intelligence-ai-chatbot-new-language-research-openai-google-a7869706.html

[13] "OCC Report Highlights Risks Facing National Banks and Federal Savings Associations", *The Office of the Comptroller of the Currency (OCC)*, July 7, 2017, accessed April 2, 2018, https://www.occ.gov/news-issuances/news-releases/2017/nr-occ-2017-78.html

[14] "Fintech Trends to Watch in 2018," Corporate Innovation Trends, July 20, 2018, accessed May 14, 2018, https://www.cbinsights.com/research/report/fintech-trends-2018/.

[15] David St. Geme, "Mortgages Are Data-rich but Insight-poor", *Blend*, accessed April 12, 2018, https://blend.com/mortgages-are-data-rich-but-insight-poor/.
[16] "SWIFT history", *SWIFT*, accessed April 12, 2018, https://www.swift.com/about-us/history.
[17] "The Global Findex Database 2017: Measuring Financial Inclusion and the Fintech Revolution", *The World Bank*, 2017, accessed May 13, 2018, https://globalfindex.worldbank.org/.
[18] Fintech trends to watch in 2018, *CB Insights*, 2018, accessed May 4, 2018, https://www.cbinsights.com/research/report/fintech-trends-2018/
[19] "Redrawing the lines: FinTech's growing influence on Financial Services", *PWC*, 2017, accessed April 2, 2018, https://www.pwc.com/jg/en/publications/pwc-global-fintech-report-17.3.17-final.pdf
[20] "Where top US banks are investing in fintech – CB Insights", *Banking Tech*, August 14, 2017, accessed April 12, 2018, https://www.bankingtech.com/2017/08/where-top-us-banks-are-investing-in-fintech-cb-insights/
[21] Harriet Taylor, "Bank of America launches AI chatbot Erica — here's what it does", *CNBC*, October 24, 2016, accessed May 12, 2018, https://www.cnbc.com/2016/10/24/bank-of-america-launches-ai-chatbot-erica--heres-what-it-does.html.
[22] Hugh Son, "JPMorgan Software Does in Seconds What Took Lawyers 360,000 Hours", *Bloomberg*, February 27, 2017, accessed May 13, 2018, https://www.bloomberg.com/news/articles/2017-02-28/jpmorgan-marshals-an-army-of-developers-to-automate-high-finance
[23] Julie Verhage and Selina Wang, "SoFi's Plan to Become the Bank of the Future Isn't Going So Well", *Bloomberg*, October 4, 2017, accessed May 5, 2018, https://www.bloomberg.com/news/articles/2017-10-04/sofi-s-plan-to-become-the-bank-of-the-future-isn-t-going-so-well.
[24] Anna Irrera, "Banks scramble to fix old systems as IT 'cowboys' ride into sunset", *Reuters*, April 10, 2017, accessed May 5, 2018, https://www.reuters.com/article/us-usa-banks-cobol/banks-scramble-to-fix-old-systems-as-it-cowboys-ride-into-sunset-idUSKBN17C0D8
[25] Dinesh Bacham and Janet Zhao, "Machine Learning: Challenges, Lessons, and Opportunities in Credit Risk Modeling", *Moody's Analytics*, July 2017, accessed May 7, 2018, https://www.moodysanalytics.com/risk-perspectives-magazine/managing-disruption/spotlight/machine-learning-challenges-lessons-and-opportunities-in-credit-risk-modeling
[26] Larry D. Wall, "Prudential Regulation, Big Data, and Machine Learning", *Federal Reserve Bank of Atlanta*, November 2016, accessed May 7, 2016,

https://www.frbatlanta.org/cenfis/publications/notesfromthevault/11-prudential-regulation-bigdata-and-machine-learning-2016-11-21.aspx

[27] Stacy Cowley, "Zelle, the Banks' Answer to Venmo, Proves Vulnerable to Fraud", *The New York Times*, April 22, 2018, accessed May 4, 2018, https://www.nytimes.com/2018/04/22/business/zelle-banks-fraud.html

[28] Alex Schmidt, "Brazilian banks lead way on biometrics", *Marketplace*, August 6, 2013, accessed May 7, 2018, https://www.marketplace.org/2013/08/06/tech/brazilian-banks-lead-way-biometrics

[29] Thomas Fox-Brewster, "Yes, Cops Are Now Opening iPhones With Dead People's Fingerprints", Forbes, March 22, 2018, accessed May 7, 2018, https://www.forbes.com/sites/thomasbrewster/2018/03/22/yes-cops-are-now-opening-iphones-with-dead-peoples-fingerprints/#16485495393e

[30] Eva Casey Velasquez, "Why children are now prime targets for identity theft", *The Hill*, February 13, 2018, accessed May 7, 2018 http://thehill.com/opinion/cybersecurity/373692-why-children-are-now-prime-targets-for-identity-theft

[31] "Combating Synthetic Identity Fraud", *US Government Accountability Office*, July 26, 2017, accessed May 8, 2018 https://www.gao.gov/products/GAO-17-708SP

[32] "Ad Fraud to cost advertisers $19 Billion in 2018," accessed May 19, 2018, Juniper Research, https://www.juniperresearch.com/press/press-releases/ad-fraud-to-cost-advertisers-$19-billion-in-2018

[33] "Money Laundering", *Association of Certified Money Laundering Specialists (ACAMS)*, accessed May 8, 2018, https://www.acams.org/aml-glossary/index-m/

[34] "Money-Laundering and Globalization", *United Nations Office on Drugs and Crime*, accessed May 11, 2018, https://www.unodc.org/unodc/en/money-laundering/globalization.html

[35] "Fighting Financial Crime", *Standard Chartered Bank*, accessed May 12, 2018, https://www.sc.com/fightingfinancialcrime/

[36] "The FinCEN Artificial Intelligence System: Identifying Potential Money Laundering from Reports of Large Cash Transactions", *Association for the Advancement of Artificial Intelligence*, 1995, accessed May 13, 2018, https://www.aaai.org/Papers/IAAI/1995/IAAI95-015.pdf

[37] "Regulatory Sandbox", *Financial Conduct Authority*, May 11, 2015, accessed May 13, 2018, https://www.fca.org.uk/firms/regulatory-sandbox

[38] Loke Kok Fai, "Singapore to increasingly use technology, data analytics to combat transnational crime: DPM Teo", *Channel News Asia*, September 28, 2017, accessed April 2, 2018,

https://www.channelnewsasia.com/news/singapore/singapore-to-increasingly-use-technology-data-analytics-to-9257988

[39] Chris Baraniuk, "Australian AI spots dodgy deals that look like money laundering", *New Scientist*, August 17, 2016, accessed April 2, 2018, https://www.newscientist.com/article/mg23130871-300-australian-ai-spots-dodgy-deals-that-look-like-money-laundering/

[40] Sara Merken, "OCC Fintech Pilot Goes On After Curry Makes Way for Trump Team", *Bloomberg BNA*, August 25, 2017, accessed May 13, 2018, https://www.bna.com/occ-fintech-pilot-n73014463724/

[41] "A New Paradigm: Redesigning the US AML/CFT Framework to Protect National Security and Aid Law Enforcement", *The Clearing House*, February 2017, accessed May 13, 2018, https://www.theclearinghouse.org/~/media/TCH/Documents/TCH%20WEEKLY/2017/20170216_TCH_Report_AML_CFT_Framework_Redesign.pdf

5. Science, Health and Medicine

[1] The Anthropocene Era is broadly discussed in Daniel Wagner and Dante Disparte's book, *Global Risk Agility and Decision-Making*, Palgrave Macmillan, 2016.

[2] "Harnessing Artificial Intelligence for the Earth", *World Economic Forum*, January 2018, accessed May 26, 2018, http://www3.weforum.org/docs/Harnessing_Artificial_Intelligence_for_the_Earth_report_2018.pdf

[3] "Along with artificial intelligence, we need shovels, bricks and common sense to build things: Ben Hammersley", *The Economic Times*, March 7, 2018, accessed May 22, 2018, https://economictimes.indiatimes.com/tech/Internet/along-with-artificial-intelligence-we-need-shovels-bricks-and-common-sense-to-build-things-ben-hammersley/articleshow/63197666.cms

[4] "The STM Report An overview of scientific and scholarly journal publishing", *International Association of Scientific, Technical and Medical Publishers*, March 2015, accessed May 22, 2018, https://www.stm-assoc.org/2015_02_20_STM_Report_2015.pdf

[5] Rose Eveleth, "Academics Write Papers Arguing Over How Many People Read (And Cite) Their Papers", *Smithsonian Mag*, March 25, 2014, accessed May 23, 2018, https://www.smithsonianmag.com/smart-news/half-academic-studies-are-never-read-more-three-people-180950222/.

[6] https://iris.ai/
[7] Nick Stockton, "IF AI CAN FIX PEER REVIEW IN SCIENCE, AI CAN DO ANYTHING", *Wired*, February 21, 2017, accessed May 24, 2018, https://www.wired.com/2017/02/ai-can-solve-peer-review-ai-can-solve-anything/
[8] David Paquette, "The Future of Scientific Publishing is 'Open'", *Synapse*, December, 5 2017, accessed May 25, 2018, https://synapse.ucsf.edu/articles/2017/12/05/future-scientific-publishing-open
[9] Matthew Hutson, "Why are AI researchers boycotting a new Nature journal—and shunning others?", *Science*, accessed May 2, 2018, http://www.sciencemag.org/news/2018/05/why-are-ai-researchers-boycotting-new-nature-journal-and-shunning-others
[10] "OpenAI," OpenAI, accessed May 02, 2018. https://openai.com/.
[11] Monya Baker, "1,500 scientists lift the lid on reproducibility", *Nature*, May 25, 2016, accessed May 25, 2018, https://www.nature.com/news/1-500-scientists-lift-the-lid-on-reproducibility-1.19970
[12] Devin Coldewey, "AI learns and recreates Nobel-winning physics experiment", *Tech Crunch*, May 16, 2016, accessed May 25, 2018, https://techcrunch.com/2016/05/16/ai-learns-and-recreates-nobel-winning-physics-experiment/
[13] "Artificial intelligence accelerates discovery of metallic glass", *Northwestern University*, April 13, 2018, accessed May 25, 2018, https://news.northwestern.edu/stories/2018/april/artificial-intelligence-accelerates-discovery-of-metallic-glass/
[14] Andy Extance, "Robot Scientist Discovers Potential Malaria Drug", *Scientific American*, February 5, 2015, accessed May 25, 2018, https://www.scientificamerican.com/article/robot-scientist-discovers-potential-malaria-drug/
[15] Nurfilzah Rohaidi, "IBM's Watson Detected Rare Leukemia In Just 10 Minutes", *Asian Scientist*, August 15, 2016, accessed May 27, 2018, https://www.asianscientist.com/2016/08/topnews/ibm-watson-rare-leukemia-university-tokyo-artificial-intelligence/
[16] "Study Suggests Medical Errors Now Third Leading Cause of Death in the U.S.", *Johns Hopkins University*, May 3, 2016, accessed May 28, 2018, https://www.hopkinsmedicine.org/news/media/releases/study_suggests_medical_errors_now_third_leading_cause_of_death_in_the_us
[17] Bertalan Mesko, "The role of artificial intelligence in precision medicine", *Expert Review of Precision Medicine and Drug Development*, July 30, 2017, accessed May 28, 2018,

https://www.tandfonline.com/doi/pdf/10.1080/23808993.2017.1380516?needAccess=true

[18] Martin Stumpe and Craig Mermel, "An Augmented Reality Microscope for Cancer Detection", *Google AI Blog*, April 16, 2018, accessed May 28, 2018, https://ai.googleblog.com/2018/04/an-augmented-reality-microscope.html

[19] "Research Shows Shortage of More than 100,000 Doctors by 2030", *Association of American Medical Colleges*, March 14, 2017, accessed June 3, 2018, https://news.aamc.org/medical-education/article/new-aamc-research-reaffirms-looming-physician-shor/

[20] "World Population Prospects: The 2017 Revision", United Nations, June 21, 2017, accessed June 18, 2018, https://www.un.org/development/desa/publications/world-population-prospects-the-2017-revision.html

[21] "Connected and coordinated: Personalized service delivery for the elderly", *PWC*, October 2015, accessed June 18, 2018, p.4, https://www.pwc.com/gx/en/healthcare/pdf/pwc-elderly-care-report.pdf

[22] Emily Mullin, "A cheap and easy blood test could catch cancer early", *MIT Technology Review*, January 18, 2018, accessed June 9, 2018, https://www.technologyreview.com/s/610008/a-cheap-and-easy-blood-test-could-catch-cancer-early/

[23] "REVOLUTIONISING CARDIOVASCULAR DIAGNOSIS", *Ultromics*, 2017, accessed June 9, 2018, http://www.ultromics.com/technology/

[24] Emily Mullin, "AI can spot signs of Alzheimer's before your family does", *MIT Technology Review*, March 19, 2018, accessed June 9, 2018, https://www.technologyreview.com/s/609236/ai-can-spot-signs-of-alzheimers-before-your-family-does/

[25] Christian De Looper, "Google's smart bathroom patent puts sensors in your toilet, tub, and mirror", *Digital Trends*, August 10, 2016, accessed June 9, 2018, https://www.digitaltrends.com/home/google-smart-bathroom-patent/

[26] Meg Tirrell, "Unlocking my genome: Was it worth it?", *CNBC*, December 10, 2015, accessed June 6, 2018, https://www.cnbc.com/2015/12/10/unlocking-my-genome-was-it-worth-it.html

[27] Jason Bardi, "Major Genetic Study of Multiple Sclerosis Reveals DNA Hot Spots of Disease Susceptibility", *University of California San Francisco*, August 12, 2011, accessed May 2, 2018, https://www.ucsf.edu/news/2011/08/10431/major-genetic-study-multiple-sclerosis-reveals-dna-hot-spots-disease

[28] Gina Kolata, "In Treatment for Leukemia, Glimpses of the Future", *The New York Times*, July 7, 2012, accessed May 2, 2018, https://www.nytimes.com/2012/07/08/health/in-gene-sequencing-treatment-for-leukemia-glimpses-of-the-future.html

[29] Rick Mullin, "Cost to Develop New Pharmaceutical Drug Now Exceeds $2.5B", *Scientific American,* November 24, 2014, accessed June 7, 2018, https://www.scientificamerican.com/article/cost-to-develop-new-pharmaceutical-drug-now-exceeds-2-5b/

[30] Megan Scudellari, "Q&A: AI Could 'Redesign' the Drug Development Process", *IEEE Spectrum*, April 27, 2018, accessed May 2, 2018, https://spectrum.ieee.org/the-human-os/biomedical/devices/qa-ai-could-redesign-the-drug-development-process

[31] Will Knight, "An AI-Driven Genomics Company Is Turning to Drugs", *MIT Technology Review*, May 3, 2017, accessed June 7, 2018, https://www.technologyreview.com/s/604305/an-ai-driven-genomics-company-is-turning-to-drugs/

[32] "Antibiotic / Antimicrobial Resistance", *Centers for Disease Control and Prevention*, March 29, 2018, accessed June 8, 2018, https://www.cdc.gov/drugresistance/index.html

[33] Alexander Sulakvelidze and Zemphira Alavidze and J. Glenn Morris, Jr., "Bacteriophage Therapy", *Antimicrobial Agents and Chemotherapy*, March 2001, accessed June 8, 2018, https://www.ncbi.nlm.nih.gov/pmc/articles/PMC90351/

[34] Emily Mullin, "Faced with failing antibiotics, scientists are using killer viruses to fight superbugs", *MIT Technology Review*, January 29, 2018, accessed June 8, 2018, https://www.technologyreview.com/s/610066/faced-with-failing-antibiotics-scientists-are-using-killer-viruses-to-fight-superbugs/

[35] Nick Stockton, "VERITAS GENETICS SCOOPS UP AN AI COMPANY TO SORT OUT ITS DNA", *Wired*, August 7, 2017, accessed June 6, 2018, https://www.wired.com/story/veritas-genomics-scoops-up-an-ai-company-to-sort-out-its-dna/

[36] "How many pages in a gigabyte", *Lexis Nexis*, 2017, accessed June 6, 2018, https://www.lexisnexis.com/applieddiscovery/lawlibrary/whitePapers/ADI_FS_PagesInAGigabyte.pdf

[37] "Genome Editing with CRISPR-Cas9", *McGovern Institute for Brain Research*, November 5, 2014, accessed June 18, 2018, https://youtu.be/2pp17E4E-O8

[38] "Researchers use AI to improve accuracy of gene editing with CRISPR", *Microsoft The AI Blog*, January 10, 2018, accessed June 18, 2018, https://blogs.microsoft.com/ai/crispr-gene-editing/

[39] "CRISPR Eliminates HIV in Live Animals", *Genetic Engineering & Biotechnology News*, May 2, 2017, accessed June 18, 2018, https://www.genengnews.com/gen-news-highlights/crispr-eliminates-hiv-in-live-animals/81254287

[40] Julia Belluz and Umair Ifran, "2 new CRISPR tools overcome the scariest parts of gene editing", *Vox*, October 25, 2017, accessed June 19, 2018, https://www.vox.com/2017/10/25/16527370/crispr-gene-editing-harvard-mit-broad

[41] Kate Charlet, "The New Killer Pathogens", *Foreign Affairs*, accessed June 19, 2018, https://www.foreignaffairs.com/articles/2018-04-16/new-killer-pathogens

[42] Christopher Mims, "The AI Doctor Will See You Now", *The Wall Street Journal*, May 20, 2018, accessed June 9, 2018, https://www.wsj.com/articles/the-ai-doctor-will-see-you-now-1526817600

[43] Michelle Z. Donahue, "How a Color-Blind Artist Became the World's First Cyborg", *National Geographic*, April 3, 2017, accessed June 12, 2018, https://news.nationalgeographic.com/2017/04/worlds-first-cyborg-human-evolution-science/

[44] Neil Harbisson, "I listen to color", *TEDGlobal*, June 20, 2012, accessed June 12, 2018, https://www.ted.com/talks/neil_harbisson_i_listen_to_color

[45] Michael Franco, "Antenna implanted in cyborg's skull gets Wi-Fi, color as sound", *CNET*, April 14, 2014, accessed June 15, 2018, https://www.cnet.com/news/cyborg-interview-hear-colors-with-antenna-in-your-skull/

[46] Daniel Wagner, *Virtual Terror: 21st Century Cyber Warfare*, CreateSpace, 2017.

[47] D.T. Max, "How Humans Are Shaping Our Own Evolution", *National Geographic*, April 2017, accessed June 12, 2018, https://www.nationalgeographic.com/magazine/2017/04/evolution-genetics-medicine-brain-technology-cyborg/

[48] Tim Hornyak, "Brain implants let paralyzed woman move robot arm", *CNET*, December 17, 2012, accessed June 16, 2018, https://www.cnet.com/news/brain-implants-let-paralyzed-woman-move-robot-arm/

[49] Brad Jones, "The US Military is Testing Mood Altering, AI-Controlled Brain Implants in Humans", *Futurism*, November 28, 2017, accessed June 12, 2018, https://futurism.com/military-testing-mood-altering-ai/

[50] Adam Jezard, "This mind-reading AI can see what you're thinking - and draw a picture of it", *World Economic Forum*, February 5, 2018, accessed June 15, 2018, https://www.weforum.org/agenda/2018/02/mind-reading-ai-creates-images-from-your-thoughts/

[51] Shilo Rea, "Beyond Bananas: CMU Scientists Harness "Mind Reading" Technology to Decode Complex Thoughts", *Carnegie Mellon University*, June 26, 2017, accessed June 16, 2018, https://www.cmu.edu/dietrich/news/news-stories/2017/june/brain-decoding-complex-thoughts.html

[52] Shilo Rea, "Brain "Reads" Sentences the Same in English and Portuguese", *Carnegie Mellon University*, November 15, 2016, accessed June 18, 2018, https://www.cmu.edu/news/stories/archives/2016/november/brain-reads-the-same.html

[53] "Brain-computer interface advance allows fast, accurate typing by people with paralysis", *Stanford University*, February 21, 2017, accessed June 16, 2018, https://med.stanford.edu/news/all-news/2017/02/brain-computer-interface-allows-fast-accurate-typing-by-people-with-paralysis.html

[54] Nick Statt, "Facebook is working on a way to let you type with your brain", *The Verge*, April 19, 2017, accessed June 18, 2018, https://www.theverge.com/2017/4/19/15360798/facebook-brain-computer-interface-ai-ar-f8-2017

6. An Evolutionary Communications Landscape

[1] "Akamai's [state of the Internet] Q1 2017 Report", *Akamai*, 2017, accessed June 30, 2018, https://www.akamai.com/us/en/multimedia/documents/state-of-the-Internet/q1-2017-state-of-the-Internet-connectivity-report.pdf

[2] Mu-Hyun Cho, "South Korea to fund 10Gb Internet push", *ZDNet*, March 19, 2018, accessed June 30, 2018, https://www.zdnet.com/article/south-korea-to-fund-10gb-Internet-push/

[3] Carol Wilson, "AI Paving the Way for 5G, IoT", *Light Reading*, August 30, 2016, accessed June 30, 2018, https://www.lightreading.com/artificial-intelligence-machine-learning/ai-paving-the-way-for-5g-iot/d/d-id/725741

[4] "Focus Group on Machine Learning for Future Networks including 5G", *International Telecommunication Union*, 2017, accessed June 30, 2018, https://www.itu.int/en/ITU-T/focusgroups/ml5g/Pages/default.aspx

[5] "Record For Fastest Data Rate Set", *UNLOC*, February 11, 2016, accessed June 30, 2018, http://www.unloc.net/index.php/news-and-events/press-releases/261-record-for-fastest-data-rate-set

[6] "Email Statistics Report, 2017-2021", *The Radicati Group*, 2017, accessed June 24, 2018, https://www.radicati.com/wp/wp-content/uploads/2017/01/Email-Statistics-Report-2017-2021-Executive-Summary.pdf

[7] Federic Lardinois, "Google says its machine learning tech now blocks 99.9% of Gmail spam and phishing messages", *TechCrunch*, May 31, 2017, accessed June 27, 2018, https://techcrunch.com/2017/05/31/google-says-its-machine-learning-tech-now-blocks-99-9-of-gmail-spam-and-phishing-messages/

[8] "Smart Reply: Automated Response Suggestion for Email", *Google*, 2016, accessed June 25, 2018, https://storage.googleapis.com/pub-tools-public-publication-data/pdf/45189.pdf

[9] Chris Johnston, "Microsoft to buy LinkedIn for $26bn", *BBC News*, June 13, 2016, accessed June 26, 2018, https://www.bbc.com/news/business-36519766

[10] Dani Deahl, "Here's how to use Gmail's new Smart Compose", *The Verge*, May 10, 2018, accessed June 27, 2018, https://www.theverge.com/2018/5/10/17340224/google-gmail-how-to-use-smart-compose-io-2018

[11] Mark Wilson, "I let Gmail's new AI write my pointless emails for me", *Co.Design*, May 16, 2018, accessed June 27, 2018, https://www.fastcodesign.com/90172296/i-let-gmails-new-ai-write-my-pointless-emails-for-me

[12] "2017 NICE inContact CX Transformation Benchmark Study - Business Wave", *NICE*, 2017, accessed June 29, 2018, https://www.niceincontact.com/call-center-resource-finder/2017-incontact-cx-transformation-benchmark-study-business-wave

[13] Lindsay Clark, "AI elevates predictive maintenance for Kone and ThyssenKrupp", *ComputerWeekly.com*, May 2018, accessed June 29, 2018, https://www.computerweekly.com/feature/AI-elevates-predictive-maintenance-for-Kone-and-ThyssenKrupp

[14] Charles Duhigg, "How Companies Learn your Secrets", *The New York Times Magazine*, February 19, 2012, accessed June 29, 2018, https://www.nytimes.com/2012/02/19/magazine/shopping-habits.html?_r=1&hp=&pagewanted=all

[15] Catherine Clifford, "You can pay for your burger with your face at this fast food restaurant, thanks to A.I.", *CNBC*, February 2, 2018, accessed June 26, 2018, https://www.cnbc.com/2018/02/02/pay-with-facial-recognition-a-i-at-caliburger-in-pasadena-california.html

[16] Sophie Fitzpatrick, "Gastrograph AI: The Company That's Digitizing Your Taste Buds", *Edgy Labs*, May 24, 2018, accessed June 30, 2018, https://edgylabs.com/gastrograph-ai-the-company-thats-digitizing-your-taste-buds

[17] Nandita Bose, "Wal-Mart's new robots scan shelves to restock items faster", *Reuters*, October 26, 2017, accessed July 3, 2018, https://www.reuters.com/article/us-usa-walmart-robots/wal-marts-new-robots-scan-shelves-to-restock-items-faster-idUSKBN1CV1N4

[18] "Google Duplex may mark the beginning of a new era of AI", *CBS News*, May 9, 2018, accessed June 18, 2018, https://youtu.be/hyw7AM7OR6U

[19] "AI Learns the Art of Debate", *IBM*, June 18, 2018, accessed July 3, 2018, https://www.ibm.com/blogs/research/2018/06/ai-debate/

[20] Dom Galeon, "An AI That Detects Deception", *Futurism*, January 9, 2018, accessed July 3, 2018, https://futurism.com/new-ai-detects-deception-bring-end-lying-know-it/.

[21] Steven Melendez, "Goodbye polygraph? New tech uses AI to tell if you're lying", *Fast Company*, May 24, 2018, accessed July 4, 2018, https://www.fastcompany.com/40575672/goodbye-polygraphs-new-tech-uses-ai-to-tell-if-youre-lying

[22] "Employee Theft Statistics", *Statistic Brain Research Institute*, 2017, accessed July 7, 2018 https://www.statisticbrain.com/employee-theft-statistics/.

[23] Matt, McFarland, "The eyes expose our lies. Now AI is noticing.", *CNN Tech*, October 4, 2017, accessed July 4, 2018, http://money.cnn.com/2017/10/04/technology/business/eyedetect-lies-polygraph/index.html

[24] "Dapiprazole", *Drugs*.com, https://www.drugs.com/cdi/dapiprazole.html

[25] "The workplace of the future", *The Economist*, March 28, 2018, accessed July 4, 2018, https://www.economist.com/leaders/2018/03/28/the-workplace-of-the-future

[26] Rosalie Chan, "The high-tech office of the future will spy on you", *The Week*, June 6, 2018, accessed July 7, 2018, http://theweek.com/articles/760582/hightech-office-future-spy.

[27] "ACCENTURE EXTENDED REALITY", *Accenture*, 2018, July 7, 2018, https://www.accenture.com/us-en/service-extended-reality

[28] Mara Faccio and John J. McConnel, "Death by Pokémon GO: The Economic and Human Cost of Using Apps While Driving", *SSRN*, November 21, 2017, accessed July 7, 2018, https://papers.ssrn.com/sol3/papers.cfm?abstract_id=3073723

[29] Dieter Bohn, "Intel is giving up on its smart glasses", *The Verge*, April 18, 2018, accessed July 6, 2018, https://www.theverge.com/2018/4/18/17255354/intel-vaunt-shut-down

[30] Vanessa Ho, "'Heritage activists' preserve global landmarks ruined in war, threatened by time", *Microsoft*, April 23, 2018, accessed July 6, 2018, https://news.microsoft.com/transform/heritage-activists-preserve-global-landmarks-ruined-in-war-threatened-by-time/?utm_source=Direct

[31] Allegra Frank, "An infamous Japanese erotic game company makes its English debut", *Polygon*, April 10, 2018, accessed July 6, 2018, https://www.polygon.com/2018/4/10/17219618/illusion-software-eroge-steam-vr-kanojo

7. The Future of Learning

[1] "U.S. High School Graduation Rates Rise to New High," The Washington Post, December 04, 2017, accessed August 26, 2018, https://www.washingtonpost.com/news/education/wp/2017/12/04/u-s-high-school-graduation-rates-rise-to-new-high/?utm_term=.ab329f3a0088

[2] Karol Markowicz, "Lowering School Standards for Diversity's Sake Hurts All Students," USA Today, June 07, 2018, accessed August 26, 2018, https://www.usatoday.com/story/opinion/2018/06/07/fairness-means-lowering-bar-education-column/673198002/

[3] "A Brief History of MOOCs," SKILLSETS, November 16, 2015, accessed August 26, 2018, https://www.mcgill.ca/maut/current-issues/moocs/history

[4] Sanjay Kapoor, "Investing in Startups and the Future of the Google Assistant," Google, May 02, 2018, accessed August 26, 2018, https://www.blog.google/products/assistant/investing-startups-and-future-google-assistant/

[5] Blair Hanley Frank and ISG, "Quizlet Raises $20 Million for AI-powered Tutoring Software," VentureBeat, February 09, 2018, accessed August 26, 2018., https://venturebeat.com/2018/02/06/quizlet-raises-20-million-for-ai-powered-tutoring-software/.

[6] "Goal 4 : Sustainable Development Knowledge Platform," United Nations, aAccessed August 26, 2018, https://sustainabledevelopment.un.org/sdg4

[7] *The World Needs Almost 69 Million New Teachers to Reach the 2030 Education Goals*, PDF, UNESCO Institute for Statistics, October 2016, accessed August 26, 2018, http://uis.unesco.org/sites/default/files/documents/fs39-the-world-needs-almost-69-million-new-teachers-to-reach-the-2030-education-goals-2016-en.pdf

[8] Dana Goldstein, "Teacher Pay Is So Low in Some U.S. School Districts That They're Recruiting Overseas," The New York Times, May 02, 2018, accessed August 26, 2018, https://www.nytimes.com/2018/05/02/us/arizona-teachers-philippines.html

[9] Boaz Weinstein, "No Ethnic Group Owns Stuyvesant. All New Yorkers Do," The New York Times, June 13, 2018, accessed August 26, 2018, https://www.nytimes.com/2018/06/13/opinion/de-blasio-stuyvesant-school.html.

[10] Leslie Brody, "New York City Mayor Alters Exam-School Admissions," The Wall Street Journal, June 04, 2018, accessed August 26, 2018. https://www.wsj.com/articles/new-york-mayor-moves-to-diversify-citys-specialized-high-schools-1527971146

[11] Jenny Anderson, "Asians Spend Seven times as Much as Americans on Tutoring to Give Their Kids an Edge," Quartz, April 27, 2017, accessed August 26, 2018. https://qz.com/970130/asians-spend-15-of-their-family-income-on-extra-education-and-tutoring-for-kids-americans-spend-it-on-cars-and-gas/

[12] Julia Angwin and Jeff Larson, "The Tiger Mom Tax: Asians Are Nearly Twice as Likely to Get a Higher Price from Princeton Review," ProPublica. accessed August 26, 2018, https://www.propublica.org/article/asians-nearly-twice-as-likely-to-get-higher-price-from-princeton-review

[13] "Transcript: Obama's State Of The Union Address," NPR, January 26, 2011, accessed August 26, 2018, https://www.npr.org/2011/01/26/133224933/transcript-obamas-state-of-union-address

[14] Saro Mohammed, "ARPA-Ed: What Would It Take?" Brookings, November 17, 2016, accessed August 26, 2018, https://www.brookings.edu/blog/brown-center-chalkboard/2016/11/17/arpa-ed-what-would-it-take/

[15] J.D. Fletcher and John E. Morrison, *Accelerating Development of Expertise: A Digital Tutor for Navy Technical Training*, PDF, INSTITUTE FOR

DEFENSE ANALYSES, November 2014, accessed August 26, 2018, http://www.dtic.mil/dtic/tr/fulltext/u2/1002362.pdf

[16] "China's Schools Are Quietly Using AI to Mark Students' Essays," South China Morning Post, May 28, 2018, accessed August 26, 2018, https://www.scmp.com/news/china/society/article/2147833/chinas-schools-are-quietly-using-ai-mark-students-essays-do

[17] "Privacy & Security | Google for Education," Google for Education: Google SVA Scholarship, accessed August 26, 2018, https://edu.google.com/k-12-solutions/privacy-security/?modal_active=none

[18] Marcelo S. Perlin, "Building and Maintaining Exams with Dynamic Content," Blog and Site, accessed August 26, 2018, https://msperlin.github.io/2017-01-30-Exams-with-dynamic-content/

[19] Makena Kelly, "France Bans Smartphone Use in Schools," The Verge, August 01, 2018, accessed August 26, 2018, https://www.theverge.com/2018/8/1/17640476/france-bans-smartphone-cellphones-school-emmanuel-macron

[20] Camilla Turner, "Headmistress Scuppered by Red Tape as She Tried to Use Mobile Phone Jamming Technology to Stop Children Using Internet in Class," The Telegraph, January 19, 2017, accessed August 26, 2018, https://www.telegraph.co.uk/education/2017/01/19/headmistress-buys-illegal-jammer-block-students-internet-use/

[21] Richard Feloni, "Consumer-goods Giant Unilever Has Been Hiring Employees Using Brain Games and Artificial Intelligence - and It's a Huge Success," Business Insider, June 28, 2017, accessed August 26, 2018, https://www.businessinsider.com/unilever-artificial-intelligence-hiring-process-2017-6.

[22] Minda Zetlin, "AI Is Now Analyzing Candidates' Facial Expressions During Video Job Interviews," Inc.com. February 28, 2018, accessed August 26, 2018, https://www.inc.com/minda-zetlin/ai-is-now-analyzing-candidates-facial-expressions-during-video-job-interviews.html

[23] "Talent Acquisition: Enter the Cognitive Recruiter," Deloitte United States, accessed August 26, 2018, https://www2.deloitte.com/insights/us/en/focus/human-capital-trends/2017/predictive-hiring-talent-acquisition.html

[24] Harvey Gavin, "'We Will Decide Who Enters Our Countries!' Hungary and Poland REJECT EU Refugee Quotas," Express.co.uk, May 15, 2018, accessed August 26, 2018, https://www.express.co.uk/news/world/960125/eu-news-hungary-poland-oppose-european-union-refugee-quota-Viktor-Orban-Morawiecki

[25] Krisztina Than, "Hungary to Arm New 'border Hunters' after Six-month Crash Course," Reuters, March 10, 2017, accessed August 26, 2018, https://www.reuters.com/article/us-europe-migrants-hungary-borderhunters/hungary-to-arm-new-border-hunters-after-six-month-crash-course-idUSKBN16G2ED

[26] Deutsche Welle. "Austria Brings Hard-line Refugee Policy to EU | DW | 02.07.2018," DW.COM, accessed August 26, 2018, https://www.dw.com/en/austria-brings-hard-line-refugee-policy-to-eu/a-44496290

[27] Stanford University, "Algorithm Improves Integration of Refugees | Stanford News," Stanford University, January 24, 2018, accessed August 26, 2018, https://news.stanford.edu/2018/01/18/algorithm-improves-integration-refugees/

8. Smarter Government

[1] Tim Dutton, "An Overview of National AI Strategies — Politics AI — Medium," Medium, June 28, 2018, accessed August 04, 2018. https://medium.com/politics-ai/an-overview-of-national-ai-strategies-2a70ec6edfd

[2] "France and Canada Collaborate on Ethical AI," Digital Innovations in Facial Recognition: Interview (Includes Interview), June 10, 2018, accessed August 04, 2018, http://www.digitaljournal.com/tech-and-science/technology/france-and-canada-collaboraten-on-ethical-ai/article/524336

[3] Real Ventures, "Canada's Artificial Intelligence Ecosystem — Montreal," Medium, April 20, 2018, accessed August 04, 2018, https://medium.com/believing/canadas-artificial-intelligence-ecosystem-4798b0517016

[4] "International Students," Ethics, Society, and Law — Futurestudents, accessed August 04, 2018, http://www.artsci.utoronto.ca/futurestudents/internationalstudents

[5] Bloomberg.com, accessed August 04, 2018, https://www.bloomberg.com/quicktake/japan-s-shrinking-population

[6] International Finance Review, "Robots: Japan Delivers 52 Percent of Global Supply," IFR International Federation of Robotics, accessed August 04, 2018, https://ifr.org/ifr-press-releases/news/robots-japan-delivers-52-percent-of-global-supply

[7] "Development of Care Robots Growing in Aging Japan," The Japan Times, accessed August 04, 2018,

https://www.japantimes.co.jp/news/2016/01/27/national/social-issues/development-care-robots-growing-aging-japan/#.W2NgDtVKipo

[8] Christopher Mims, "Why Japanese Love Robots (And Americans Fear Them)," MIT Technology Review, October 22, 2012, accessed August 04, 2018, https://www.technologyreview.com/s/421187/why-japanese-love-robots-and-americans-fear-them/

[9] "Robots Have Positive Effect on Seniors at Nursing Care Facilities: Study - The Mainichi," May 31, 2017, accessed August 04, 2018, https://mainichi.jp/english/articles/20170531/p2a/00m/0na/006000c

[10] *Innovation at DARPA*, PDF, DARPA, July 2016, accessed August 4, 2018, https://www.darpa.mil/attachments/DARPA_Innovation_2016.pdf

[11] Arjun Kharpal, "Flying Cars Will Be in the Air within Five Years, CEO of Larry Page-backed Firm Says," CNBC, February 12, 2018, accessed November 04, 2018, https://www.cnbc.com/2018/02/12/flying-car-in-air-in-5-years-kitty-hawk-ceo-sebastian-thrun.html

[12] "Don't Be Silly: Lawmakers 'Rarely' Read Legislation and Oftentimes Don't Understand It . . . But That's Okay," Penn State Law Review RSS, accessed August 04, 2018, http://www.pennstatelawreview.org/penn-statim/dont-be-silly-lawmakers-rarely-read-legislation-and-oftentimes-dont-understand-it-but-thats-okay/

[13] John Mannes, "Primer Helps Governments and Corporations Monitor and Understand the World's Information" TechCrunch, October 24, 2017, accessed August 04, 2018. https://techcrunch.com/2017/10/24/primer-helps-governments-and-corporations-monitor-and-understand-the-worlds-information/

[14] Gallup, Inc, "Confidence in Institutions," Gallup.com, accessed August 04, 2018, https://news.gallup.com/poll/1597/confidence-institutions.aspx

[15] *Corruption: Costs and Mitigating Strategies*, PDF, INTERNATIONAL MONETARY FUND, May 2016, accessed August 4, 2018, https://www.imf.org/external/pubs/ft/sdn/2016/sdn1605.pdf

[16] "Singapore to Build Capabilities in Artificial Intelligence," South China Morning Post, November 22, 2017, accessed August 04, 2018, https://www.scmp.com/news/asia/southeast-asia/article/2092584/singapore-build-capabilities-artificial-intelligence

[17] "Download Entire World Economic Outlook Database, October 2017," Picture This -- Girl Power -- Finance & Development, March 2017, accessed August 04, 2018, http://www.imf.org/external/pubs/ft/weo/2018/01/weodata/download.aspx

[18] "Land & Liveability National Innovation Challenge," RIE2020 Plan, accessed August 04, 2018, https://www.nrf.gov.sg/programmes/national-innovation-challenges/land-liveability-national-innovation-challenge

[19] Ran Goldblatt, "Artificial Intelligence for Smart Cities: Insights from Ho Chi Minh City's Spatial Development" Governance for Development, March 12, 2018, accessed August 04, 2018, https://blogs.worldbank.org/opendata/artificial-intelligence-smart-cities-insights-ho-chi-minh-city-s-spatial-development

[20] Jamie Smyth, "Small Satellites and Big Data: A Commercial Space Race Hots up" Financial Times, January 24, 2018, Accessed August 04, 2018, https://www.ft.com/content/32d3f95e-f6c1-11e7-8715-e94187b3017e

[21] "World Urbanization Prospects - Population Division," United Nations, accessed August 04, 2018, https://esa.un.org/unpd/wup/

[22] "Super Smart City - Happier Society with Higher Quality | Deloitte China | Public Sector," Deloitte United States, June 29, 2018, accessed August 04, 2018, https://www2.deloitte.com/cn/en/pages/public-sector/articles/super-smart-city.html

[23] Lucas Laursen, "City Saves Money, Attracts Businesses with Smart City Strategy," MIT Technology Review, December 01, 2014, accessed August 04, 2018, https://www.technologyreview.com/s/532511/barcelonas-smart-city-ecosystem/

[24] *Drive.ai Announces On-Demand Self-Driving Car Service on Public Roads in Texas*, PDF, Frisco, Texas: Drive.ai, May 7, 2018, accessed August 4, 2018, https://s3.amazonaws.com/www-staging.drive.ai/content/uploads/2018/05/06164346/Press-Release_Drive.ai-Texas-Deployment.pdf

[25] *Adding Value to Parcel Delivery*, PDF, Accenture, September 29, 2015, accessed August 4, 2018, https://www.accenture.com/t20170227T024657Z__w__/ph-en/_acnmedia/Accenture/Conversion-Assets/DotCom/Documents/Global/PDF/Dualpub_23/Accenture-Adding-Value-to-Parcel-Delivery.pdf

[26] Chris Velazco, "Estonia Is First in the EU to Let Cute Delivery Bots on Sidewalks" Engadget, June 15, 2017, Accessed August 04, 2018, https://www.engadget.com/2017/06/15/estonia-welcomes-delivery-robots-to-sidewalks/.

[27] Leo Benedictus, "Chinese City Opens 'phone Lane' for Texting Pedestrians," The Guardian, September 15, 2014, accessed August 05, 2018, https://www.theguardian.com/world/shortcuts/2014/sep/15/china-mobile-phone-lane-distracted-walking-pedestrians

[28] "UAS Integration Pilot Program," FAA Seal, May 07, 2018, accessed August 04, 2018, https://www.faa.gov/uas/programs_partnerships/uas_integration_pilot_program/

[29] *Estimated U.S. Energy Consumption in 2017: 97.7 Quads*, PDF, Lawrence Livermore National Laboratory, 2017, accessed August 4, 2018, https://flowcharts.llnl.gov/content/assets/docs/2017_United-States_Energy.pdf

[30] Diane Cardwell, "Utility Helps Wean Vermonters From the Electric Grid," The New York Times, July 29, 2017, accessed August 04, 2018, https://www.nytimes.com/2017/07/29/business/energy-environment/vermont-green-mountain-power-grid.html

[31] Steve Lohr, "Homes Try to Reach Smart Switch," The New York Times, December 21, 2017, accessed August 04, 2018, https://www.nytimes.com/2015/04/23/business/energy-environment/homes-try-to-reach-smart-switch.html

[32] Daniel Wagner, *Virtual Terror*, CreateSpace, 2017, p. 220.

[33] John Kosowatz, "Using AI to Manage the Grid," ASME, January 2018, accessed August 04, 2018, https://www.asme.org/engineering-topics/articles/energy/using-ai-manage-the-grid.

[34] "Visualization and Analytics of Distribution Systems with Deep Penetration of Distributed Energy Resources (VADER)," Grid Modernization Laboratory Consortium, accessed August 04, 2018, https://gridmod.labworks.org/projects/si-1756

[35] Raffi Avo Sevlian, et al, *VADER: Visualization and Analytics for Distributed Energy Resources*, PDF, ArXiv, August 30, 2017, accessed August 4, 2018, https://arxiv.org/pdf/1708.09473.pdf

[36] *AI-augmented Government: Using Cognitive Technologies to Redesign Public Sector Work*, PDF, Deloitte University Press, 2017, accessed August 4, 2018, https://www2.deloitte.com/content/dam/insights/us/articles/3832_AI-augmented-government/DUP_AI-augmented-government.pdf

[37] "High-Priority Programs," PaymentAccuracy.gov, accessed August 04, 2018, https://paymentaccuracy.gov/high-priority-programs/

[38] "How to Reduce Improper Payments in Medicaid," Georgetown University Health Policy Institute, Georgetown.edu., accessed August 04, 2018, https://ccf.georgetown.edu/2018/04/19/measuring-medicaids-mistakes-estimating-improper-payments/

[39] Office, U.S. Government Accountability, "Medicare and Medicaid: CMS Needs to Fully Align Its Antifraud Efforts with the Fraud Risk Framework,"

U.S. Government Accountability Office (U.S. GAO), December 05, 2017, accessed August 04, 2018, https://www.gao.gov/products/GAO-18-88

[40] "The Challenge of Health Care Fraud," The National Healthcare Antifraud Association, accessed August 04, 2018, https://www.nhcaa.org/resources/health-care-anti-fraud-resources/the-challenge-of-health-care-fraud.aspx

[41] "National Health Care Fraud Takedown Results in Charges Against Over 412 Individuals Responsible for $1.3 Billion in Fraud Losses," The United States Department of Justice, December 11, 2017, accessed August 04, 2018, https://www.justice.gov/opa/pr/national-health-care-fraud-takedown-results-charges-against-over-412-individuals-responsible

[42] "National Science Foundation - Where Discoveries Begin," NSF - National Science Foundation, accessed August 04, 2018, https://www.nsf.gov/news/news_summ.jsp?cntn_id=137848

[43] "Using AI to Trace Leaking Pipes," The Economist, April 26, 2018, accessed August 04, 2018, https://www.economist.com/science-and-technology/2018/04/26/using-ai-to-trace-leaking-pipes

[44] David L. Chandler and MIT News Office, "Finding Leaks While They're Easy to Fix," MIT News, July 17, 2017, accessed August 04, 2018. http://news.mit.edu/2017/robot-finds-leaks-water-pipes-0718

[45] Christopher Woody, "Mexico's Oil Company Is Losing More than a Billion Dollars a Year to Cartels - and Its Own Employees Are Helping Them out," Business Insider, April 13, 2018, accessed August 04, 2018, https://www.businessinsider.com/cartels-mexico-oil-theft-pemex-2018-4

[46] "Bureau of Alcohol, Tobacco, Firearms and Explosives," National Firearms Act Handbook | Bureau of Alcohol, Tobacco, Firearms and Explosives, May 01, 1970, accessed August 04, 2018, https://www.atf.gov/resource-center/fact-sheet/fact-sheet-national-tracing-center

[47] "The ATF's Nonsenical Non-Searchable Gun Databases, Explained," The Trace, April 20, 2018, accessed August 04, 2018, https://www.thetrace.org/2016/08/atf-non-searchable-databases/

[48] "In Your Face: China's All-seeing State," BBC News, December 10, 2017, accessed August 04, 2018, https://www.bbc.com/news/av/world-asia-china-42248056/in-your-face-china-s-all-seeing-state

[49] "China's Facial Recognition Catches Another Fugitive Jacky Cheung Fan," South China Morning Post, May 22, 2018, accessed August 04, 2018, https://www.scmp.com/news/china/society/article/2147245/chinas-facial-recognition-cameras-apprehend-third-fugitive-jacky

[50] Mariel Myers, "Here's the Tech China's Using to Monitor, Shame and Rate Citizens," CNET, April 26, 2018, accessed August 04, 2018,

https://www.cnet.com/news/china-turns-to-tech-to-monitor-shame-and-rate-citizens/
[51] Ben Hammersley, "Concerned about Brexit? Why Not Become an E-resident of Estonia," WIRED, September 29, 2017, accessed August 04, 2018, https://www.wired.co.uk/article/estonia-e-resident.
[52] "Independent Report on E-voting in Estonia," Independent Report on Evoting in Estonia, accessed August 04, 2018, https://estoniaevoting.org/findings/summary/
[53] Damien McGuinness, "How a Cyber Attack Transformed Estonia," BBC News, April 27, 2017, accessed August 04, 2018. https://www.bbc.com/news/39655415
[54] "Interoperability Services," E-Estonia, accessed August 04, 2018, https://e-estonia.com/solutions/interoperability-services/.
[55] PricewaterhouseCoopers, "PwC's Global Artificial Intelligence Study: Sizing the Prize," PwC, accessed August 04, 2018, https://www.pwc.com/gx/en/issues/data-and-analytics/publications/artificial-intelligence-study.html.

9. The AI/Cyber Nexus

[1] Marc Goodman, *Future Crimes: Inside the Digital Underground and the Battle for Our Connected World,* (2nd ed.) (USA: Anchor, 2015), p. 406.
[2] Careers, VEON, and VEON Careers, "Dogs, Wolves, Data Science, and Why Machines Must Learn Like Humans Do," Medium, Augmenting Humanity, June 9, 2017, accessed March 6, 2018, https://medium.com/veon-careers/dogs-wolves-data-science-and-why-machines-must-learn-like-humans-do-213b08036a10.
[3] Neil Gerschenfeld and Isaac L. Chuang, "Quantum Computing with Molecules," *Scientific American*, June 1998, accessed July 25, 2017, http://cba.mit.edu/docs/papers/98.06.sciqc.pdf
[4] "Quantum Computing," *Wikipedia*, accessed July 25, 2017, https://en.wikipedia.org/wiki/Quantum_computing
[5] Shannon Bond, "Artificial intelligence and quantum computing aid cybercrime fight", *Financial Times*, May 24, 2017, accessed July 26, 2017, https://www.ft.com/content/1b9bdc4c-2422-11e7-a34a-538b4cb30025.
[6] Divesh Aggarwal et al., *Quantum Attacks on Bitcoin, and How to Protect against Them*, PDF, ArXiv, October 28, 2017, accessed July 26, 2018, https://arxiv.org/pdf/1710.10377.pdf.
[7] Chantel McGee, "The man who built a virtual nervous system explains how humans will interact with machines in ten years," *CNBC*, May 21,

2017, accessed July 26, 2017, http://www.cnbc.com/2017/05/21/mark-sagar-how-humans-will-interact-with-machines-in-ten-years.html

[8] Dan Falk, "Godlike 'Homo Deus' Could Replace Humans as Tech Evolves," *NBC*, May 31, 2017, accessed July 26, 2017, https://www.nbcnews.com/mach/technology/godlike-homo-deus-could-replace-humans-tech-evolves-n757971

[9] "First robotic cop joins Dubai police," *Yahoo*, accessed July 26, 2017, https://www.yahoo.com/tech/first-robotic-cop-joins-dubai-police-092135132.html

[10] Knight Scope, accessed July 26, 2017, http://www.knightscope.com/

[11] Carina Kolodny, "Stephen Hawking Is Terrified of Artificial Intelligence," *The Huffington Post*, May 5, 2014, accessed July 26, 2017, http://www.huffingtonpost.com/2014/05/05/stephen-hawking-artificial-intelligence_n_5267481.html

[12] Linda Sui, "44% of World Population Will Own Smartphones in 2017," *Strategy Analytics*, December 21, 2016, accessed July 26, 2017, https://www.strategyanalytics.com/strategy-analytics/blogs/smart-phones/2016/12/21/44-of-world-population-will-own-smartphones-in-2017#.WVATGGjyvic

[13] Rachel Metz, "Google Glass Is Dead; Long Live Smart Glasses," *MIT Technology Review*, November 26, 2014, accessed July 26, 2017, https://www.technologyreview.com/s/532691/google-glass-is-dead-long-live-smart-glasses/

[14] Christopher Mims, "In 10 Years, Your iPhone Won't Be a Phone Anymore," *The Wall Street Journal*, June 15, 2017, accessed July 26, 2017, https://www.wsj.com/articles/in-10-years-your-iphone-wont-be-a-phone-anymore-1498395600.

[15] Nina Larson, "AI 'good for the world'...says ultra-lifelike robot," *Phys.Org*, June 8, 2017, accessed March 6, 2018, https://phys.org/news/2017-06-ai-good-world-ultra-lifelike-robot.html

[16] Heungseok Koh, "Despite Assumption of Bias NYTimes, WSJ Photos Neutral," *Newspaper Research Journal* 33, no. 4 (2012): 102-16, accessed March 6, 2018. doi:10.1177/073953291203300409

[17] Derek Manky, "Artificial Intelligence: Cybersecurity Friend or Foe?" *Dark Reading*, May 11, 2017, accessed July 26, 2017 http://www.darkreading.com/threat-intelligence/artificial-intelligence-cybersecurity-friend-or-foe-/a/d-id/1328838.

[18] Cesar Cerrudo and Lucas Apa. "Hacking Robots Before Skynet," *IOActive*, 2017, accessed July 25, 2017, https://ioactive.com/pdfs/Hacking-Robots-Before-Skynet.pdf

[19] Georgia Diebelius, "Sex Robots Could Kill You If They Are Hacked, Expert Warns," Metro, Metro.co.uk, September 9, 2017, accessed March 6,2018, http://metro.co.uk/2017/09/09/sex-robots-could-kill-you-if-they-are-hacked-expert-warns-6915172/

[20] Georgia Diebelius, "Sex Robots Could Kill You If They Are Hacked, Expert Warns," Metro, Metro.co.uk, September 9, 2017, accessed March 6, 2018, http://metro.co.uk/2017/09/09/sex-robots-could-kill-you-if-they-are-hacked-expert-warns-6915172/

[21] Nick Ismali, "The importance of creating a cyber security culture," *Information Age*, April 8, 2017, accessed July 26, 2017, http://www.information-age.com/importance-creating-cyber-security-culture-123465778/

[22] Tereza Pultarova, "IBM's artificially intelligent Watson computer set to tackle cybercrime," *Engineering and Technology*, February 13, 2017, accessed July 26, 2017, https://eandt.theiet.org/content/articles/2017/02/ibms-artificially-intelligent-watson-computer-set-to-tackle-cybercrime/

[23] Andrew Lerner, "Intent-based Networking," *Gartner*, February 7, 2017, accessed July 26, 2017, http://blogs.gartner.com/andrew-lerner/2017/02/07/intent-based-networking/

[24] Derek Manky, "Extreme Makeover: AI & Network Cybersecurity," *Dark Reading*, May 10, 2017, accessed July 26, 2017, http://www.darkreading.com/threat-intelligence/extreme-makeover-ai-and-network-cybersecurity-/a/d-id/1328837

[25] "Artificial Intelligence and Life in 2030: One Hundred Year Study on Artificial Intelligence: Report of the 2015 Study Panel," *Stanford*, September 2016, accessed July 26, 2017, https://ai100.stanford.edu/sites/default/files/ai100report10032016fnl_singles.pdf

[26] *Machine Learning: A Higher Level of Automation*, PDF, DarkTrace, 2016, accessed July 26, 2017, https://www.ciosummits.com/Online_Asset_Darktrace_Whitepaper-Machine_Learning.pdf

[27] Research, P&S Market, 2017, "Artificial Intelligence in Cyber Security Market to Reach $18.2 Billion by 2023: P&S Market Research," GlobeNewswire News Room, "GlobeNewswire", November 6, 2017, accessed March 6, 2018, https://globenewswire.com/news-release/2017/11/06/1174755/0/en/Artificial-Intelligence-in-Cyber-Security-Market-to-Reach-18-2-Billion-by-2023-P-S-Market-Research.html

[28] Ryan Kh and Google, "How AI Is the Future of Cybersecurity," Infosecurity Magazine, December 1, 2017, accessed March 6, 2018

https://www.infosecurity-magazine.com/next-gen-infosec/ai-future-cybersecurity/

[29] "AI and Machine Learning in Cyber Security – Towards Data Science." Towards Data Science, Towards Data Science, January 1, 2018, accessed March 18, 2018, https://towardsdatascience.com/ai-and-machine-learning-in-cyber-security-d6fbee480af0

[30] "AI in Cybersecurity: Where We Stand & Where We Need to Go," n.d., Dark Reading, accessed May 1, 2018, https://www.darkreading.com/threat-intelligence/ai-in-cybersecurity-where-we-stand-and-where-we-need-to-go/a/d-id/1330787

[31] *The Malicious Use of Artificial Intelligence: Forecasting, Prevention, and Mitigation*, PDF, University of Oxford, February 2018, accessed March 6, 2018, https://www.eff.org/files/2018/02/20/malicious_ai_report_final.pdf, pp. 16-22.

[32] *The Malicious Use of Artificial Intelligence: Forecasting, Prevention, and Mitigation*, PDF, University of Oxford, February 2018, accessed March 6, 2018, https://www.eff.org/files/2018/02/20/malicious_ai_report_final.pdf, pp. 33-41.

[33] *The Malicious Use of Artificial Intelligence: Forecasting, Prevention, and Mitigation*, PDF, University of Oxford, February 2018. accessed March 6, 2018. https://www.eff.org/files/2018/02/20/malicious_ai_report_final.pdf, pp. 51-52, 64-65.

[34] "Albert Einstein Quotes." *Brainy quote*, accessed July 26, 2017, https://www.brainyquote.com/quotes/quotes/a/alberteins121993.html

[35] "Winning the Cyber War with AI & Cognitive Computing: The Experts' View," *AI Business*, September 9, 2016, accessed July 26, 2017, http://aibusiness.org/winning-the-cyber-war-with-ai-cognitive-computing-the-experts-view/

[36] *The Malicious Use of Artificial Intelligence: Forecasting, Prevention, and Mitigation*, PDF, University of Oxford, February 2018, accessed March 6, 2018, https://www.eff.org/files/2018/02/20/malicious_ai_report_final.pdf, pp. 3-7.

10. Spying and Fighting

AI SUPREMACY

[1] Greg Allen and Taniel Chan, *Artificial Intelligence and National Security*, PDF, Harvard Kennedy School, Belfer Center, July 2017, accessed May 1, 2018 https://www.belfercenter.org/sites/default/files/files/publication/AI%20NatSec%20-%20final.pdf

[2] "Artificial Intelligence," n.d., RealClearDefense, accessed May 1, 2018, https://www.realcleardefense.com/articles/2018/02/06/artificial_intelligence_an_offset_to_the_ics_information_overload_113020.html

[3] "For Artificial Intelligence to Thrive, It Must Explain Itself," The Economist, The Economist Newspaper, February 15, 2018, accessed May 1, 2018, https://www.economist.com/news/science-and-technology/21737018-if-it-cannot-who-will-trust-it-artificial-intelligence-thrive-it-must

[4] "About – CosmiQ Works," n.d., CosmiQ Works, accessed May 1, 2018, http://www.cosmiqworks.org/about

[5] Simonite, Tom, "Why Amazon and the CIA Want Algorithms to Understand Satellite Photos," MIT Technology Review, MIT Technology Review, August 25, 2016, accessed May 1, 2018, https://www.technologyreview.com/s/602239/amazon-and-the-cia-want-to-teach-ai-to-watch-from-space/

[6] Dave Gershgorn, "Facebook Made Detailed Maps Of 20 Countries For Its Internet Drones," Popular Science, February 22, 2016, accessed May 01, 2018, https://www.popsci.com/facebooks-new-population-maps-will-guide-its-Internet-drones

[7] Max de Haldevang, "Russia's Election Is so Predictable We Already Know Many of the Numbers," Quartz, Quartz, March 17, 2018, accessed May 1, 2018, https://qz.com/1231401/russia-election-putins-victory-is-so-predictable-we-already-know-many-of-the-numbers/

[8] Blair Hanley Frank and ISG, "In-Q-Tel Invests in SigOpt to Help U.S. Intelligence Agencies Optimize AI," VentureBeat, VentureBeat, February 12, 2018, accessed May 1, 2018, https://venturebeat.com/2018/02/12/in-q-tel-invests-in-sigopt-to-help-u-s-intelligence-agencies-optimize-ai/

[9] Jenna McLaughlin, "Artificial Intelligence Will Put Spies Out of Work, Too," Foreign Policy, Foreign Policy, June 9, 2017, accessed May 1, 2018, http://foreignpolicy.com/2017/06/09/artificial-intelligence-will-put-spies-out-of-work-too/

[10] "Artificial Intelligence: Putting Machines to Work for the Spies," n.d., U.S. News & World Report, U.S. News & World Report, accessed May 1, 2018, https://www.usnews.com/news/national-news/articles/2017-08-10/artificial-intelligence-putting-machines-to-work-for-the-spies

[11] Frank Konkel, "The CIA Says It Can Predict Social Unrest as Early as 3 to 5 Days Out," Defense One, October 5, 2016, accessed May 1, 2018, http://www.defenseone.com/technology/2016/10/cia-says-it-can-predict-social-unrest-early-3-5-days-out/132121/?oref=search_Digital Directorate

[12] Osonde A. Osoba and William Welser, *The Risks of Artificial Intelligence to Security and the Future of Work*, PDF, Rand Corporation, 2017, accessed May 1, 2018, https://www.rand.org/content/dam/rand/pubs/perspectives/PE200/PE237/RAND_PE237.pdf

[13] Jackie Snow, "Can AI Win the War against Fake News?" MIT Technology Review, MIT Technology Review, December 13, 2017, accessed May 1, 2018, https://www.technologyreview.com/s/609717/can-ai-win-the-war-against-fake-news/

[14] Will Knight, "AI Algorithms Are Creating a Frighteningly Realistic Fake Future," MIT Technology Review, MIT Technology Review, May 9, 2017, accessed July 19, 2018, https://www.technologyreview.com/s/604270/real-or-fake-ai-is-making-it-very-hard-to-know/

[15] Mahesh Saptharishi, "The New Eyes of Surveillance: Artificial Intelligence and Humanizing Technology," Wired, Conde Nast, August 7, 2015, accessed May 1, 2018, https://www.wired.com/insights/2014/08/the-new-eyes-of-surveillance-artificial-intelligence-and-humanizing-technology/

[16] Jenna McLaughlin, "Shake-Up at Pentagon Intelligence Agency Sparks Concern," Foreign Policy, Foreign Policy, January 12, 2018, accessed May 1, 2018, http://foreignpolicy.com/2018/01/12/shake-up-at-pentagon-intelligence-agency-sparks-concern/

[17] Jeremy Hsu, "AI Can Help Hunt Down Missile Sites in China," Wired, Conde Nast, November 21, 2017, accessed May 1, 2018, https://www.wired.com/story/ai-can-help-hunt-down-missile-sites-in-china/

[18] M.L. Cummings, *Artificial Intelligence and the Future of Warfare*, PDF, Chatham House: The Royal Institute of International Affairs, January 2017, accessed May 1, 2018, https://www.chathamhouse.org/sites/files/chathamhouse/publications/research/2017-01-26-artificial-intelligence-future-warfare-cummings-final.pdf, pp. 1, 12-13.

[19] Ruth A. David and Paul Nielsen, *Defense science board summer study on autonomy*, Defense Science Board Washington United States, 2016.

[20] Swarm technology remains in an evolutionary state, but early applications—whether for military or commercial use—point to a need to proceed with caution, for it remains an inaccurate science with plenty of room for error.
[21] *Department of Defense: Artificial Intelligence, Big Data, and Cloud Taxonomy*, PDF, Govini, 2011, accessed May 1, 2018, https://en.calameo.com/read/0000097792ddb787a9198
[22] Cade Metz, "Artificial Intelligence Is Now a Pentagon Priority. Will Silicon Valley Help?" The New York Times, The New York Times, August 26, 2018, accessed May 1, 2018, https://www.nytimes.com/2018/08/26/technology/pentagon-artificial-intelligence.html?action=click&module=Top Stories&pgtype=Homepage
[23] Patrick Tucker, "The Pentagon Is Nervous about Russian and Chinese Killer Robots," Defense One, December 14, 2015, accessed May 1, 2018, http://www.defenseone.com/threats/2015/12/pentagon-nervous-about-russian-and-chinese-killer-robots/124465/?oref=search_Russian chinese killer robots
[24] Amir Husain, "Putin Says Russia's New Weapons Can't Be Beat. With AI and Robotics, They Can," Defense One, March 13, 2018, accessed May 1, 2018, http://www.defenseone.com/ideas/2018/03/putin-says-russias-new-weapons-cant-be-beat-ai-and-robotics-they-can/146631/
[25] "USNI Logo," n.d., SparkCognition Inc, accessed May 1, 2018, https://www.sparkcognition.com/2017/07/on-hyperwar/
[26] Mark Pomerleau, "DoD Stands up Team to Take on PED/Intel Problem," C4ISRNET, C4ISRNET, August 8, 2017, accessed May 1, 2018, https://www.c4isrnet.com/intel-geoint/isr/2017/04/28/dod-stands-up-team-to-take-on-ped-intel-problem/
[27] Mark Pomerleau, "Automation 'Big as the Introduction of Nuclear Weapons,' Says Intel Official," C4ISRNET, C4ISRNET, August 8, 2017, accessed May 1, 2018, https://www.c4isrnet.com/intel-geoint/isr/2017/06/20/automation-big-as-the-introduction-of-nuclear-weapons-says-intel-official/
[28] "Project Maven Brings AI to the Fight against ISIS," Bulletin of the Atomic Scientists, June 28, 2018, accessed May 1, 2018, https://thebulletin.org/project-maven-brings-ai-fight-against-isis11374
[29] Brandon Knapp, "Here's Where the Pentagon Wants to Invest in Artificial Intelligence in 2019," C4ISRNET, C4ISRNET, February 21, 2018, May 1, 2018, https://www.c4isrnet.com/intel-geoint/2018/02/16/heres-where-the-pentagon-wants-to-invest-in-artificial-intelligence-in-2019/

[30] "Coming Soon to a Battlefield near You: Pentagon's Robot Vehicles," n.d., RT International, RT, accessed May 1, 2018, https://www.rt.com/usa/425613-pentagon-battle-robots-darpa/

[31] "An AI-Powered Network Could Save the US Navy Billions of Dollars," Futurism, Futurism, December 26, 2017, accessed May 1, 2018, https://futurism.com/ai-powered-network-save-us-navy-billions-dollars/

[32] Patrick. Tucker, "The Pentagon Is Nervous about Russian and Chinese Killer Robots," Defense One, December 14, 2015, accessed May 1, 2018, http://www.defenseone.com/threats/2015/12/pentagon-nervous-about-russian-and-chinese-killer-robots/124465/?oref=search_Russian chinese killer robots

[33] Kyle Mizokami, "How Russia's New Doomsday Torpedo Works," Popular Mechanics, Popular Mechanics, March 7, 2018, accessed May 1, 2018, https://www.popularmechanics.com/military/weapons/a19160734/how-russias-new-doomsday-torpedo-works/

[34] "In Space and Cyber, China Is Closing in on the United States," SpaceNews.com, January 10, 2018, accessed May 1, 2018, http://spacenews.com/in-space-and-cyber-china-is-closing-in-on-the-united-states/

[35] "Red Robots Rising," n.d., RealClearDefense, accessed May 1, 2018, https://www.realcleardefense.com/articles/2017/12/12/red_robots_risin_112770.html

[36] Jeff Kimmons, Ronald Makuta, and Graham Gilmer, *Remaking Intelligence, Processing, Exploitation and Dissemination*, PDF, Booz Allen Hamilton, 2017, accessed May 1, 2018, http://www.defenseone.com/media/ped-thought-piece-presentedby-booz-allen.pdf.

[37] "Meaningful Human Control, Artificial Intelligence and Autonomous Weapons," n.d., Article36, accessed May 1, 2018, http://www.article36.org/autonomous-weapons/roff-moyes-fli-paper/

[38] Patrick Tucker, "Report: Weapons AI Increasingly Replacing, Not Augmenting, Human Decision Making," Defense One, September 26, 2016, accessed May 1, 2018, http://www.defenseone.com/technology/2016/09/report-weapons-ai-increasingly-replacing-not-augmenting-human-decision-making/131826/

[39] Patrick Tucker, "Report: Weapons AI Increasingly Replacing, Not Augmenting, Human Decision Making," Defense One, September 26, 2016, accessed May 1, 2018, http://www.defenseone.com/technology/2016/09/report-weapons-ai-increasingly-replacing-not-augmenting-human-decision-making/131826/, pp. 1-5, 78.

11. China's Quest for AI Supremacy

[1] Patrick Tucker, "Report: Weapons AI Increasingly Replacing, Not Augmenting, Human Decision Making," Defense One, September 26, 2016, accessed May 1, 2018, http://www.defenseone.com/technology/2016/09/report-weapons-ai-increasingly-replacing-not-augmenting-human-decision-making/131826/, pp. 1-5, 78.

[2] Joshua Philipp, "Extensive Network of Secret Chinese Military Units Attack US on Daily Basis", *The Epoch Times*, November 21, 2014, accessed July 21, 2017, http://www.theepochtimes.com/n3/1094262-chinas-silent-war-on-the-us/

[3] Joshua Philipp, "EXCLUSIVE: How Hacking and Espionage Fuel China's Growth", *The Epoch Times*, September 10, 2015, accessed July 21, 2017, http://www.theepochtimes.com/n3/1737917-investigative-report-china-theft-incorporated/

[4] Margi Murphy, "Chinese Facial Recognition Company Becomes World's Most Valuable AI Start-Up," Yahoo! News, Yahoo! April 9, 2018, accessed July 19, 2018, https://www.yahoo.com/news/chinese-facial-recognition-company-becomes-105823939.html

[5] Christina Larson, "China's Massive Investment in Artificial Intelligence Has an Insidious Downside" Science | AAAS. American Association for the Advancement of Science, February 8, 2018, accessed March 4, 2018, http://www.sciencemag.org/news/2018/02/china-s-massive-investment-artificial-intelligence-has-insidious-downside

[6] "Big Data Reshapes China's Approach to Governance | Mercator Institute for China Studies," n.d., Mercator Institute for China Studies (MERICS), accessed March 4, 2018, https://www.merics.org/en/blog/big-data-reshapes-chinas-approach-governance

[7] *State Council Notice on the Issuance of the Next Generation Artificial Intelligence Development Plan*, (translated by New America Foundation), PDF, China's State Council, July 20, 2017, accessed March 4, 2018, file:///C:/Users/Admin/Downloads/translation-fulltext-8.1.17.pdf

[8] *State Council Notice on the Issuance of the Next Generation Artificial Intelligence Development Plan*, (translated by New America Foundation), PDF, China's State Council, July 20, 2017, accessed March 4, 2018, file:///C:/Users/Admin/Downloads/translation-fulltext-8.1.17.pdf, pp. 4-5.

[9] "China's Plan to 'Lead' in AI: Purpose, Prospects, and Problems, " n.d., New America, accessed March 4, 2018, https://www.newamerica.org/cybersecurity-initiative/blog/chinas-plan-lead-ai-purpose-prospects-and-problems/

[10] Jeffery Ding, *Deciphering China's AI Dream*, PDF, Future of Humanity Institute & University of Oxford, March 2018, accessed March 4, 2018, https://www.fhi.ox.ac.uk/wp-content/uploads/Deciphering_Chinas_AI-Dream.pdf

[11] "China's Plan to 'Lead' in AI: Purpose, Prospects, and Problems," n.d., New America, accessed March 4, 2018, https://www.newamerica.org/cybersecurity-initiative/blog/chinas-plan-lead-ai-purpose-prospects-and-problems/

[12] "World News, Politics, Economics, Business & Finance," n.d., The Economist, The Economist Newspaper, accessed March 4, 2018, https://www.economist.com/news/business/21725018-its-deep-pool-data-may-let-it-lead-artificial-intelligence-china-may-match-or-beat-.america?zid=291&ah=906e69ad01d2ee51960100b7fa502595

[13] "Espacenet," n.d., Espacenet - Bibliographic Data. Accessed September 5, 2018, https://worldwide.espacenet.com/

[14] Based on keyword searches of title and abstract.

[15] "China's Surveillance State: AI Startups, Tech Giants Are At The Center Of The Government's Plans," Corporate Innovation Trends, March 27, 2018, accessed July 19, 2018, https://www.cbinsights.com/research/china-surveillance-ai/

[16] "Research and Development (R&D) - Gross Domestic Spending on R&D - OECD Data," n.d., The OECD, accessed September 6, 2018, https://data.oecd.org/rd/gross-domestic-spending-on-r-d.htm

[17] Will Knight, "Inside the Chinese Lab That Plans to Rewire the World with AI," MIT Technology Review, MIT Technology Review, March 10, 2018 accessed July 19, 2018, https://www.technologyreview.com/s/610219/inside-the-chinese-lab-that-plans-to-rewire-the-world-with-ai/

[18] Sarah Zhang, "China's Artificial-Intelligence Boom," The Atlantic, Atlantic Media Company, February 27, 2017, accessed March 4, 2018, https://www.theatlantic.com/technology/archive/2017/02/china-artificial-intelligence/516615/

[19] "Why China's AI Push Is Worrying," The Economist, The Economist Newspaper, July 27, 2017, accessed March 4, 2018, https://www.economist.com/news/leaders/21725561-state-controlled-corporations-are-developing-powerful-artificial-intelligence-why-chinas-ai-push

[20] "Central Huijin Investment," Wikipedia, Wikimedia Foundation, August 30, 2018, accessed September 5, 2018, https://en.wikipedia.org/wiki/Central_Huijin_Investment.
[21] "State Administration of Foreign Exchange," Wikipedia, Wikimedia Foundation, August 30, 2018, accessed September 5, 2018, https://en.wikipedia.org/wiki/State_Administration_of_Foreign_Exchange
[22] "China Investment Corporation," 2018. Wikipedia. Wikimedia Foundation. August 26, 2018. Accessed September 5, 2018, https://en.wikipedia.org/wiki/China_Investment_Corporation
[23] Paul Mozur and Jane Perlez, "China Bets on Sensitive U.S. Start-Ups, Worrying the Pentagon," The New York Times, The New York Times, March 22, 2017, accessed March 4, 2018, https://www.nytimes.com/2017/03/22/technology/china-defense-start-ups.html
[24] Tom Simonite, "China Targets Nvidia's Hold on Artificial Intelligence Chips," Wired, Conde Nast, November 20, 2017, accessed March 4, 2018, https://www.wired.com/story/china-challenges-nvidias-hold-on-artificial-intelligence-chips/
[25] Phil Stewart, "U.S. Weighs Restricting Chinese Investment in Artificial Intelligence," Reuters, Thomson Reuters, June 14, 2017, accessed March 4, 2018, https://www.reuters.com/article/us-usa-china-artificialintelligence/u-s-weighs-restricting-chinese-investment-in-artificial-intelligence-idUSKBN1942OX
[26] "Silicon Valley Gets Queasy about Chinese Money," The Economist, August 09, 2018, accessed August 11, 2018, https://www.economist.com/business/2018/08/09/silicon-valley-gets-queasy-about-chinese-money.
[27] Tom Simonite, "The Trump Administration Can't Stop China From Becoming an AI Superpower," Wired, Conde Nast, June 30, 2017, Accessed March 4, 2018. https://www.wired.com/story/america-china-ai-ascension/
[28] Gregory Allen and Elsa B. Kania, "China Is Using America's Own Plan to Dominate the Future of Artificial Intelligence," Foreign Policy, Foreign Policy, September 8, 2017, accessed March 4, 2018, http://foreignpolicy.com/2017/09/08/china-is-using-americas-own-plan-to-dominate-the-future-of-artificial-intelligence/
[29] *Artificial Intelligence: Implications for China*, PDF, McKinsey Global Institute, April 2017, accessed March 4, 2018, https://www.mckinsey.com/~/media/McKinsey/Featured%20Insights/China/Artificial%20intelligence%20Implications%20for%20China/MGI-Artificial-intelligence-implications-for-China.ashx

[30] *Findings of the Investigation Into China's Acts, Policies, and Practices Related to Technology Transfer, Intellectual Property, and Innovation Under Section 301 of the Trade Act Of 1974*, PDF, Executive Office of the President of the United States, March 22, 2018, accessed April 4, 2018, https://ustr.gov/sites/default/files/Section%20301%20FINAL.PDF, p. 67.

[31] Echo Huang, "Chinese Investment in the US Skyrocketed Last Year," Quartz, Quartz, January 3, 2017, accessed March 4, 2018, https://qz.com/876693/chinese-investment-in-the-us-skyrocketed-in-2016/

[32] "Why Does Everyone Hate Made in China 2025?" n.d., Council on Foreign Relations, Council on Foreign Relations, accessed March 4, 2018, https://www.cfr.org/blog/why-does-everyone-hate-made-china-2025

[33] Paul Mozur, "Inside China's Dystopian Dreams: A.I., Shame and Lots of Cameras," The New York Times, The New York Times, July 8, 2018, accessed July 19, 2018, https://www.nytimes.com/2018/07/08/business/china-surveillance-technology.html?rref=collection/spotlightcollection/china-reach

[34] Patrick Tucker, "Thanks, America! How China's Newest Software Could Predict, Track, and Crush Dissent," Defense One, March 7, 2016, accessed July 19, 2018, http://www.defenseone.com/technology/2016/03/thanks-america-china-aims-tech-dissent/126491/

[35] Elsa B. Kania, "China Is On a Whole-of-Nation Push for AI. The US Must Match It," Defense One, December 8, 2017, accessed March 4, 2018, http://www.defenseone.com/ideas/2017/12/us-china-artificial-intelligence/144414/

[36] Elsa B. Kania, "Chinese Sub Commanders May Get AI Help for Decision-Making," Defense One, February 12, 2018, accessed March 4, 2018, http://www.defenseone.com/ideas/2018/02/chinese-sub-commanders-may-get-ai-help-decision-making/145906/

[37] "Chinese Satellite Uses Quantum Cryptography for Secure Video Conference between Continents," MIT Technology Review, MIT Technology Review, January 30, 2018, accessed March 4, 2018, https://www.technologyreview.com/s/610106/chinese-satellite-uses-quantum-cryptography-for-secure-video-conference-between-continents/

12. International Relations

[1] "Robots, Immune to Fear or Favor, Are Making China's Foreign Policy," South China Morning Post, South China Morning Post, July 31, 2018, accessed August 4, 2018,

https://www.scmp.com/news/china/society/article/2157223/artificial-intelligence-immune-fear-or-favour-helping-make-chinas

[2] Michael Chui, James Manyika, and Mehdi Miremadi, "The Countries Most (and Least) Likely to Be Affected by Automation," Harvard Business Review, September 20, 2017, accessed March 4, 2018, https://hbr.org/2017/04/the-countries-most-and-least-likely-to-be-affected-by-automation

[3] Alena Kudzko, "Future Now: How AI Is Already Changing the Global and Military Landscape," GLOBSEC, February 06, 2018, accessed March 04, 2018, https://www.globsec.org/future-now-ai-already-changing-global-military-landscape/.

[4] Daveed Gartenstein-Ross, "Terrorists Are Going to Use Artificial Intelligence," Defense One, May 3, 2018, accessed July 19, 2018, https://www.defenseone.com/ideas/2018/05/terrorists-are-going-use-artificial-intelligence/147944/

[5] Allegedly. No public proof has been provided by the US government.

[6] Ben Scott, Stefan Heumann, and Philippe Lorenz, *Artificial Intelligence and Foreign Policy*, PDF, Stiftung Neue Verantwortung, January 2018. Accessed March 4, 2018. https://www.stiftung-nv.de/sites/default/files/ai_foreign_policy.pdf

[7] "Open Letter on Autonomous Weapons," n.d., Future of Life Institute, The FLI Team, accessed March 4, 2018, https://futureoflife.org/open-letter-autonomous-weapons/

[8] "Climate Change and Human Rights," n.d., Global Policy Journal, accessed March 4, 2018, https://www.globalpolicyjournal.com/blog/14/04/2016/lethal-artificial-intelligence-problem

[9] "Climate Change and Human Rights," n.d., Global Policy Journal, accessed March 4, 2018, https://www.globalpolicyjournal.com/blog/14/04/2016/lethal-artificial-intelligence-problem

[10] Denise Garcia, "Future Arms, Technologies, and International Law: Preventive Security Governance," Cambridge Core, Cambridge University Press, January 27, 2016, accessed March 4, 2018, https://www.cambridge.org/core/journals/european-journal-of-international-security/article/future-arms-technologies-and-international-law-preventive-security-governance/7AEE439FBCC7ADC9EF3703D5822DF1F0/core-reader

[11] Edward Geist and Andrew J. Lohn, *How Might Artificial Intelligence Affect the Risk of Nuclear War?* PDF, RAND Corporation, 2018, accessed March 4, 2018, file:///C:/Users/Admin/Downloads/RAND_PE296.pdf

[12] *One Alliance: The Future Tasks of the Adapted Alliance*, PDF, GLOBSEC, November 2017, accessed March 4, 2018, https://www.globsec.org/wp-content/uploads/2017/11/GNAI-Final-Report-Nov-2017.pdf

[13] . Bloomberg.com. Bloomberg, accessed March 4, 2018, https://www.bloomberg.com/news/articles/2018-03-27/european-technology-irrelevance-feared-as-u-s-china-dominate

[14] Philip Kaleta et al., "Macron's €1.5 Billion Plan to Drag France into the Age of Artificial Intelligence," POLITICO, April 14, 2018, accessed March 04, 2018, https://www.politico.eu/article/macron-aims-to-drag-france-into-the-age-of-artificial-intelligence/

[15] "#FranceIA: the National Artificial Intelligence Strategy Is Underway," n.d., Gouvernement.fr., accessed March 4, 2018, http://www.gouvernement.fr/en/franceia-the-national-artificial-intelligence-strategy-is-underway

[16] James Temperton, "Samsung Developing Robots to Replace Cheap Chinese Labour," WIRED, WIRED UK, October 4, 2017, accessed March 4, 2018, http://www.wired.co.uk/article/samsung-south-korea-robots-cheap-labour

[17] Abishur Prakash, "Forget The Markets, Robots Are China's New Worry," Forbes, Forbes Magazine, January 28, 2016, accessed March 4, 2018, https://www.forbes.com/sites/realspin/2016/01/28/forget-the-markets-robots-are-chinas-new-worry/#292a39543cb7

[18] Abishur Prakash, "Three Geopolitical Predictions for Artificial Intelligence," Medium, Augmenting Humanity, April 21, 2016, accessed March 4, 2018, https://medium.com/next-geopolitics/three-geopolitical-predictions-for-artificial-intelligence-94359c03559f

[19] Ben Scott, Stefan Heumann, and Philippe Lorenz, *Artificial Intelligence and Foreign Policy*, PDF, Stiftung Neue Verantwortung, January 2018, accessed March 4, 2018, https://www.stiftung-nv.de/sites/default/files/ai_foreign_policy.pdf, pp. 2-4, 31-32.

[20] "UNICRI Centre for Artificial Intelligence and Robotics," UNICRI, accessed March 04, 2018, http://www.unicri.it/in_focus/on/UNICRI_Centre_Artificial_Robotics

[21] David Gosset, "Artificial Intelligence (AI) And Global Geopolitics," The Huffington Post, TheHuffingtonPost.com, June 30, 2017, accessed March 4, 2018, https://www.huffingtonpost.com/david-gosset/artificial-intelligence-a_2_b_10710612.html

[22] John R. Allen and Amir Husain, "The Next Space Race Is Artificial Intelligence," Foreign Policy, Foreign Policy, November 3, 2017, accessed March 4, 2018, http://foreignpolicy.com/2017/11/03/the-next-space-race-is-artificial-intelligence-and-america-is-losing-to-china/

[23] "Text - H.R.5356 - 115th Congress (2017-2018): National Security Commission Artificial Intelligence Act of 2018," Congress.gov. May 22, 2018, accessed July 19, 2018, https://www.congress.gov/bill/115th-congress/house-bill/5356/text

[24] Jack Corrigan, "How Much Does Artificial Intelligence Threaten National Security?" Nextgov.com, Nextgov, March 21, 2018, accessed March 4, 2018, https://www.nextgov.com/policy/2018/03/how-much-does-artificial-intelligence-threaten-national-security/146844/

[25] "UAE Strategy for Artificial Intelligence - The Official Portal of the UAE Government," n.d., Cancer - The Official Portal of the UAE Government. Accessed March 4, 2018. https://government.ae/en/about-the-uae/strategies-initiatives-and-awards/federal-governments-strategies-and-plans/uae-strategy-for-artificial-intelligence

[26] "An inside Look at the World's First Nation with a Minister for Artificial Intelligence," Futurism, Futurism, December 11, 2017, accessed March 4, 2018, https://futurism.com/uae-minister-artificial-intelligence/

13. Legal Safeguards

[1] James Ovenden, "Why The Legal Profession Is Turning To Machine Learning | Articles | Analytics," Articles | Finance | Innovation Enterprise, April 18, 2017, accessed March 22, 2018, https://channels.theinnovationenterprise.com/articles/why-the-legal-profession-is-turning-to-machine-learning

[2] "ROSS and Watson Tackle the Law – Watson," IBM Cognitive Advantage Reports, IBM Corporation, May 5, 2017, accessed March 22, 2018, https://www.ibm.com/blogs/watson/2016/01/ross-and-watson-tackle-the-law/

[3] "Meet 'Ross,' the Newly Hired Legal Robot," The Washington Post, WP Company, May 16, 2016, accessed March 22, 2018, https://www.washingtonpost.com/news/innovations/wp/2016/05/16/meet-ross-the-newly-hired-legal-robot/?utm_term=.d830ba735f7c

[4] Steve Lohr, "A.I. Is Doing Legal Work. But It Won't Replace Lawyers, Yet," The New York Times, The New York Times, March 19, 2017, accessed March 22, 2018, https://www.nytimes.com/2017/03/19/technology/lawyers-artificial-intelligence.html?_r=0

[5] Jason Koebler, "Rise of the Robolawyers," The Atlantic. Atlantic Media Company, March 14, 2017, accessed March 22, 2018,

https://www.theatlantic.com/magazine/archive/2017/04/rise-of-the-robolawyers/517794/

[6] In 2017 the European Commission published a strategy for integrating what it refers to as "cooperative intelligent transportation systems", which establishes an approach to developing a standardized intelligent transport infrastructure allowing vehicles to communicate with each other, centralized traffic management systems and other highway users. See: *Result of C-ITS Platform Phase II*, PDF, European Commission, December 2017, accessed March 22, 2018, https://ec.europa.eu/transport/sites/transport/files/c-its_security_policy_release_1.pdf

[7] "Artificial Intelligence: the Real Legal Issues," n.d., Osborne Clarke, accessed March 4, 2018, http://www.osborneclarke.com/insights/artificial-intelligence-the-real-legal-issues-an-article-by-john-c-buyers-osborne-clarke-llp/

[8] Such as is the case with OpenAI. See: "OpenAI," OpenAI, accessed May 02, 2018, https://openai.com/

[9] Jeremy Elman and Abel Castilla. "Artificial Intelligence and the Law," TechCrunch, TechCrunch, January 28, 2017, accessed March 4, 2018, https://techcrunch.com/2017/01/28/artificial-intelligence-and-the-law/.

[10] "Tesla Model X Was in Autopilot before Fatal Crash in California," n,d., CNNMoney. Cable News Network, accessed March 4, 2018, http://money.cnn.com/2018/03/31/technology/tesla-model-x-crash-autopilot/index.html

[11] Edward Helmore, "Emergency Brake Was Disabled on Self-Driving Uber That Killed Woman," The Guardian, Guardian News and Media, May 24, 2018, accessed June 4, 2018, https://www.theguardian.com/technology/2018/may/24/emergency-brake-was-disabled-on-self-driving-uber-that-killed-woman

[12] "Liability and Legal Questions Follow Uber Autonomous Car Fatal Accident," Insurance Journal, March 21, 2018, accessed March 22, 2018, https://www.insurancejournal.com/news/national/2018/03/20/483981.htm

[13] Mark Harris, "Exclusive: Arizona Governor and Uber Kept Self-Driving Program Secret, Emails Reveal," The Guardian, Guardian News and Media, March 28, 2018, https://www.theguardian.com/technology/2018/mar/28/uber-arizona-secret-self-driving-program-governor-doug-ducey

[14] Ian Bogost, "Can You Sue a Robocar?" The Atlantic. Atlantic Media Company, March 20, 2018, accessed March 22, 2018,

https://www.theatlantic.com/technology/archive/2018/03/can-you-sue-a-robocar/556007/

[15] "Volvo CEO: We Will Accept All Liability When Our Cars Are in Autonomous Mode," n.d., Fortune, Fortune, accessed March 4, 2018, http://fortune.com/2015/10/07/volvo-liability-self-driving-cars/

[16] *Artificial Intelligence, Ethics and Enhanced Data Stewardship*, PDF, The Information Accountability Foundation, September 20, 2017, accessed March 4, 2018, https://www.privacyconference2017.org/eng/files/ai.pdf

[17] "Privacy - Approach to Privacy," n.d., Apple, accessed March 4, 2018, https://www.apple.com/privacy/approach-to-privacy/

[18] Kate Conger and Natasha Lomas, "What Apple's Differential Privacy Means for Your Data and the Future of Machine Learning," TechCrunch, TechCrunch, June 14, 2016, accessed March 4, 2018, https://techcrunch.com/2016/06/14/differential-privacy/

[19] "General Data Protection Regulation," Wikipedia, September 01, 2018, accessed March 04, 2018, https://en.wikipedia.org/wiki/General_Data_Protection_Regulation

[20] Evan Taparata, "President Trump, how is letting Internet providers sell consumers' browsing data in the public interest?" *PRI*, April 16, 2017, accessed July 26, 2017, https://www.pri.org/stories/2017-04-16/president-trump-how-letting-Internet-providers-sell-consumers-browsing-data

[21] "GDPR Key Changes", *EUGDPR*, accessed July 26, 2017, http://www.eugdpr.org/key-changes.html

[22] Bryce Goodman and Seth Flaxman, *European Union Regulations on Algorithmic Decision-making and a "right to Explanation"*, PDF, ArXiv, August 31, 2016, accessed March 4, 2018, https://arxiv.org/pdf/1606.08813.pdf

[23] "Artificial Intelligence and Copyright," n.d., Oman: Basic Law of the Sultanate of Oman (Promulgated by the Royal Decree No. 101/96), accessed March 4, 2018, http://www.wipo.int/wipo_magazine/en/2017/05/article_0003.html

[24] Louise Matsakis, "Copyright Law Makes Artificial Intelligence Bias Worse," Motherboard, October 31, 2017, accessed March 4, 2018, https://motherboard.vice.com/en_us/article/59ydmx/copyright-law-artificial-intelligence-bias

[25] Alex Campolo, Madelyn Sanfilippo, Meredith Whittaker, and Kate Crawford, *AI Now 2017 Report*, PDF, AI Now, 2017, accessed March 4, 2018, https://assets.contentful.com/8wprhhvnpfc0/1A9c3ZTCZa2KEYM64Wsc2

a/8636557c5fb14f2b74b2be64c3ce0c78/_AI_Now_Institute_2017_Report_.pdf, pp. 30-31.

[26] Jason Tashea, "Courts Are Using AI to Sentence Criminals. That Must Stop Now," Wired, Conde Nast, February 2, 2018, accessed March 4, 2018, https://www.wired.com/2017/04/courts-using-ai-sentence-criminals-must-stop-now/

[27] "Should Artificial Intelligence Be Regulated?" The Huffington Post, TheHuffingtonPost.com, July 27, 2017, accessed March 4, 2018, https://www.huffingtonpost.com/entry/should-artificial-intelligence-be-regulated_us_597a452de4b09982b737630c

[28] Sandra Wachter, Brent Mittelstadt, and Luciano Floridi, "Transparent, Explainable, and Accountable AI for Robotics," Science Robotics, May 31, 2017, accessed March 4, 2018, http://robotics.sciencemag.org/content/2/6/eaan6080.full

[29] "New York City Aims to Bring Accountability to Algorithms with New Bill," Immuta, January 19, 2018, accessed March 4, 2018, https://www.immuta.com/new-york-city-aims-to-bring-accountability-to-algorithms-with-new-bill/

[30] David Beyer, Artificial Intelligence and Machine Learning in Industry: Perspectives from Leading Practitioners, O'Reilly Media, First Edition, March 2017, pp. 12-17, accessed March 14, 2018.

[31] "AI for Law," n.d., Legal Robot, accessed March 4, 2018, https://www.legalrobot.com/

[32] Catherine Baksi, "The Digital Shake up of Law," Raconteur, Raconteur Media Ltd., March 27, 2017, accessed March 4, 2018, https://www.raconteur.net/business/the-digital-shake-up-of-law

[33] *2016 ILTA/InsideLegal Technology Purchasing Survey*, PDF, International Legal Technology Association, 2016, accessed March 4, 2018, https://insidelegal.typepad.com/files/2016_ILTA_InsideLegal_Technology_Purchasing_Survey.pdf

[34] "Artificial Intelligence," Canadian Lawyer Mag, April 3, 2017, accessed July 19, 2018, http://www.canadianlawyermag.com/article/artificial-intelligence-3585/

[35] "AI Principles," n.d., Future of Life Institute, The FLI Team, https://futureoflife.org/wp-content/uploads/2015/10/FLI_logo-1.png. Accessed July 19, 2018. https://futureoflife.org/ai-principles/

14. An AI Future to be Feared or Embraced?

[1] Vlad Savov, "Google's Selfish Ledger Is an Unsettling Vision of Silicon Valley Social Engineering," The Verge, The Verge, May 17, 2018, accessed July 19, 2018, https://www.theverge.com/2018/5/17/17344250/google-x-selfish-ledger-video-data-privacy

[2] Mark Bergen, "Google Is Training Machines to Predict When a Patient Will Die - Tech News | The Star Online," Nation | The Star Online, June 19, 2018, accessed July 19, 2018, https://www.thestar.com.my/tech/tech-news/2018/06/19/google-is-training-machines-to-predict-when-a-patient-will-die/

[3] Bergen, Mark. 2018. "Google Is Training Machines to Predict When a Patient Will Die." Bloomberg Quint. Bloomberg Quint. June 18, 2018. https://www.bloombergquint.com/business/2018/06/18/google-is-training-machines-to-predict-when-a-patient-will-die.

[4] Jack Corrigan, "The Pentagon Wants to Bring Mind-Controlled Tech To Troops," Nextgov.com, Nextgov, July 17, 2018, accessed July 19, 2018, https://www.nextgov.com/emerging-tech/2018/07/pentagon-wants-bring-mind-controlled-tech-troops/149776/

[5] ""China Developing Robotic Subs to Launch New Era of Sea Power," South China Morning Post, South China Morning Post, July 23, 2018, accessed July 19, 2018, https://www.scmp.com/news/china/society/article/2156361/china-developing-unmanned-ai-submarines-launch-new-era-sea-power

[6] "Robots Grow Mini-Organs from Human Stem Cells," ScienceDaily, ScienceDaily, May 17, 2018, accessed July 19, 2018, https://www.sciencedaily.com/releases/2018/05/180517123300.htm

[7] "Child Sex Dolls, Human-Bot Babies and the End of Society: RT Attends London's Sex Robot Conference," n.d., RT International, accessed July 19, 2018, https://www.rt.com/uk/413762-sex-love-robot-conference/

[8] "Digital Immortality? Funeral Agency Wants to 'Revive' the Dead in AI Chatbot Project," n.d., RT International, accessed July 19, 2018, https://www.rt.com/news/419706-digital-immortality-funeral-agency/

[9] "Black Mirror in Action: Swedish Scientists to Revive the Dead in Robots with AI," Sputnik International, February 23, 2018,accessed July 19, 2018, https://sputniknews.com/science/201802231061926036-ai-sweden-revive-dead-robots/

[10] Moa Petersén, "The Reason Thousands of Swedish People Are Inserting Microchips into Themselves," Quartz, August 29, 2018, accessed July 19, 2018, https://qz.com/1313537/biohacking-in-sweden-why-thousands-are-inserting-microchips-into-themselves/

[11] Monica Torres, "One Person Was Able to 'Shoplift' from Amazon's New Convenience Store," Ladders | Business News & Career Advice. Ladders |

Business News & Career Advice, January 23, 2018, accessed July 19, 2018, https://www.theladders.com/career-advice/shoplift-from-amazon-go-convenience-store

[12] Monica Torres, "Amazon Wants Alexa to Be Your New Coworker," Ladders | Business News & Career Advice. Ladders | Business News & Career Advice, December 4, 2017, accessed July 19, 2018, https://www.theladders.com/career-advice/amazon-wants-alexa-to-be-your-new-co-worker

[13] Hannah Devlin, "Scientists Genetically Engineer Pigs Immune to Costly Disease," The Guardian, Guardian News and Media, June 20, 2018, accessed July 19, 2018, https://www.theguardian.com/science/2018/jun/20/scientists-genetically-engineer-pigs-immune-to-costly-disease

[14] Josh Horwitz, "Alibaba Is Using Artificial Intelligence to Help Raise Pigs in China," Quartz, Quartz, February 15, 2018, accessed July 19, 2018, https://qz.com/1202142/alibaba-is-using-use-artificial-intelligence-to-help-raise-pigs/

[15] "China Construction Bank," Wikipedia, Wikimedia Foundation, September 5, 2018, accessed July 19, 2019, https://en.wikipedia.org/wiki/China_Construction_Bank

[16] Helen Roxburgh, "Inside Shanghai's Robot Bank: China Opens World's First Human-Free Branch" The Guardian, Guardian News and Media, May 14, 2018, accessed July 19, 2018, https://www.theguardian.com/cities/2018/may/14/shanghai-robot-bank-china-worlds-first-human-free-branch-construction

[17] "Come on, Barbie, Let's Go Party: Europe's First Sex Doll Brothel Opens in Barcelona (PHOTOS)," n.d. , RT International, RT, accessed July 19, 2018, https://www.rt.com/viral/378799-sex-doll-brothel-barcelona/

[18] Mark Branagan, "Sex Robot Will Turn down Steamy Romps If It's Not in the Mood," Mirror, Mirror.co.uk, June 17, 2018, accessed July 19, 2018, https://www.mirror.co.uk/news/weird-news/sex-robot-turns-down-steamy-12726878

[19] "Creepy AI Can Predict Your Moves in Advance," Daily Mail Online, Associated Newspapers, June 15, 2018, accessed July 19, 2018, http://www.dailymail.co.uk/sciencetech/article-5847767/Creepy-AI-predict-moves-advance-lead-level-Big-Brother-surveillance.html

[20] *The Malicious Use of Artificial Intelligence: Forecasting, Prevention, and Mitigation*, PDF, University of Oxford, February 2018, accessed July 19, 2018, https://www.eff.org/files/2018/02/20/malicious_ai_report_final.pdf

[21] Philip Gerbert, Jan Justus, and Martin Hecker, "Competing in the Age of Artificial Intelligence," January 16, 2017, BCG Henderson Institute, accessed July 19, 2018, https://www.bcg.com/publications/2017/competing-in-age-artificial-intelligence.aspx

[22] Martin Reeves, Daichi Ueda, Philipp Gerbert, and Ralf Dreischmeier, "The Integrated Strategy Machine: Using AI to Create Advantage," April 19, 2016, BCG Henderson Institute, accessed July 19, 2018, https://www.bcg.com/publications/2016/strategy-technology-digital-integrated-strategy-machine-using-ai-create-advantage.aspx

[23] *2018 AI Predictions: 8 Insights to Shape Business Strategy*, PDF, PWC, 2018, accessed July 19, 2018, https://www.pwc.com/us/en/advisory-services/assets/ai-predictions-2018-report.pdf

[24] Watson (Computer)," Wikipedia, Wikimedia Foundation, August 31, 2018, https://en.wikipedia.org/wiki/Watson_(computer)

[25] Timothy Green, "Here's IBM's Blueprint for Winning the AI Race," The Motley Fool, June 2, 2018, accessed July 19, 2018, https://www.fool.com/investing/2018/06/02/heres-ibms-blueprint-for-winning-the-ai-race.aspx

[26] Evangelos Power and Brad Simoudis, "The 5 Things IBM Needs to Do to Win at AI," Harvard Business Review, April 24, 2017, accessed July 19, 2018, https://hbr.org/2016/02/the-5-things-ibm-needs-to-do-to-win-at-ai

[27] From Michael Polanyi's book, *The Tacit Dimension* (University of Chicago Press, 1966).

[28] "The Business of Artificial Intelligence," Harvard Business Review, accessed July 19, 2018, https://hbr.org/cover-story/2017/07/the-business-of-artificial-intelligence

[29] "Survey: 80 Percent of Enterprises Investing in AI, but Cite Significant Challenges Ahead," n.d. ,Teradata Latest Press Releases | A Leader in Data and Analytics, accessed July 19, 2018, https://www.teradata.com/Press-Releases/2017/Survey-80-Percent-of-Enterprises-Invest-in-AI

[30] "Gartner Says AI Technologies Will Be in Almost Every New Software Product by 2020," n.d., Hype Cycle Research Methodology | Gartner Inc. Gartner, Inc., accessed July 19, 2018, https://www.gartner.com/newsroom/id/3763265

[31] "3 Keys to Winning the Great Artificial Intelligence (AI) War! – InFocus Blog | Dell EMC Services," InFocus Blog | Dell EMC Services, December 12, 2017, accessed July 19, 2018, https://infocus.dellemc.com/william_schmarzo/3-keys-to-winning-the-great-ai-war/

[32] See Dante Disparte and Daniel Wagner, "Do you Know What your Company's Data is Worth?", Harvard Business Review, September 16, 2016, accessed July 19, 2018, https://hbr.org/2016/09/do-you-know-what-your-companys-data-is-worth

[33] Henry A. Kissinger, "How the Enlightenment Ends," The Atlantic, Atlantic Media Company, May 16, 2018.accessed July 19, 2018, https://www.theatlantic.com/magazine/archive/2018/06/henry-kissinger-ai-could-mean-the-end-of-human-history/559124/

[34] "The Need for a Digital Geneva Convention - Microsoft on the Issues," The Official Microsoft Blog, May 15, 2018, July 19, 2018, https://blogs.microsoft.com/on-the-issues/2017/02/14/need-digital-geneva-convention/

[35] Tad Friend, "How Frightened Should We Be of A.I.?" The New Yorker. The New Yorker, May 31, 2018, accessed July 19, 2018, https://www.newyorker.com/magazine/2018/05/14/how-frightened-should-we-be-of-ai

AI SUPREMACY

Index

A

ABI (Activity-Based Intelligence), 259
abundance, 61, 200, 301, 403; economics of, 61
accountability, 80, 197, 332, 334, 362, 379–80
accuracy, 10, 22, 24, 26, 28, 45, 79, 263, 268, 270, 272, 377, 384, 393, 401; complete, 414; increasing human, 270
ACORD data standards, 47
activities, abnormal, 109, 140, 208, 211, 260, 270
actors, 20, 163, 242–44, 253, 255, 300, 332, 357, 401–2; global, 324; international, 324; malevolent, 267; malicious, 6, 246, 249, 256, 401
ACTUV (ASW Continuous Trail Unmanned Vessel), 282
ADHD (Attention Deficit Hyperactivity Disorder), 31
advancement, 13, 17, 122, 138, 401
adversarial, 240; applied, 246
Africa, 200
agencies, 207, 213–15, 259, 265–66, 271, 339, 396, 414; bureaucratic, 299; international, 347
agents, 92, 156, 211, 267, 361; adaptive, 215; human, 266; intelligent, 246
AGI, 416; self-improving, 416
agility, 85, 233, 297, 326, 338, 403; programmatic, 279
agreements, 67, 333, 359, 382–83; indemnification, 359; non-disclosure, 10
agriculture, 56, 89, 298, 301, 314, 398
AI-based defenses, 248–49, 254

AI-based defense systems, 249
AI-based learning platforms, 185–86
AI-based surveillance and sensor technology, 190
AI-based technology, 90, 93
AI-based tools, 80
AI-cybersecurity nexus, 247
AI-dominated world, 61, 215, 413–14
Airbnb, 60
Airbus, 66
aircraft, 66, 81, 287–88; commercial, 81; piloted, 285; stealth, 195; unmanned, 284
airlines, 82, 355
alerts, 131, 157; crisis management, 344; false positive, 116
Alexa, 30–31, 134, 160, 168, 229, 397–98
Algorithmic Accountability Bill, 380
Algorithmic Warfare Cross-Functional Team (AWCFT), 276–77
algorithms, 18–20, 25, 44, 81, 89–97, 105–9, 125–28, 151–54, 199–201, 208–10, 262–64, 268–69, 354–57, 371–75, 377; best, 81; categorization, 154; decision-making, 380; deploying, 118; modifying, 277; optimization, 407; predictive, 372; supervised, 241; test, 262; winning, 127
Alibaba, 40, 295, 302–6, 410
Alipay, 302, 304–5
ALIS (Autonomous Learning Investment Strategies), 50
allies, 233, 326, 328
AlphaGo, 21, 25–26
Alzheimer, 133, 436
Alzheimer's disease, 132
Amazon, 30–31, 40, 72, 134, 160, 165, 172, 220, 262, 305–6, 397–99, 404, 407, 410, 415
analysts, 49, 196–97, 208, 260–61, 264, 266–67, 270–71, 286, 312, 317; data management role, 259; human, 242, 277; human imagery, 271–72; political risk, 343; predictive, 30; satellite imagery, 272; top-notch, 271
analytics, 42, 67, 205, 266, 277; advanced, 81, 404; behavioral, 246; drive train, 83; predicative, 29;

predictive, 27, 29, 81, 101, 106, 403
anomalies, 111, 234, 240–41, 248; genetic, 136; identifying, 241
Anthropocene Era, 119, 434
anthropomorphizing, 13, 162
anti-money laundering, robust, 112
Apple, 22–23, 109–10, 167, 221–22, 361, 363–65, 411, 415
appliances, 30, 73; household, 6; mundane, 6; smart, 134
applications: basic, 68; chatbot, 35; competitive, 317; consumer, 406; customer-facing, 118; disruptive, 251; flagship, 155; hyperwar, 275; illegal, 112; industrial, 76; intelligent, 59; mission-critical, 408; mobile banking, 99; predictive, 233; social media, 150
AR. *See* augmented reality
architecture, 151, 260, 331; global legal/political, 334; inadequate security, 213

arena, 4, 25, 34, 54, 101, 195, 283–84, 326, 328, 344, 369, 389, 392–93, 405, 413–15; cyber, 247, 254, 261, 326–27, 401
armaments, 333; regulating, 333
arms race, 25–26, 331, 387; biological, 139; global, 5; lethal autonomous, 330, 332; metaphorical, 273
army, 102, 217, 280
ARPA-Ed, 180, 183
Artificial Intelligence, x, 12, 419–21, 426–27, 433–35, 450–51
Asia, 179, 200
Asian Infrastructure Investment Bank, 41
assassinations, 242, 271, 326, 330
assessments, 185, 369; cognitive, 81; complete security, 230; screening, 188–89
assets, 49, 83, 92, 100, 229, 296, 411; balance sheet, 411; low-cost, 274; personal, 198; strategic, 407
assets under management. *See* AUM
assistant, 28, 32, 143; artificial, 35; automated intelligent,

383; robotic lab, 124–25; voice-activated, 229
ATMs (automated teller machine), 69–70, 109, 399
attackers, 227–29, 236, 239, 242–43, 245–49, 327, 401–2; automated, 237; cyber, 226, 236, 238, 254; human, 224; skilled, 248
attorneys, 10, 351–53, 378, 381, 383, 385; intelligent, 352
augmented reality (AR), x, 7, 128, 167–69, 172, 183–84, 221–24, 435
AUM (assets under management), 92–93
AUSTRAC (Australian Transaction Reports and Analysis Centre), 116
authentication, 225, 230, 304; biometric, 109; two-factor, 108, 111, 247
authority: central, 103; decision-making, 320; strong, 317
automation, 9, 11, 46, 50, 70–74, 77, 79, 84–87, 206, 233, 237, 242, 318, 325–26, 420; demand-flexible, 51; extreme, 50–52, 427

autonomy, 146, 195, 251, 273–74, 287, 293, 376, 416; degrees of, 251, 318
AVATAR system, 163
AWCFT (Algorithmic Warfare Cross-Functional Team), 276–77

B

BAE Systems, 281
Baidu, 294–95, 302–3, 307, 314, 410
balances: crafted, 216; delicate, 111; global, 326; risk/return, 406
banks, 51, 69, 99–101, 103, 107–9, 117, 196, 200, 431; large, 147
barriers, 59, 114, 409–10; critical, 127; lowering, 355; regulatory, 284; technical, 74, 78; virtual, 337
battle, 239, 253, 275–76, 319, 325, 449; behind-the-scenes, 236; silent, 30
battlefields, 236, 265, 274, 281, 283, 285, 287–88, 318, 320, 325–26; roboticized, 283; virtual, 328
BCI, 142–44, 146
behaviors: conformist, 167; predicted, 92; rule-abiding, 296

Beijing, 297, 307–8, 310, 313, 315–16, 318–19, 341
beneficiaries, 207
biases, 3, 106, 125, 162, 166, 187, 223, 234–35, 267, 343, 375, 389, 408–9; algorithmic, 380; cognitive, 127, 162, 267; encoded, 375; hidden, 408; historical, 162; human, 235; social, 329
Big Data, 23–24, 50, 54, 98, 105–6, 266, 272, 274–75, 277, 294, 296, 299, 414–16, 422, 432
bills, 46, 114, 196–97, 348, 365, 380; freight, 46–47; restaurant, 294; scans, 46
Bitcoin, 100–102, 219
Black Mirror, 396
blockchain, 34, 47–48, 51–52, 57, 101–3, 305, 426
blockchain concept, 101
blockchain policies, 52
blockchain schemes, 219
Bloomberg, 99, 430–31
Bluetooth, 141, 225
bodies, 5–6, 123, 131, 133–34, 137–38, 140–41, 147, 162, 169, 172, 184, 219–21, 394, 397, 399; human, 3, 15, 132, 136, 140, 389; human-like, 162; international, 333, 347; physical, 15, 32, 142, 160, 167; smart, 221
borders, 163, 189; national, 40
boundaries, 2, 18, 40, 142, 170, 238, 324, 357, 366; clear, 131; ethical, 170; legal, 366
BP (British Petroleum), 66
brains, 13–16, 18, 32, 50, 129, 140, 142–45, 171, 219, 275, 394, 439; animal, 12; human, 14, 21–22, 144, 165, 217, 377
brothel, robot, 400
budget, 180, 207, 275, 381; annual, 58; federal, 54; national, 53
businesses, 38–40, 45–49, 61–62, 67–68, 72–73, 84–87, 100, 158–59, 161–62, 170–72, 216–18, 229, 231, 397–98, 406–7; construction, 64; country risk, 344–45; drug development, 136; global, 43, 425; high-risk, 100; information-based, 117; international, 38, 46, 324, 365; modern, 238; online food, 50; online lending, 90;

private, 212, 293, 377;
shifting, 232;
transforming, 85
business models, 36, 42, 81, 86; country-based, 40; legacy, 66

C

Cambridge Analytica, 27–29, 376, 424
Cambridge University, 27–28
cameras, 6, 220, 228–29, 234, 242, 270, 317–18, 397; closed circuit television, 212, 303; depth-sensing, 269; digital, 77
campaigns, 27, 29, 198, 218, 279, 424; misinformation, 27; phishing, 108; political, 244, 268; presidential, 27, 328; public relations, 113
Canada, 54, 163, 193–94, 384
cancer, 127, 132, 135–36, 436; advanced kidney, 136; blood, 126; curing, 138; diagnose skin, 18; late-stage breast, 393
CANES (Consolidated Afloat Networks and Enterprise Services), 282

capabilities, 13–14, 64, 66, 163, 242, 246, 259–61, 275–76, 278–79, 281, 317–18, 320, 326, 334–37, 408–10; algorithmic breaker, 206; anti-fraud, 207; autonomous, 281; censorship, 345; cognitive, 220; computing, 412; counter-AI, 319–20; facial recognition, 26; foreign, 261; human, 42, 239, 244–45, 401, 416; industrial, 309; nuclear, 335; offense, 11; prediction, 255; technological, 38, 325
capacity, 6, 42, 57, 93, 96, 202, 259, 273, 276, 299, 404, 412; computing, 105; human, 336; intelligence collection, 260; memory, 86
CARA (Case Analysis Research Assistant), 381–82
cars, 20, 60, 82, 86, 237, 359, 361–62; autonomous, 307, 358, 360–61; self-driving, 21, 201, 303, 361; semi-autonomous, 310; smart, 221
cash, 33, 99, 293, 301, 339, 358

CCP (Chinese Communist Party), 291, 296, 303, 315
cells, 127–29; bacterial, 137; blood vessel, 396; damaged skin, 396; healthy, 136; human skin, 396; living, 138, 170; potential cancer, 407; stem, 396
Central Intelligence Agency. *See* CIA
CFIUS (Committee on Foreign Investment in the United States), 299, 311–12
chatbots, 97, 99, 101, 121, 169, 171, 219, 302, 355, 396
China, 25–26, 41, 52–53, 60, 180–81, 200–201, 283–85, 291–302, 304–19, 321–24, 326, 338, 341, 399–400, 455–57
Chinese, 17, 53, 292, 295, 297, 299–302, 306–9, 311, 313, 315, 317, 324, 339
Chinese companies, 303, 306, 310–11, 314, 317, 321; private, 316
Chinese government, 181, 292, 294–95, 300–301, 308, 313, 315, 317, 321, 324

CIA (Central Intelligence Agency), 262, 266
cities, 10, 102, 134, 178, 199–201, 206, 210, 234, 260, 277, 303, 305, 317, 326, 361; ancient, 170; coastal, 301; enemy, 102; smart, 56, 199–201, 221, 301, 310, 348, 406; vulnerable, 205
citizens, 36, 44–45, 52, 62, 190, 200, 208, 212, 214–15, 302, 304, 307, 317–18, 365, 367; law-abiding, 212; ordinary, 253
classes, 184, 237, 246, 271, 395; benign, 238; distinct, 238; middle, 71
climate change, 35, 39, 61, 119, 323
Clinton, Hillary, 26, 269
cloud, 2, 129, 133, 140, 217, 225, 233, 240, 263, 274–75, 278, 304, 306, 351, 369
CMS, 207–8, 434
CMS anti-fraud framework, 207
CNN, 421
COBOL (Common Business-Oriented Language), 104
Coca Cola, 411
code, 103–4, 159, 208, 224, 247, 371; bar, 80,

159; binary, 246; card verification, 159; confirmation, 108; postal, 371; standard, 121; unknown, 236
cognification, 5–7, 101, 206, 419
COiN, 103
Cold War, 292
collaboration, 28, 68, 194, 201, 235, 306, 405; brain-computer, 299; foreign, 294; human-machine, 319
color, 7, 140–41, 272; person's skin, 28
commanders, 276; strategic, 276; submarine, 395
commands, 222, 246, 296, 320, 334, 401; central, 30, 203; external, 225; infecting, 328; point-and-click, 146; verbal, 232
community, 162, 261, 307, 334, 381; decision-making, 35; international, 345; legal, 378; local, 157; scientific, 121; venture capital, 262
companies, 23–25, 31, 36, 44–46, 76–80, 82–85, 115, 155–60, 166–67, 221–22, 305–10, 363–68, 390–93, 404–7, 410–11; antivirus, 247, 402; client, 406; commercial, 221; domestic, 300, 310; external, 99; foreign, 311; global, 369; industrial, 66, 83; non-European, 365; nonfinancial, 98; opaque, 115; pharmaceutical, 125; phone, 366; shell, 114, 164; state-owned, 308
COMPAS, 223, 355, 377
competition, 45, 52, 60, 91, 121, 127, 139, 147, 179, 188, 261, 306; strategic, 319
competitiveness, 85, 117, 338; global, 325, 340
components, 73, 77, 121, 125, 182, 211, 239, 250, 275, 284, 287, 323, 327, 383; basic electronic, 119; critical, 298; fundamental, 145; robot ecosystem, 225; satellite, 284
computers, xi, 13–14, 23, 64, 66, 71, 73, 143, 145, 221, 223, 228–30, 374, 394–95, 397; air-gapped, 244; desktop, 301; high powered, 123; onboard, 237; personal, 79; quantum, 218–19, 321

Computer Science and Artificial Intelligence Laboratory, 20
conflicts, 66, 68, 197–98, 285–86, 332, 336, 368; inter-state, 250; sociocultural, 329; violent, 332
congestion, 150, 201; managing traffic, 200
connections, 25, 95, 286, 352, 377, 408; direct, 308; hidden, 123; neural-like, 21
connectivity, 50, 57, 345, 427; extreme, 51; massive, 30
consciousness, 14–16, 18, 144; collective, 12; public, 102
constraints, 169, 252, 371; budgetary resource, 286; self-imposed, 340
consumers, 30–32, 61–62, 64–65, 98–101, 105, 149–50, 158–59, 161, 170, 172, 202–4, 365, 385, 392, 410–11; average, 364; connected, 32
context, 13, 145, 236, 241, 243, 245, 318, 328–29, 331, 333, 362, 366, 381, 404, 411; high-stakes, 376; information security, 238; religious, 12

contracts, 46, 103, 308–9, 370, 382–83; catalog, 45; futures, 94; legal, 10
control, 4, 6, 15, 17, 52–53, 142–43, 146, 148, 204–6, 216, 226, 228, 386–87, 392, 394–95; human, 52, 252, 296, 455
conversations, 14, 161, 229, 364; global, 59; guiding, 156; open, 166; perennial, 276
Corporate Transparency Act, 114
corporations, 39, 74, 188, 190, 235, 293, 422; public, 422
corruption, 197–98; global, 164
costs, 22, 33, 45–46, 50–51, 61, 64–65, 74–79, 87, 103–4, 130, 135–36, 288–89, 292–93, 340, 381; administrative, 52, 80; company's liability, 359; high labor, 100; human labor, 10; industrial robots, 73; operating, 149, 160, 195; systems-engineering, 75; theft, 292; total, 74–75, 78; transaction, 92
counter-terrorism, 337

countries, 40–41, 52–55, 59, 114–16, 149, 189–91, 193–97, 199–202, 212–16, 263–66, 284–87, 293–96, 320–27, 340–43, 345–48; developing, 38–39, 99, 177, 325; industrialized, 41, 73
CQW (CosmiQ Works), 262
creation, 42, 54–59, 108, 110, 176–77, 232, 234, 289, 307, 310, 332–33, 376, 378, 409, 412
creativity, 61, 71, 85, 183–84, 236
credit, 48, 54, 106, 110, 118, 322, 369
crimes, 109, 170, 211–14, 220, 234, 316, 366, 377, 433, 449; combat, 114; financial, 113, 116, 214; organized, 116
crisis, financial, 94, 118, 348, 430
CRISPR (Clustered Regularly Interspaced Short Palindromic Repeats), 138, 219, 438
CRM (customer relationship management), 157

culture, 7, 85, 145, 170, 179, 386, 406, 408; risk-taking, 196
CVC (card verification code), 159
cyberattacks, 18, 21, 118, 148, 204–5, 213, 224, 227, 230, 239–40, 256, 265, 325, 327, 421
cybersecurity, 57, 203, 205, 225, 230–34, 236, 239–41, 245–47, 254–55, 257, 366, 384

D

DAMO, 305
dangers, 17, 73, 144, 168, 327, 336, 357, 359, 381, 410, 412
DARE system, 163
Dark Web, 34, 108, 170, 269, 326
DARMS, 235
DARPA, 58, 146, 180, 195, 261–62, 275, 282, 285, 394, 454
data: browsing, 365, 461; crunches, 342; domestic, 300; laser, 357; objective, 403; private, 339; text-based, 357
databases, 26, 111, 120, 238, 268, 276, 286–87, 294, 304, 397, 399; biased, 375; corporate, 115;

modern, 211; national, 318; searchable, 211
datasets, 235, 318, 371; large, 247–48, 402; updated, 35
DDC, 46
Deep Learning. *See* DL
DeepMind, 11, 21, 25, 128, 420, 422
Deep Neural Network, 318
defense, 5, 27, 111, 239, 243, 248–50, 253, 263, 273–74, 298, 402–3; antisubmarine, 335; boundary, 236; enemy, 335–36; high-altitude missile, 283; national, 58, 299; system-level, 248
deliveries, 46, 83, 251; courier, 202; crowd sourcing, 202; parcel, 202; public service, 57; time-sensitive, 203
democracies, 256, 328; liberal, 52–53, 345
Department of Defense. *See* DoD
departments, 247, 276, 279, 292, 295, 402; compartmentalized, 118; corrections, 377; legal, 383
deploy, 33, 42, 56, 234, 237, 245, 250, 254, 283, 286, 292, 296, 399, 401–2, 405

deployment, 50, 67, 76, 176, 248, 298, 319, 330, 371; ethical, 392; fast, 228; large-scale, 250; mass, 74; weapon, 327
design, 41, 77, 81, 83, 233, 243, 329, 358, 368, 371; computer-aided, 77; top-level, 296; uncoordinated, 285
designers, human, 327
detection, 49, 225, 239, 247, 249, 267, 401; advanced submarine, 284; early, 35, 131, 210; evading, 224
developers, 82, 165, 230, 235, 278–79, 306, 357, 377, 386, 394, 424
devices, 6, 32, 131, 134, 167–68, 205, 209, 221–22, 224, 232–33, 237–38, 241, 302, 304, 363–64; brain-computer interface, 142; digital, 140, 146; external, 228; implantable, 139, 146; mobile, 224, 294; personalized, 390; physical, 172; smart, 23, 30, 204–5, 220–22; smart home, 204; technological, 140; wearable, 133

Differential Privacy. *See* DP
DigitalGlobe, 272; satellite operator, 262
digitization, 40, 97, 296; complete, 101; increased, 117
Director of National Intelligence, 318
discoveries, 13, 36, 79, 123, 126, 305, 347, 353; electronic, 354; scientific, 123, 162, 414
discrimination, 44, 329, 371, 373
disinformation, 27, 44, 244, 267–68, 328–29; dispelling, 329
disinformation campaigns, 328
DL (Deep Learning), 22, 24, 34–35, 80, 112, 127, 136, 145, 154, 193, 236, 308, 313, 323, 405
DL algorithms, 127–28, 145, 272, 377; trained, 271
DNA, 132, 135, 138, 212–13, 398, 437; human immunodeficiency virus, 138; tumor, 132
DNS (Domain Name System), 241
doctors, 16, 51, 61, 126–27, 129, 131, 134, 140, 142, 147, 170, 207, 393, 436; human, 9, 129; medical, 126; routine, 129
DoD (Department of Defense), 26, 274–82, 308–9, 312, 423, 453–54
DOE (Department of Energy), 124, 203, 205
domains, 2, 19, 21, 56–57, 147, 221, 246, 254, 325, 329, 333, 337, 353, 401–2, 407; cyber, 253; legal, 352; legitimate, 246, 401; offense/defense, 289; physical, 32; political, 254; public, 26, 68, 375; social, 254
DoNotPay, 354–55
DP (Differential Privacy), 363–65
dream, 201, 429; lucid, 172
drivers, 5, 10, 154, 201, 282, 300, 358–59, 406; central, 403; human, 10, 201, 361; human back-up, 359; truck, 9–10, 12
drones: aerial, 249; anti-submarine, 282; autonomous, 242; cheap hobbyist, 244; human piloted, 250; nuclear-powered undersea, 335; regulated, 34; weaponizing, 33

drugs, 33–34, 95, 125, 133, 136–37, 164, 172, 303, 377, 433, 437; experimental, 133; psychedelic, 15
DTRA (Defense Threat Reduction Agency), 293
Dubai, 220, 348
dystopian, 2, 257, 397

E

East China University of Science and Technology, 292
Economic Times, 434
economic value, 19, 70, 112, 348, 411
economies, 9, 19, 34, 36, 60–61, 70–71, 79, 139, 147, 175, 191, 248, 297, 342, 348; advanced, 269; associated, 248; digital, 118; global, 11, 25, 40, 52, 79, 191, 342, 407, 409; healthier, 65; national, 78; reconfigure, 215
ecosystem, 6, 57, 118, 141, 300, 411; changing, 215; complex, 33; digital assistant, 176; dynamic, 339; investor, 48; mature cybersecurity, 240; open, 347; personal, 30

edge, 36, 61, 91, 205, 261, 265, 276, 316, 320, 339, 394, 402, 407; decisive, 276
EDR (Endpoint Detection and Response), 246
education, 56–57, 175, 177–82, 185, 193, 230, 379; mathematics, 59; technology-guided, 180; transforming, 180
effectiveness, 188, 198, 240, 245, 325, 328, 407; improved equipment, 83; operational, 279; relative, 249
Einstein, Albert, 255, 416
election, 27, 198, 269, 401; presidential, 26, 29, 246, 256, 267, 269
emergency, 359; medical, 354; vehicle's, 359
employees, 9, 36, 42, 72, 84–85, 87, 95, 103, 164–67, 172, 342, 375, 398, 405; campaign, 27; customer service, 156; full-time, 406; skilled, 56; tracking, 318
encryption, 226, 230, 321, 363; channel, 226; end-to-end, 364; weak, 225
energy, 34, 124, 200, 203–4, 206, 294, 320;

distributed, 34; renewable, 206
energy grids, distributed, 34
engineers, 33–34, 66, 230, 252, 255, 273, 378, 416, 425; well-trained computer, 264
engines, 82, 93, 309, 421; airplane, 81; combustion, 411; diesel-electric, 395; operating, 81; rocket, 308
Enron, 375
ENTC (Explosives National Tracing Center), 211
entities, 16, 103, 114–15, 207, 241, 249, 293, 359; global, 289; government-owned, 308; intelligent, 15, 220; legal, 114–15; physical, 169, 321
environments, xi–xii, 60, 76, 112, 119, 141, 146, 167, 203, 206, 224, 254, 281, 287, 317; combat, 260; controlled, 201; cyber-contested, 280; dangerous natural, 144; digital, 172; monitored, 270; off-road, 281; regulatory, 101, 150, 302, 365; strategic, 276; uncontrolled, 250; unpredictable, 152; urban, 361
epidemics, 96, 147, 254; global, 139
era, 28, 71, 172, 191, 212, 223, 237, 275, 340, 345, 362, 401, 442; digital, 85; early industrial, 86; emerging, 36
errors, xi, 46, 208, 267, 271, 409, 435, 454; human, 270, 358, 362; instant, 383; potential, 186
Estonia, 202, 213–14
ethics, 3, 55–58, 142, 170, 299, 307, 327, 338, 371, 376, 386, 413; democratic, 346
Europe, 43, 77, 189, 215, 312, 321, 328, 338–40, 361, 379, 400
Everbright, 308
evolution, 16, 62, 121, 138, 191, 335, 408, 422, 439; rapid, 80
exams, 134, 185–86; annual physical, 131; physical, 131; standardized, 175
exchange, 41, 96, 100, 129, 135, 147–48, 151, 191, 214, 354, 361; foreign, 308, 399

expense, 178, 215, 320, 354; legal, 383; operating, 105, 293
experiment, 97, 123–24, 126, 147, 181, 297, 410; historical, 124; obscure physics, 124; scientific, 124
experts, 2, 127, 213, 263, 265, 272, 275, 289, 309, 346, 357; digital authoritarianism, 53; human, 20–21, 421; technical, 252
exploitation, 243, 247, 260, 274, 276, 327, 329; enhanced, 261; opportunistic, 327
EyeDetect, 164
eyewear, 168, 186; smart, 26, 168–69, 186

F

Facebook, 23, 27, 146, 263, 305, 313, 364, 366, 376, 410, 415, 422, 424, 430, 439
Facebook-Cambridge Analytica data scandal, 146
faces, 18, 22, 66, 221, 317–18, 358, 385, 399; categorizing, 271
factories, 75, 77–78, 80, 82–84, 230; engine, 84; virtual, 80; wind generator, 83

FAGMA (Facebook, Apple, Google, Microsoft, and Amazon), 23
fake news, 268–69, 328, 337, 376
FANUC Intelligent Edge Link and Drive, 84
FCC (Federal Communications Commission), 365
FDA (Food and Drug Administration), 136, 377
Federal Energy Regulatory Commission, 375
Federal Reserve Bank, 432
Federal Savings Associations, 430
feedback, 7–8, 20–22, 76, 113–15, 140, 181, 183, 196, 205, 208–9, 279, 297; client, 80; instant, 177; interactive, 184; rapid, 156; real-time, 81–82
Financial Crimes Enforcement Network. *See* FinCEN
Financial institutions. *See* FIs
financial products, 101, 104, 110; algorithm-based, 94; traditional, 104
financial services, 89–90, 98–100, 112, 117–18, 305, 431

Financial Times, 423, 426, 430, 449
FinCEN (Financial Crimes Enforcement Network), 113–14, 117
FinCEN Artificial Intelligence System, 433
fingerprints, 6, 109, 111, 120, 244, 432
FIR (Fourth Industrial Revolution), 6–7, 50, 64, 119, 139, 419, 425, 427
firms, 43, 50, 80, 84, 86, 90–91, 93, 98–99, 147–48, 221–22, 302, 383–84, 405, 409–12, 415; money transfer, 51; scaling, 407; tech-savvy, 407; telecommunication, 363; venture capital equity, 98
FIRRMA (Foreign Investment Risk Review Modernization Act), 312–13
First Industrial Revolution, 5
FIs (Financial institutions), 46, 89–91, 98, 100, 103–18, 121, 164
Fitbit, 133, 356
food, 13, 136, 159–60, 229, 411; healthy, 134

footage, 277; security camera, 212
force, 36, 46, 77, 103, 127, 184, 212, 275, 284, 286, 298, 307, 331–32, 339, 412–13; armed, 259, 285–86, 395; dominant, 410; global, 167; lethal, 251; retaliatory, 335
foreign policy, 323, 327, 329, 345–46; common, 328
formats, 121, 268, 363; electronic, 367; machine-readable, 375; non-traditional, 46; structured, 187
France, 142, 186, 235, 338–39
fraud: fast-moving, 118; financial, 112; identity, 111; takeover, 108; vendor, 164
freedoms, 53, 299, 345, 367, 370, 386; personal, 216
functions: automatable, 85; back-office, 46; bodily, 131, 134; checkup, 129; delayed reward, 91; fundamental, 89

G
Gallup poll, 197

game, 14, 21, 25, 43, 95–96, 153, 170, 189, 215, 242, 280, 291, 303, 322, 354; erotic, 170; neuroscience-based, 187; virtual, 109; zero-sum, 152
GAO (Government Accountability Office), 110, 207, 432
gaps, 7, 38, 74, 149, 239, 273, 305, 315, 390, 422; coverage, 20; cybersecurity skills, 233; growing, 250, 402; information security, 201; plug, 391
Gates, Bill, 17
GDP (global domestic product), 11, 24, 215, 348
GDP, global, 112, 197
GDPR (Global Data Protection Regulation), 44, 253, 365–71, 373, 379
Gemological Institute, 47
gene-editing methods, 219
gene mutations, 132
General Data Protection Regulation, 253
General Electric, 51
generation, 9, 11, 93, 139, 142, 150, 274, 276, 283, 289, 298, 317, 319, 324, 390;

advanced natural language, 246
genes, 51, 136–38, 219; malfunctioning, 136; mutant, 138; transferring, 398
genome, 51, 135–39, 436; human, 135, 137, 142, 391
Germany, 55, 74, 339
GitHub, 68
Glassdoor, 419
Global Findex Database, 431
globalization, 38–41, 43, 60, 62, 425, 433; political, 52
Gmail, 153–55, 182, 441
Goldman Sachs, 90, 101, 429
Google, 23, 25, 128, 133, 153–56, 182–83, 221, 305–6, 361, 364, 366, 390–93, 401–2, 435–36, 440
Google algorithm, 393
governance: better, 301; collective, 340; global data, 43; shared, 44
governments, 3, 9, 43–44, 52–59, 181–82, 195, 198–202, 205–9, 211–19, 231, 288–89, 297–304, 313–15, 326–28, 341–43; central, 299; digital, 213–14; domestic, 53; effective, 213; federal,

130, 207; local, 295–96, 299–300, 312; subnational, 299; tribal, 203
GPS, 351
GPS location data, 150
Great Recession, 71, 98, 381
GRID projects, 205
grids, 205–6; autonomous, 204; electric, 203–6
groups, 7, 15, 20, 38, 55, 102, 124–25, 159, 170, 180, 188, 215, 250–51, 326, 372; criminal, 236; ethnic, 178, 295; political, 245; terrorist, 251, 326; violent, 235
growth, 39–41, 49, 55, 73–74, 85, 87, 93, 99, 153, 162, 171, 215, 236, 239, 412; aggressive, 136; economic, 11, 38–39, 41, 55–56, 116, 191, 194, 298; export-led, 41; population, 200; skin, 89; sustainable, 42

H

habits, 3, 8, 160, 316–17, 372, 391; eating, 134; information consumption, 268
hackers, 50, 226–28, 231, 239, 363, 397

hacking, 26, 141, 182, 228, 247; automated, 254, 256
Haiyin Capital, 309
Harbisson, Neil, 140–41
hardware, 242, 244, 248, 250, 275, 300, 309, 313, 363, 380, 402; custom, 412; graphics, 269; snooping-oriented, 317
Harvard Business Review, 425, 429, 464
Hawking, Stephen, 17, 221, 416
health, 6–7, 44, 57–58, 119, 131, 134, 172, 222, 263, 370–71, 387, 391, 398, 434; smart, 134
healthcare, 5, 42, 51, 56, 89, 128, 130–31, 137, 147–48, 208, 237, 393; accessible, 129; effective, 147; improving, 200
hedge funds, 25, 91–93, 429
highlights, 122, 128, 151, 158, 228, 372; algorithmic decisions, 370
history, 13, 65, 72, 110, 182, 193, 288, 387, 413, 420; accident, 19; browsing, 112; criminal, 355; economic, 72; human,

119; metaphysical, 15; political, 29
Homeland Security, 163
Hong Kong, 46, 343, 374
hospitals, 16, 42, 126–28, 130, 147–48, 339, 393
HRL (Human rights law), 332–33
HSBC, 48, 116
Huffington Post, 426, 450
humanitarian law, 333; international, 332
humans, 8–9, 12–13, 17–18, 72–73, 119, 138–39, 231, 233–34, 244, 270–71, 278, 335–37, 394–97, 403–4, 408–9; intelligent, 50; outperform, 17, 126; outperformed, 16, 163; well-intentioned, 223
HWCE (Hyper War Center of Excellence), 337
hyperwar, 275–76

I

IAEA (International Atomic Energy Agency), 272, 347
IAIA (International AI Agency), 346–47
IARPA (Intelligence Advanced Research Projects Activity), 145, 262
IBM, 47, 218, 232, 305–6, 314, 405–7, 426, 442, 451

IBM's Watson, 127, 218–20, 232, 352–53, 405–7
IBNS (intelligence banknote neutralization system), 232–33
ICTAIs (International Center for Transformational AI), 56
ID cards, 399; electronic, 304; physical, 304
Identity Theft Act, 110
IHL (international humanitarian law), 332–33
illnesses, 135, 146, 208–9; common, 126; foodborne, 208; potential, 32; rare, 126
imagery analysis, 272, 278; migrating drone, 279
images, 15, 18–19, 22, 127, 129, 141, 145, 262–63, 268, 271, 278, 317, 356–57, 391, 396; computer-generated, x; digital, 168, 221; facial, 26; high-resolution, 200, 262; holographic, 7; medical, 303; storing, 304; three-dimensional, xii
implants, 139–40, 142, 146; neural lace, 143

implications, 3, 11, 51, 65, 223–24, 264, 268, 275, 345–46, 362, 365, 373, 378, 386, 389; economic, 128, 144; global, 274; legal, 54, 56, 193; moral, 386; political, 17; potential negative, 244

incentive, 10, 190, 269, 307, 341; economic, 361

inclusion, 99, 335, 368, 370; financial, 57, 99; social, 55

income, 36, 71, 87, 117, 179, 210, 372; median, 71; middle-class, 39; universal basic, 11, 420

India, 11, 38, 41–42, 55–56, 83, 200, 325, 341, 374

indicators, 133; early, 132; key performance, 80, 384; static, 344; vocal pattern, 163

individuals, 162–63, 167, 170, 205, 207, 211, 214, 243, 245, 269, 272–73, 295–96, 365–67, 369, 378–79; high profile, 17; high-risk, 198; imitate, 244; impersonate, 268; micro-target, 29; supplant, 10; tracking, 295

industrial revolution, 52, 64, 72; the Fourth, 427

industries, 5, 51–52, 58–59, 68, 70, 73–78, 84, 89–90, 150, 262, 264, 294–95, 311–12, 383, 385; accuracy-critical, 263; advertising, 19; auto, 143; automotive, 74; autonomous vehicle, 359; banking, 51; chip, 310; cyber security, 238; electrical-equipment manufacturing, 75; exchange-traded fund, 49; fabricated-metals, 76; global, 74; healthcare, 129; heavy, 82; high-tech commercial, 294; multi-billion-dollar, 343; primary-metals, 73; quantitative fund, 118; recruiting, 154; satellite manufacturing, 284; saturated, 202; scientific publishing, 122; strategic, 311; transformed, 5

inefficiencies, 47, 83, 103, 110, 204; bureaucratic, 286

inequality, 39, 86, 128, 371; economic, 39;

global socioeconomic, 323
information technology, 6, 76, 295, 404
infrastructure: computational, 278; critical, 204; developing, 193; heavy, 34; hybrid cloud, 46; pipeline, 209; public, 199; risk management, 99; smart, 34; social, 175; technical, 403; urban, 199, 263
innovation, 39, 41, 45, 47–48, 84–85, 87, 98–99, 116–17, 179, 195–96, 235–36, 238, 298, 338–39, 347; digital, 44, 266; disruptive, 116; educational, 180; engineering, 280; groundbreaking, 213; technological, 119, 279, 296–97, 300
instability, 267, 316; social, 314
institutions, 41, 116–17, 214, 313, 327; democratic, 308, 329, 345; educational, 175; financial, 46, 89, 121, 164; post-war, 41
instructions, 17, 33, 77, 97, 168, 357; human, 412; obeying, 259; relay, 102

insurance, 66, 305, 358, 385, 426; auto, 362; marine, 47; political risk, 343
integration, 3, 34, 41, 55, 64, 124, 190, 239, 297, 325, 384, 389; agile, 325; heavy, 284; renewable energy, 205
intelligence, 10, 12–14, 16, 255, 267–68, 270, 275, 283, 286, 416, 420, 452, 457, 460, 464; actionable, 36, 105, 277; emotional, 13, 188; forward-looking, 405; offensive, 224; persistent tactical, 260; security attack, 232
intelligence agencies, 34, 259–60, 269–70, 308
intelligence community, 196, 259, 264, 271, 289
interfaces, 35, 151, 395; interactive visual, 120; invasive, 394; neural, 394; non-invasive, 394; standardize, 152; touch, 221; visual, 121
International Monetary Fund, 197
Internet, 19, 23, 27, 42–43, 53, 141, 143, 149, 186, 190, 195, 218, 220, 225, 227

Internet of Things. *See* IoT
investments, 54–55, 57, 59, 71, 74–75, 92, 293, 295, 300, 312–13, 315–16, 341, 343, 406–7, 427; early-stage, 311; equity, 298; foreign, 312–13; foreign direct, 316; high-tech, 299; state-guided, 309
investors, 50, 308, 312, 343–44, 406; foreign, 312; human, 94, 343; minority, 312; retail, 94; third-party, 354
IoT (Internet of Things), 6, 30, 35, 54, 64, 82, 146, 150, 233, 440
IP (intellectual property), 182, 292, 407, 457
IQT (In-Q-Tel), 263
Iran, 48, 53, 260, 265, 271, 324
Islamic State, 251
ISPs (Internet Service Providers), 31, 134, 150, 366, 372, 391
ISR (Intelligence, Surveillance and Reconnaissance), 260
Israel, 288
Italy, 73
ITTI, 48
ITU (International Telecommunication Union), 150, 440

J
Japan, 57, 74, 126, 144, 194–95, 400
JEDI, 338
Jobs, Steve, 167
Johns Hopkins University, 127, 131, 435
Joint European Disruption Initiative, 337
journals, 121, 295, 426; scientific, 120
JP Morgan Chase, 101, 103
jurisdictions, 49, 117, 370, 383
Justice Research Institute, 346

K
Kaggle, 81
Kaspersky Lab, 32, 424
Kenya, 325
keywords, 120–21; extracting, 120–21
Kim, Jong Un (Marshal of the Republic), 270
Kissinger, Henry, 412
knowledge, 120, 122–23, 126, 128, 141, 176, 178, 183, 190–91, 194, 231, 377, 390–91, 401, 407; enterprise, 84; expert, 368; human, 32, 190, 408; impart, 191; intuitive, 408; physician's, 131; scientific, 120–21, 124
KPMG, 49

L

labels, 20–22, 111, 113, 115, 140, 154, 208, 212–13, 238, 278, 411; classification, 200
labor, 77, 86, 112, 248, 260, 340; administrative, 51; expensive engineering, 81; human, 5, 74–75, 255; manual, 74; skilled, 86, 243
landscape, 20, 38, 120, 149, 255, 262, 273, 311, 410, 413; competitive, 149; global educational, 177; operational, 84; regulatory, 118, 366; strategic, 246
Latin America, 164
law, 111, 113, 196, 266, 297–98, 321, 332, 334, 351–58, 362–63, 365–66, 370–71, 374, 379, 382–86; anti-spoofing, 97; copyright, 355, 373–75; cyber security, 307; emerging, 373; federal, 377; zoning, 200
Law Enforcement Act, 114
Lawrence Livermore National Laboratory, 203
lawyers, 10, 197, 223, 343, 351–54, 357, 382–84, 426, 431; human, 10, 419; traditional, 384
leaders, 26, 40, 42, 52, 54, 68, 84, 148, 276, 283, 291, 305, 310, 332, 415–16; effective, 188; elected, 297; global, 59, 194, 213, 283, 315
leadership, 42, 54, 253, 406
leaks, 209–10, 228, 363; tracing, 209
ledger, 102, 390; distributed, 102; irrevocable timestamped, 101
ledger system, open, 102
Legal Innovation Center, 384
legislation, 182, 196–97, 213, 231, 311, 370; bipartisan, 110; protective, 313
lenders, 343–44
lending, 89, 105–6; mortgage, 98; personal online, 100
leukemia, 126–27, 437
LexisNexis, 353–54
limb, 15; missing, 14–15; moving robotic, 147; prosthetic, 394
limitations, 13, 20, 111, 120, 238, 273, 318, 320

LinkedIn, 153–55, 441
Litecoin, 100
Little Dragon, 399
Liveability National
  Innovation Challenge,
  199
loans, 90, 100, 103, 106,
  305, 372; mortgage,
  106
locations, 105, 108, 138,
  160, 190, 205, 209–
  10, 241, 270, 302,
  357, 370, 391;
  geographic, 19, 179,
  190; mortar, 104; on-
  shore, 282
Lockheed Martin's Convoy
  Active Safety
  Technology system,
  281–82
losses, 21, 25, 106, 110–
  11, 229, 342, 362,
  389; annual, 110;
  financial, 230;
  monetary, 207;
  temporary job, 8

M

machines, x–xi, 2–5, 12–
  14, 77–80, 163–64,
  217, 231, 236–37,
  271, 374, 382, 400–
  401, 403–5, 408–9,
  413–14; automated
  teller, 69; autonomous,
  333; deployable, 265;
  fastest, 288;
  intelligent, 261, 337,
416; remote, 394;
  well-oiled, 404
Macron, Emmanuel
  (President), 194, 338–
  39
maintenance, 81–83, 275,
  332–33, 403;
  predictive, 65, 441;
  preventive, 81, 83
malicious, 238, 246, 402;
  potential, 252, 254–55
malware, 224, 228, 233–
  34, 240, 244, 246,
  267–68, 401;
  autonomous, 224–25;
  custom, 227; human-
  authored, 246;
  traditional, 254
management, 8, 24, 172,
  187, 207, 368;
  pension, 301;
  regulating waste, 200;
  upper, 166; water
  level, 304
managers, 68, 147, 157,
  166, 180, 188, 191,
  410; effective, 87;
  frontline, 68; hiring,
  188–89
manipulation, 90, 95, 320;
  efficient, 267; remote,
  252; social, 256
manufacturing, 60, 64, 69–
  72, 80, 82–84, 86–87,
  194, 204, 264, 285,
  301, 310, 315, 340,
  427–28; furniture, 76;
  global, 65, 340;

intelligent, 80, 298;
robot-enhanced, 341;
semiconductor, 321;
smart, 64, 82, 87
maps, 133, 210, 262–63, 271, 453
Maritime Operations Centers, 282
marketplace, 11, 56, 251, 273, 348, 362, 432; global, 50, 82, 340–42; global cyber security, 239; global digital, 54; growing commercial, 274
markets, 34, 50–51, 77, 91–96, 107, 136, 233, 341, 344–45, 347, 415, 422, 426; black, 164, 330; competitive, 273; consumer, 146; end-use, 42; financial, 91–92, 94–95, 118; futures, 96; global, 40; job, 3, 144, 194; public security, 318; spoofing, 95; stock, 206
MaRS, 194
McGovern Institute, 438
McKinsey Global Institute, 11, 420, 457
Medicaid, 207
medical diagnoses, 148, 296; inaccurate, 127
Medicare, 207
medication, 6, 207; non-habit-forming, 32

medicine, 58, 119–20, 126, 136, 148, 298, 395, 408, 434; modern, 131
members, 27, 100, 104, 108, 148, 154, 162, 172, 196, 245, 253, 382, 401; national, 41
Mexico, 33, 57, 116, 210, 325
microchips, 315, 397
microphones, 220, 228
microscope, 128; compound light, 127; modified, 128
Microsoft, 23, 154–55, 170, 302, 305–6, 314, 374, 413, 438, 441, 443
military, 189, 193, 251, 261, 271, 273–74, 277, 285–87, 289, 293, 297, 318–20, 327, 338, 394–95
military robotics, 285–86; unmanned, 285
military robots, 308
Mind Mapping, 265
Mind Reader, 144
MindSphere, 83
missiles, 274, 395; ballistic, 270; defensive, 287; drone-carried Hellfire, 282
misuse, 44, 252, 362, 386; potential, 139, 158
MIT (Massachusetts Institute of Technology), 132, 210

mobility, 57, 146; economic, 71; robot's, 229; robust, 281; smart, 56

models, 19–20, 28–29, 60, 65, 67, 118, 124, 126, 194, 197, 204, 240, 246, 263, 358; billable hour, 353; classic statistical, 106; climate, 35; data exchange, 214; delivery, 202; developing analytical, 260; digital, 170; economic, 40–41, 60; global, 297; hybrid, 68; mathematical, 125; non-predatory, 194; open-access, 185; predictive, 29, 81, 380; supply chain, 80; weather, 35

modes, 61, 66; autonomous, 360–61; dummy, 400; fixed distribution, 206

Modi, Narendra (Prime Minister), 341

money, 33, 39, 53–54, 89, 94–95, 125, 128, 175, 183, 308–9, 313, 339, 342, 344, 353

money laundering, 48, 112–16, 432

MOOCs (massively open online course), 176

Moorfields Eye Hospital NHS Foundation Trust, 128

Morocco, 325

Moscow, 26, 285–86

movements, 112, 132–33, 165–66, 172, 229, 282, 299, 302, 345, 370, 390; forecast, 90; forecasting market, 342; global social, 274; weapon, 272

MRIs, 79

MSRA (Microsoft Research Asia), 46

MST (Ministry of Science and Technology), 295

Musk, Elon, 17, 143, 416, 421

MyPersonality, 27

N

NAFTA (North American Free Trade Agreement), 44

NASA, 309

nations, 14, 38, 41, 43, 45, 129–30, 180, 182, 213–15, 286–89, 309, 311, 340, 342–43, 347–48; advanced, 298, 301, 324; competitive, 324; developing, 129, 148; nuclear, 334; wealthier, 342

NATO (North Atlantic Treaty Organization), 284, 336–37
Nest, 204
networks, 20, 22–23, 102–3, 105, 108, 146, 150–52, 154, 209–10, 228, 230–33, 238, 241, 248, 250; advertising, 305; cloud infrastructure, 278; criminal, 211; high-speed, 152; industrial, 227–28; mobile, 150; pipeline, 209–10; private, 47; registered, 108; self-governing, 102; social, 227, 326, 363, 369
neural net algorithms, 278
New Paradigm, 433
New York, 178, 355, 380
New York Times Magazine, 158, 441
Nextlaw Labs, 354
NGA, 271
NGAV (next-generation anti-virus), 246
NGOs (non-governmental organizations), 196, 198, 253
NIC (National Innovation Challenge), 199
NITI Aayong, 56
NLP (natural language processing), xi, 19, 99, 154, 197, 405

NLP algorithms, 24–25, 115
NMI (Nature Machine Intelligence), 122
North Korea, 24, 48, 261, 271, 324
Northwestern University, 124
NSA (National Security Agency), 261
NTSB (National Transportation Safety Board), 359
nuclear weapons, 288, 330; space-based, 330

O

Obama, Barack (President), 179–80, 269, 365
Obama administration, 59, 264, 313–14
obsolete, 231, 237, 353, 381
OCC (Office of the Comptroller of the Currency), 97, 105, 116, 430
Ohio State Board, 175
Ohio State University, 396
oil, 23, 82, 410, 422
online, 10, 79, 113, 161, 176, 187, 191, 212–13, 267, 294, 301, 304, 366; government services, 215; posted, 363
online activity, 166

online chat rooms, 297
online education, 176
online loan, 66
online payments, 102
OOCL, 46
OpenAI, 460
operators, 81, 270, 282, 320, 360; human, 262, 275, 327, 360; military drone, 244; remote, 282
O'Reilly Media Inc, 421–22
organizations, 41–43, 65–69, 73–74, 84–85, 87, 90, 112–13, 158, 236–37, 292–93, 364–65, 390–92, 404–5, 409, 414–15; extremist, 325; global, 42, 410; international, 152, 332; medical, 128–29; non-governmental, 196; proxy, 27
orientation, 2, 53, 67, 87, 338; futuristic, 104; interactive, 72; reactive, 347; sexual, 28, 371, 380; weapon, 280
ownership, 51, 67, 115, 181, 373, 403
Oxford Hospital, 132

P

pace, 3–4, 68, 75, 79, 85, 110, 122–23, 150, 173, 183, 239, 264, 284, 289, 335–36; fast, 219; frantic, 215; rapid, 5, 71, 286
pains, 15, 59, 366; economic, 314; phantom limb, 15
Pakistan, 42, 271
Paris, 337, 339
participants, 2, 10, 24, 47, 92, 177, 248, 262, 410; paralyzed, 145
partnership, 253, 314, 379; government-wide, 57
passwords, 108, 225, 227; administrator's, 228; complex, 247; correct, 226; unique, 231
patents, 133, 304, 347, 436; smart bathroom, 133, 436
patients, 126–32, 140, 146, 408; elderly, 130–31; infected, 137; pacemaker, 51
Patriot missile system, 320
payments, 89, 99, 101–2, 108, 207, 248, 303, 367; cross-border, 99; digital, 295; fraudulent, 108; improper, 207; on-time, 46; real-time, 108
payoff, 91; potential economic, 76
PDF malware classifiers, 246

peace, 317, 330–33; global, 334
PED, 276
Pemex, 210
Pentagon, 275, 280, 309, 311
permission, 31, 103, 214, 226, 363, 366; explicit, 365; grant, 396; parental, 185
personal data, 27, 134, 150, 302, 363, 365–68, 370–71, 387; protecting, 44
perspectives, 104, 151, 153, 346, 369, 383, 397; historical, 267; legal, 380; operational, 353; social, 56
phases, 40–41, 57, 95, 101, 136–37, 187, 265, 278, 425; initial, 101, 277
phones, 6, 32–33, 109, 177, 187, 222, 310, 390, 450; ordinary cell, 221; smart, 70, 171, 202, 244, 302, 356, 397, 416
physics, 185, 321; quantum, 320–21
pigs, 398–99; engineered, 398; guinea, 172
pilot, 116, 201, 275, 320, 395; airline, 2
pipelines, 149, 203, 209; global, 128

PLA (People's Liberation Army), 293, 318–19
plants, 13, 15, 35, 83; digitizing power, 42; nuclear power, 408; power generation, 205; steel, 83; virtual power, 34
platforms, 21, 46–47, 99–100, 128–29, 151, 154, 196, 198, 208, 224–25, 299, 301, 303, 305–6, 343–44; autonomous aerial, 275; blockchain insurance, 426; cognitive computing, 218; common communication, 203; core banking, 104; digital, 40, 98, 101, 103–4; drone imagery, 279; drug-development, 136; financial, 103; global ISR, 260; industrial IoT, 84; messaging, 154; mobile, 275; open, 303; prediction, 323; threat intelligence, 344
players, 14, 25, 170, 300, 303, 340, 342, 347
Polanyi's Paradox, 408, 464
policymakers, 43–45, 59, 177, 182, 185, 189,

191, 252, 255, 259–61, 288, 323
politics, 8, 30, 301, 340, 344; corporate, 188; global, 340–41; international, 324
pollution, 119; reducing, 200
populations, 36, 38, 100, 129, 189, 194, 200, 215, 301, 328, 390; aging, 130, 147, 199, 342; ethnic, 179; global, 153; local, 265; subduing, 330
portfolios, 90, 100, 289, 305, 383, 406
Portuguese, 30, 145, 439
poverty, 390–91
power, 93–94, 105, 147, 149, 152, 178, 198, 291, 294, 296, 298, 324, 326, 338, 340; enforcement, 297; intuitive, 344; nuclear, 272; political, 339; steam, 5; unchecked, 223; voting, 41
precedent, 249, 377, 381; legal, 359–60
predictions, 20–21, 24, 124, 126, 128–29, 137, 140, 158–59, 162, 165, 372–73, 382, 384, 389, 391
Predix, 82–83
PRI, 461

price, 75, 78, 92, 94, 96, 154, 160, 168, 179, 183, 200, 204, 206, 284, 394; artificial, 96; base, 78; stock, 25
PriceWaterhouseCoopers, 11
privacy, 43–44, 54, 56, 148, 155, 157–58, 166, 211–12, 216, 223, 231, 235–36, 302, 363–64, 368; erosion of, 3, 389; personal, 8, 397; protecting, 45, 365
private sector, 25, 36, 55, 195, 203, 252, 265, 289, 293, 313, 315, 413
productivity, 8, 38–39, 41, 57, 71–72, 79, 85; advanced, 39; digital, 42; human, 416; worker, 9, 73
professionals, 155, 246, 256, 378, 402; aid security, 249; legal, 353, 385; medical, 207, 393; skilled, 61
profiles, 27, 49, 120, 376; behavioral, 95, 161; digital, 32, 160, 171–72; genetic, 212; global, 324; historical, 113; international, 54; personality, 27; physical, 30, 172; psychometric, 28–29;

risk/return, 406;
technological, 289;
virtual, 30
profiling, 369–71;
algorithmic, 371, 373
profits, 23, 25, 65, 94–97,
182, 292, 354;
declining, 406; illicit,
94; quarterly, 42
programmers, 212, 217,
223, 234, 374; human,
217; skilled, 104
Project Debater, 162
Project Maven, 276–79
projects, 59, 67, 146, 232,
235, 275, 278–79,
291, 293–94, 304–5,
308, 310, 319, 338–
39, 341; ambitious,
235; construction,
201; cooperative, 128;
cutting-edge, 306;
foreign investment,
323; long-term, 196;
nation-wide
surveillance, 303;
smart city, 200, 304–5
proliferation, 244, 327,
331; global, 249;
rapid, 250
proof, 102, 269; clear, 279;
public, 458; spoof,
109
propaganda, 27, 319;
divisive, 329; terrorist,
170
properties, 76, 125, 205;
illicitly-derived, 112;
intellectual, 292, 407,
457; physical, 108;
private, 303; shared,
60
ProPublica, 179, 355
protection, 151, 225, 227–
28, 237, 302, 362,
374; border, 203; civil
liberty, 302; legal,
347; robust privacy,
235
protocols, 151; standard,
121, 127
providers, 153, 171, 204,
207, 305; health care,
135, 208
provisions, 43, 284, 296,
311, 370; legal, 375
PSG (Preventive Security
Governance), 333–34
Putin, Vladimir
(President), 26, 284,
335, 423
PWC, 11, 108, 215, 419,
425, 431, 436

Q
QSAR (quantitative
structure– activity
relationship), 125

R
ransomware, 228
Rapid Prototype
Development
program, 280

R&D, 57–58, 180, 199, 285, 305, 313; private, 59; public, 59
recognition, 229, 441; facial, 16, 126, 213, 217, 244, 295, 297, 303–4, 356, 399
recommendations, 55–57, 66, 158–60, 261, 274, 324, 336; basic, 230; smartphone-based, 134
Redesign, 437
regimes, 331, 367; authoritarian, 297; autocratic, 308; international legal, 330, 334; regulatory, 116, 347
regulations, 18, 48, 196, 202, 212, 231, 235, 298, 331–32, 361–62, 365, 378, 380; arms, 331; environmental, 297; international, 333, 357; strict, 302; stringent privacy, 201
regulators, 96, 113–14, 118, 231, 257, 357, 405; national securities, 343; voltage, 205
Renaissance Institutional Equities fund, 91
repository, 114, 157, 185, 286; currency, 60; digital ID, 304; massive, 156, 208, 212
researcher, 121, 128; human, 232; legal, 352
resolution, 225, 262, 332; civil law, 379; high, 280
resources, 53, 61–62, 67, 69, 243, 245, 285, 288, 301, 303, 315, 318, 326, 403–4, 417; economic, 333; engineering, 77; fiscal, 60; foreign innovation, 298; global innovation, 298; grid, 206; managing, 157; monetary, 325; natural, 14; technological, 415
responsibility, 7, 220, 223, 235, 252–53, 334, 357, 359, 362–63, 378, 382, 386
restrictions, 44–45, 155, 228, 284, 333, 343; legal, 34
revolution, 5, 64–65, 85, 87, 101, 114, 118, 122, 237, 314, 389, 403; algorithmic, 93; digital, 54, 158, 202
rewards, xi, 148, 407, 411
risk assessments, 207, 346, 378; algorithmic, 378; criminal justice, 378
risk management, 35, 324

risks, 90–91, 96–98, 132–34, 143–44, 148, 167, 172–73, 184–86, 196, 202, 240, 319–20, 377–79, 383, 407–9; cyber, 118, 134, 323; cybersecurity, 240; evaluating credit, 106; legal, 46, 375; liability, 146; lower, 202; managed, 47; national security, 129; non-commercial, 343; predicting, 106; strategic, 98
RL (reinforcement learning), xi, 19, 21, 84, 105, 246–47, 357, 402
RMIT University, 116
RNS (responsive neurostimulation system), 140
robotics, 8, 19, 50, 64–65, 76, 194, 202, 219, 221, 244, 250–51, 275, 346, 376, 379–80; advanced, 74; automated, 281; deploying, 75; industrial, 84
robotics systems, 74–75, 77, 251–52, 281, 380, 395
robots, 72–78, 84, 160, 194–95, 202, 210, 219–20, 225–31, 244, 249–52, 281, 340–41, 396–97, 399–400, 414; business, 229; hacked, 227, 229; sex, 226, 400
ROS (Robot Operating System), 226
Russia, 26, 53, 264, 274, 283–86, 324, 328, 335, 338

S

safeguards, 165, 173, 223, 240, 253, 366, 370, 380, 397
Samsung, 341
satellite data, 200, 263
satellite imagery, 262–63, 272; raw, 272
satellite images, 24, 200, 204, 210
Saudi Arabia, 42
SCADA systems, 204–5
scalability, 245, 401
scan, 26, 124, 134, 140, 211, 381; facial, 295; radiological, 393; visual, 356
scanning, 160, 271; text-message, 356
scenarios, 9, 17, 20, 48, 96, 107–8, 111, 152–53, 203, 208, 234, 377; combat, 275; extreme, 32; hypothetical, 114; negotiation, 48
science fiction, 60, 138, 143, 219, 321, 356, 391

scores, 159, 164, 178, 181, 188, 212, 218, 247, 294; high risk, 377; honesty, 165; risk assessment, 223; screening, 188
screen, xii, 115, 134, 146, 164, 168, 270, 277; touch, 395
search engines, 93, 120, 202; optimizing, 120
Second Industrial Revolution, 5
security, 56, 92, 96, 205, 218, 231–32, 234–35, 247, 251, 255, 330–34, 337, 340, 344–45, 403; airport, 235; automated, 233; cyber, 237, 411; digital, 253; internal, 317; national, 260, 262, 311, 318, 345, 348, 364; national, 111; physical, 253; political, 253; public, 299, 301, 304; water, 119
security tools, 224, 247, 402; evading, 247, 402
sensors, xii, 6, 23, 35, 75–76, 82–83, 133, 141, 195, 201, 204, 301, 308, 397, 400; biometric, 220; external, 394; people-counting, 24; perception, 273

sentient, 17, 93, 397
Sentient Technologies, 93
sequencing, 135–37, 391; genetic, 135–36, 148, 213
service providers, 66, 302, 343–44, 363, 410–11; external, 368
Sex Robot Conference, 400
Siemens, 83–84
Siemens' Electronic Works Amberg, 80
Siemens gas turbines, 83
SIF (Synthetic Identity Fraud), 110, 432
signals, 125, 187, 276, 306, 319; brain's motor, 395
SigOpt, 263–64
Silicon Alley, 179
Silicon Valley, 33, 298, 307, 313–14, 325, 354, 364, 420
simulation, 15, 77, 235, 280; computer-generated, xii; geopolitical environment, 323
Singapore, 57–58, 116–17, 199
situational awareness, 276, 281; heightened, 337; sharing, 281
skills, 9, 11, 55–57, 72, 85, 87, 154–55, 190, 245, 250, 342, 356, 384, 405, 408; administrative, 180;

basic management, 87; decision-making, 280; digital, 54; math, 175; soft, 188; worker, 80
SMEs (Size Enterprises), 48, 54–55, 58
SML (supervised machine learning), xi, 19–21, 105–6, 108, 112, 196, 200, 208, 231, 238, 241
SML algorithms, 19–21, 111, 113, 115, 260
SMSS (Squad Mission Support System), 282
social media, 23–24, 27–28, 30, 155, 226, 235, 246, 268, 325, 351, 356–57, 396
societies, 3, 5, 12, 52, 61, 71, 86, 337, 340, 345, 371, 373, 376, 378, 387; advanced digital, 213; black box, 372; civil, 253, 346; smart, 301; values, 198
Socratic, 177–78
SoFI, 103–5
software, 60, 217, 219, 224, 242, 249–51, 263, 268–69, 352, 354–55, 357, 377–78, 380–81, 399–400, 402; advanced, 105; commercial antivirus, 248; facial recognition, 159, 212, 220; lifecycle management, 233; networking, 233; open source, 313; scheduling, 222; traffic optimization, 303; word, 374
South Korea, 24, 60, 74, 149, 341, 440
space, 86, 138, 149, 183, 199, 201, 252–53, 283–84, 294, 298, 309, 313–14, 338, 344, 347; digital, 221; human occupied, 252; open, 329; public, 317, 363; storage, 137–38; underground, 199
speeds, 152, 232; high, 77; increased data, 150
spoofing, algorithmic, 96
spying, 259, 262, 452
SSA (Social Security Administration), 110–11
SSNs (social security numbers), 110–11
SSRL (system–Stanford Synchrotron Radiation Lightsource), 124
stability, 97, 317, 330; economic, 210; enhancing strategic, 336; maintaining political, 296; maintaining strategic nuclear, 335; social, 299

stakeholders, 188, 200, 362, 386; citizen, 214
standards, 18, 58, 99, 121, 124, 151–52, 190, 195, 216, 329, 357, 378, 383, 413; common, 162; data privacy, 348; financial, 98; legal, 256; risk management, 91; universal, 171
Stanford, 176, 190, 195, 451
Stanford University, 145, 269, 439
startups, 20, 72, 136, 166, 176, 179, 194–95, 262, 298, 302, 304, 338, 407; ambitious, 196; chip, 310; domestic, 300; high-tech, 166; well-funded, 147
start-ups, 55, 57–58, 309, 311–12, 317, 406; foreign-based, 325
Statistic Brain Research Institute, 442
strategy, 53–58, 92–93, 131, 193–94, 204–5, 215, 275, 292, 299, 339, 348, 354, 404, 410, 417; bioweapons, 289; competitive, 319; coordinated, 316; cyberattack, 205; defensive, 27; deployable, 53; effective, 278; industrial, 59; information warfare, 26; legal, 354; long-term, 26, 129, 406; multilateral, 340; non-proliferation, 328; nuclear, 334, 336; post-war, 335; well-crafted, 294
stress, 83, 107; financial, 107; post-traumatic, 244
submarines, 282, 319–20, 335, 395; nuclear, 319–20; quiet diesel-electric, 282; semi-submersible, 33; submersible, 33
supercomputers, 136; energy-hungry, 35
superintelligence, 2, 17, 320, 335, 387
surveillance, 52, 167, 228, 234, 256, 317–18; constant, 161, 222; enhanced government, 345; offline, 304
surveillance cameras, 303–4, 317; multiple closed-circuit, 265
Sweden, 397
Syria, 283

T

tablet, 51, 89, 164, 187; clay, 89

tactics, 247, 275, 283, 291–92, 402
Taiwan, 58
tariffs, 341; tit-for-tat, 341
teams, 28, 69, 124–25, 146, 153, 195, 210, 213, 230, 263, 277, 305, 314, 358, 414; cross-functional, 279; multi-disciplinary, 58; strong, 407
tech companies, 155, 304, 310, 361, 364
technologies: antiquated, 117; blockchain, 48; blocking, 187; chip, 300; computer, x, 167; core, 155; cutting-edge, 308, 311; cutting-edge military, 329; cyber security, 238; disruptive, 262; domestic and foreign, 292; driverless, 358; embedded, 7; emerging, 6, 34, 85, 305, 354, 376, 392; financial, 98; foreign, 315; foundational, 304; gene editing, 138–39; gene-editing, 398; innovative, 325; laser, 294; lethal, 347; marine, 294; military-related, 309; mobile, 150, 152; national security, 260, 289; neural interface, 395;

research, 381; self-driving, 201; sensory, 236; wearable, 186
terrorism, 114, 323
terrorist attacks, 323; lone wolf, 250, 402; thwart, 265
terrorists, 139, 235, 253, 325, 330
Tesla, 6, 303, 358, 361
Third Industrial Revolution, 6
ThreatMetrix, 108
threats, 128–29, 205, 224, 227, 232–33, 236–40, 243–45, 249, 254–56, 282, 284, 296, 337, 380, 384; advanced, 246; cyber, 237, 239; detected, 232; detecting, 239–40; emerging, 237, 252; incoming, 287; potential, 144, 249, 270, 311, 317; unidentified, 238
TOEFL (Test of English as a Foreign Language), 176
tools, 66, 68, 81–82, 225, 227, 232, 234–35, 261–62, 265, 306–8, 344–45, 353, 373, 377–78, 381–82; advanced analytical, 122; customer scoring, 66; molecular, 398; national security, 265;

neural network, 323, 393; payload, 224; portable leak detection, 209; prediction, 235; predictive, 343; risk-assessment, 378; training, 280
ToS (Terms of Service), 31, 222, 248, 366, 391
TPP (Trans-Pacific Partnership), 43–44, 48
track, 64, 83, 131–33, 165, 172, 183, 199–200, 203, 205, 218, 224, 282, 302, 354, 394
trade, 40, 43–45, 48, 89, 91–93, 95–96, 103, 313, 325, 341, 343, 426; cross-border, 43, 46, 343; drug, 33; free, 193; illegal, 430; profitable, 95
traders, 92, 94–96, 343–44, 430; digital, 93; human, 91; independent, 95; stock, 91, 342
train, 58, 122, 126, 145, 151, 153, 156, 180, 182, 189, 191, 197, 262–63, 354, 357
transactions, 89, 98, 102–3, 111, 113, 115, 158–59, 208, 224, 309, 311, 353, 356; acquisition, 24;

automated ledger, 47; cross-border, 89, 312; early-stage venture, 312; financial, 103, 218; non-sensitive, 311; overseas, 300; recurring, 113
transparency, 42, 55, 103, 167, 197, 234, 254, 256, 324, 331, 333–34, 377, 380, 386, 413; increased, 372–73
Transparency International's Corruption Perceptions Index, 197
treaties, 327, 331, 334, 357
Trump, Donald (President), 269, 312, 365, 461
Trump administration, 59, 309, 311, 313–14
truth, 61, 76, 163, 264, 301, 311, 364, 392, 407, 414
TSA, 235
Turing, Alan, 14
tutors, 176–77, 180; digital, 180; traditional, 176
Twitter, 27, 209

U

UAE (United Arab Emirates), 117, 348
UAS (Unmanned Aircraft Systems), 203

UAVs (unmanned aerial vehicle), 7, 33–35, 195, 287
Uber, 60, 150, 358–60, 390
UBS White Paper, 427
UK government, 141
UML (unsupervised machine learning), xi, 19–21, 105, 109, 112, 120, 208, 238–39, 241
UNESCO (United Nations Educational, Scientific and Cultural Organization), 169, 177
United Nations, 41, 332, 436
UR5 robotic arm, 78
US Air Force, 308
US Congress, 110, 114, 197, 284
US Dodd-Frank Act, 107
users, 30–32, 167–70, 204–5, 226, 228, 231, 240–41, 247–48, 263–64, 278–79, 362–65, 374, 390–91, 394–95, 405; authorized, 214, 230; enabled, 27; normal, 249; preventing unauthorized, 214; voice-activated, 30
US military, 271, 274, 279, 281, 283, 286, 312, 319, 439
UUVs (unmanned underwater vehicles), 33–34, 319

V
VADER, 205–6
VCs (venture capitalists), 344, 406
vehicles, 82, 195, 220, 270–71, 281–82, 285, 301, 358–61, 366, 460; armed, 210; autonomous driverless, 250; autonomous unmanned aerial, 7; remote-controlled, 281; self-driving, 359–60; self-driving Uber, 358; sport utility, 70; tracked, 281; unmanned, 264, 285; unmanned underwater, 33
Venezuela, 284
veterans, 15; disabled, 394
video, 7, 19, 61, 161, 234, 262, 266, 268, 270, 276, 286, 295, 390, 400; courtroom trial, 163; pre-recorded, 400
Vietnam, 22, 53, 83, 200
virtual assistants, 6–8, 30, 32–33, 81–82, 121, 168, 171; voice-controlled, 398
VR (virtual reality), x, xii, 60, 65, 167, 169–70, 172, 183–84, 276, 280
vulnerabilities, 225, 227, 229–30, 232, 234,

239, 242–45, 247–49, 252, 254–56, 268, 351; human, 256; potential, 204, 219; systemic, 236, 267

W

wages, 10, 74, 86, 99, 332
Wall Street, 24, 33, 118, 147, 407, 422, 430
war, 3, 5, 33, 36, 67, 93, 147, 169, 264–65, 273, 276, 287, 292, 330–32, 336–37
warfare, 27, 277, 288, 291–92, 318, 327, 332, 336–37; unrestricted, 291, 294
water, 35, 119, 200, 311; contaminated, 119
weapons, 223, 271–74, 282–83, 287, 291–92, 327, 330–33, 345, 394, 458; autonomous military, 283; biological, 139, 330; hypersonic, 284; intelligent, 287; laser, 331; standoff, 287
websites, 112, 198, 232
WeChat, 304
WeChat Pay, 302
WeChat wallet, 294
WePay, 101
wind turbines, 51
Wireless Telegraphy Act, 187

workers, 8, 11–12, 70–73, 75–78, 80, 85–86, 98, 191, 215; displaced, 9; efficient human, 61; foreign, 194; highly-skilled, 87; knowledge-based, 10; manual, 10; warehouse, 165
workforce, 56, 70–72, 80, 85, 260, 314, 405; global, 11; tech-savvy, 348
World War II, 41, 72, 213
WTO (World Trade Organization), 43, 47

X

X-Road, 214

Y

Yahoo, 450
YouTube, 429
Yulchon, 48–49

Z

Zambia, 53
Zelle, 108, 111, 432
Zimbabwe, 53

About the Authors

Daniel Wagner

Daniel Wagner is the founder and CEO of Country Risk Solutions and has three decades of experience managing cross-border risk in the private and public sectors.

Daniel began his career at AIG in New York and subsequently spent five years as Guarantee Officer for the Asia Region at the World Bank Group's Multilateral Investment Guarantee Agency in Washington, DC. During that time, he was responsible for underwriting political risk insurance for projects in a dozen Asian countries. After serving as Regional Manager for Political Risks for Southeast Asia and Greater China for AIG in Singapore, Daniel moved to Manila, Philippines, where he served in a variety of capacities in the Asian Development Bank's Office of Co-financing Operations, including as Senior Guarantees and Syndications Specialist. He then became Senior Vice President of Country Risk at GE Energy Financial Services.

Daniel has published more than 600 articles on current affairs and risk management and is a regular contributor to the *South China Morning Post, Sunday Guardian*, and *The National Interest*, among many others. He is also the author of four previous books—*Virtual Terror, Global Risk Agility and Decision-Making, Managing Country Risk*, and *Political Risk Insurance Guide*.

He holds master's degrees in International Relations from the University of Chicago and in International Management from the American Graduate School of International Management (Thunderbird), in Phoenix. Daniel received his bachelor's degree in Political Science from Richmond College, in London.

AI SUPREMACY

Contact Information:

E-mail: daniel.wagner@countryrisksolutions.com
LinkedIn: www.linkedin.com/in/danielwagnercrs
Twitter: www.twitter.com/countryriskmgmt

Keith Furst

Keith Furst is the Managing Director of Data Derivatives, a consulting firm focused on implementing, validating, and fine-tuning financial crime systems. He has many years of extensive consulting experience working for a wide variety of financial institutions focusing on the documentation, design, model development, implementation, calibration, and validation of financial crime systems.

Keith has broad exposure and hands-on involvement in all phases of the project life cycle for transaction monitoring, know your customer (KYC), customer due diligence (CDD), sanctions, trade compliance, and anti-fraud systems. He has worked for banks in New York City under extreme regulatory scrutiny and helped guide them through the complexity of transforming their compliance system infrastructure to meet industry best practices and guidelines.

He has also optimized financial crime detection models and recommended tactical configuration changes to thresholds and parameters by using data mining and other statistical techniques. Through multiple project engagements, he has developed a stellar reputation as the "go-to" person when mission critical defects or failures emerge, as he has proven his ability to be one of the few individuals who can lead the effort to quickly identify a problem, drive executive buy-in for the proposed solution, execute the plan, and validate the final results.

Keith has also spoken at industry conferences, private executive dinners, and via live video webinars around the world, including in London, Paris, New York, Toronto, and Singapore. He has trained other financial crime professionals and developed a majority of the content for a sanctions screening module.

He writes thought pieces frequently and has been published in the *American Banker*, *The Asian Banker*, the *Association of Certified Financial Crime Specialists*, and *The South China Morning Post* on a diverse range of intricate topics such as financial crime, artificial intelligence, and risk management.

Keith holds an MBA from the Baruch College Zicklin School of Business and a BA in Philosophy from Queens College, both in New York.

Contact Information:

E-mail: kfurst@dataderivatives.com
LinkedIn: www.linkedin.com/in/keithfurst
Twitter: www.twitter.com/keithfurst

Made in the USA
San Bernardino, CA
22 November 2018